T0300124

Millimeter Wave Communications in 5G and Towards 6G

This book explores different facets of millimeter wave systems, which form a central part of 5G communication systems. It explains how these systems serve as a foundational building block of 5G-Advanced/6G as these systems evolve.

Millimeter Wave Communications in 5G and Towards 6G focuses on millimeter wave channel modeling, radio frequency (RF) and antenna level constraints imposed on beamforming, beamforming design for link level incorporating the RF/antenna constraints and the channel structure, as well as system level deployment considerations. With significant academic and industrial experience, the authors are well-equipped in explaining how the millimeter wave research developed, the fundamental principles/math beneath the technology, and in explaining precisely the "Why?" behind the "What?" that make the 5G-NR specifications. The authors examine point-to-point systems at a single link level and show how the traditional sub-7 GHz-based beamforming procedures simplify to a simplistic signal processing approach of directional beam scanning. This book examines the foundational background that led to specific choices in the millimeter wave part of the 5G-NR spec as well as chart out the roadmap in terms of future research and development activities in this arena. The book ends by providing a scope of future research in this area.

This book is geared towards both introductory as well as advanced researchers in industry and academia working in the areas of 5G, 5G-Advanced and 6G communications. It would also be useful for senior undergraduate and graduate students in universities focusing on wireless communications topics.

Millimeter Wave Communications in 5G and Towards 6G

Vasanthan Raghavan
Junyi Li
Ashwin Sampath
Ozge H. Koymen
Tao Luo

CRC Press is an imprint of the
Taylor & Francis Group, an **informa** business

Designed cover image: ssuaphotos/Shutterstock

First edition published 2025
by CRC Press
2385 NW Executive Center Drive, Suite 320, Boca Raton FL 33431

and by CRC Press
4 Park Square, Milton Park, Abingdon, Oxon, OX14 4RN

CRC Press is an imprint of Taylor & Francis Group, LLC

ISBN: 978-1-032-70374-9 (hbk)
ISBN: 978-1-032-70377-0 (pbk)
ISBN: 978-1-032-70375-6 (ebk)

DOI: 10.1201/9781032703756

Typeset in Nimbus Roman
by KnowledgeWorks Global Ltd.

Dedicated to my parents, wife, brother and his family, children and in-laws

Vasanthan Raghavan

Dedicated to my children Jessica and William Li

Junyi Li

Dedicated to my family, parents, teachers and all my colleagues over the years

Ashwin Sampath

Dedicated to my parents, siblings, wife and children for their unconditional love and unwavering support

Ozge H. Koymen

Dedicated to my parents Zhongcheng Luo and Geng Fangyang

Tao Luo

Contents

Foreword

Qualcomm Technologies, Inc., a global leader in wireless technologies, has been at the forefront of innovation and product development since the early days of 3G—the third generation of cellular technology, which began with CDMA technology. As a member of this remarkable organization, I have had the privilege of participating and witnessing our industry-leading contributions firsthand. For over two decades, I have proudly called the authors of this book my colleagues. Their impact on the wireless industry is immeasurable, as they have played a pivotal role in shaping the specifications of cellular technologies—from 3G to 5G, and soon even 6G. Their pioneering work, particularly in millimeter wave (mmWave) technology for 5G networks, has left an indelible mark.

I vividly recall a visit to our New Jersey office a few years ago, during a period when the company was deeply immersed in 5G research and the definition of critical physical layer concepts. Our goal was to contribute these innovations to the emerging 5G specification. The authors, my esteemed colleagues, were laser-focused on prototyping a mmWave wireless system. Their mission? To demonstrate that this cutting-edge technology could effectively support vehicular mobility.

The prototype system was strategically positioned, its antennas pointed toward the parking lot and the surrounding streets near the Qualcomm Technologies research building. The stakes were high, and the excitement palpable. The team achieved what seemed improbable: a seamless connection between a moving van equipped with a mmWave modem and the prototype system. As I stood there, witnessing this breakthrough, I could feel nothing but awe. This demonstration solidified our belief that mmWave technology could indeed thrive within a cellular network, paving the way for the future of wireless communications.

A few years later, during Super Bowl 58 in Las Vegas, NV, fans enjoyed an unparalleled user experience on their devices. Those with mmWave-capable devices witnessed downlink speeds exceeding 4 Gbps and uplink speeds of 700 Mbps, enabling them to seamlessly run applications and enhance their football game-watching experience. Credit is also due to the service providers who meticulously planned and ensured robust mmWave coverage at the venue.

The fifth generation of cellular technology, 5G, remains in its infancy even six years after the initial network launches in the US and China. While 5G primarily operates in the sub-7 GHz frequency bands worldwide, it also supports deployments in the mmWave bands, spanning from 24.25 to 52.6 GHz in its initial phase. Although only a handful of mmWave networks exist globally at the time of writing, our optimism lies in the inevitable adoption of this technology.

Why? The answer is simple: **spectrum scarcity.** As data demands continue to surge exponentially, network operators will eventually need to harness these higher frequency bands. These new use cases are driven by recent advancements, including an exciting enabler: **AI technology.**

This book delves into the intricacies of 5G mmWave technology, shedding light on its unique features and advantages compared with sub-7 GHz offerings. Specifically, it explores these key points:

1. **High Bandwidth:** MmWave technology boasts exceptionally high bandwidth, enabling lightning-fast downloads and data transfers when compared with traditional wireless technologies.
2. **Ultra-Low Latency:** The minimal delay provided by mmWave connections is critical for applications like augmented reality and specific industrial use cases. It is also a perfect fit for densely populated urban areas and enterprise environments.
3. **Localized Cellular Coverage:** MmWave ensures efficient connectivity in crowded city spaces. Imagine a football stadium—the ideal setting for showcasing its capabilities.
4. **Abundant Spectrum:** Operating in frequency bands ranging from 24.25 to 52.6 GHz, mmWave technology offers extreme capacity due to the ample spectrum available.
5. **Innovative Applications:** From industrial automation to immersive education and training simulations, mmWave supports a wide range of innovative use cases. Businesses can leverage mmWave to enhance productivity and explore new possibilities.

In summary, mmWave cellular networks combine high speeds, low latency, and localized coverage, positioning them as a powerful choice for next-generation communications. This book dives into the foundational aspects of this technology, unraveling the "what" and "why" behind the key performance indicators.

The authors of this book have consistently demonstrated their mastery at Qualcomm Technologies. They excel in defining innovative concepts through white papers, building prototypes to validating these ideas, and ultimately guiding our development teams to create the best realizations of these concepts in Qualcomm Technologies' modem products.

I envision this book as a valuable reference for practicing engineers seeking to grasp the foundational principles behind mmWave technology. Additionally, I eagerly anticipate witnessing how this technology evolves into 6G, fueled by the expertise and innovative minds of my esteemed colleagues and the authors of this book.

Dr. Baaziz Achour
Deputy Chief Technical Officer (CTO), Qualcomm Technologies, Inc.

Preface

The focus of this book is on explaining different facets of millimeter wave (mmWave) systems. These systems form one of the core components of Fifth Generation (5G) cellular networks today and will remain a central pillar of the emerging Sixth Generation (6G) systems and beyond. Strictly speaking, the mmWave regime stands for systems that operate at a frequency regime corresponding to millimeter wavelengths (30–300 GHz). In practice, at the Third Generation Partnership Project (3GPP) level, the mmWave regime is denoted as *Frequency Range 2* (FR2) and stands for the 24.25–71 GHz regime. Extensions to FR2 include *Frequency Range 4* (FR4) focusing on 71–114.25 GHz and *Frequency Range 5* (FR5) focusing on beyond 114.25 GHz frequencies. Different design aspects of FR2 systems also impact the design of *Frequency Range 3* (FR3) systems focusing on 7.125–24.25 GHz, which are also of broad importance as 6G evolves from an available spectrum perspective.

FR2 spectrum and more broadly mmWave systems did not attract attention from practicing wireless engineers till the beginning of the 5G standardization process due to a number of reasons. Some of these reasons included:

- Lack of licensed spectrum in FR2 for cellular operations
- Device level design complexities of radio frequency (RF) components and their high cost, area and power consumption profiles as well as the thermal management associated with higher power consumption
- Poor link budgets that implied a small cell operation corresponding to 200–500 meter inter-site distances (ISDs) in contrast to the 1–3 km coverage for Third/Fourth Generation (3G/4G) systems
- Increased capital expenses (Capex) and operational expenses (Opex) associated with the deployment of these small cells
- Necessity of multi-antenna beamforming at the RF level to bridge the link budget, requiring advances in antenna array design, radio frequency integrated circuit (RFIC) chip design and packaging technologies
- Lack of a systematic physical layer design that would take advantage of the unique features associated with these frequencies and combat their impairments.

Over the last ten years, mmWave wireless systems have seen a massive revolution starting from white-papers and vision documents leading to experimental prototypes and pre-5G standardization efforts followed by the Fifth Generation-New Radio (5G-NR) standardization process for Release 15 (Rel. 15) at 3GPP. This led to the design, development, and evolution of 5G-NR compliant modems such as the Snapdragon®[1] X50, X55, X60, X65, X70 and X75 modem families from Qualcomm Technologies.

---

[1]Snapdragon and Qualcomm branded products are products of Qualcomm Technologies, Inc. and/or its subsidiaries.

These developments have in parallel led to the evolution of the 3GPP specifications in Rel. 16 and beyond. Given this background, this book addresses, explores and *demystifies* the fundamental issues behind mmWave technology. We will see that the evolution can be seen as a *continuum* that starts with the multiple-input multiple-output (MIMO) revolution that forms the backbone of 3G and 4G systems, as well as the Institute of Electrical and Electronics Engineers (IEEE) 802.11 series of standards for WiFi systems.

At this stage, there are a number of books [1, 2, 3, 4] that have been published which address topics such as 5G, massive MIMO and mmWave communications including details on the 5G-NR standardization specifications and agreements. This book differs from the above class of books in the following aspects. This book is written with the express goal of explaining how the mmWave research developed and evolved. It explains the fundamental principles beneath the signal processing and technology determinants and more precisely, the *Why?* behind the *What?* that make the 5G-NR specifications. It is also written by theoreticians firmly grounded in the wireless industry and with significant academic and industrial experience.

The subject matter of this book concerns the tradeoffs between different possible mmWave MIMO precoding and beamforming schemes, and how they are uniquely impacted by device level constraints (at the RF and antenna levels) such as cost, complexity, area, energy consumption, channel structure, performance gains, etc. The aim of this book is to help the reader understand the scope and scale of complexities involved in form factor user equipment (UE) and customer premises equipment (CPE) design, how the standardization process tries to address these complexities, and the efficacy of the solutions addressed by the standardization efforts. Upon reading this book, one can understand how mmWave technology concepts are applied in mobile and fixed wireless access systems. Further, one can understand how they perform with a focus on practical deployment considerations.

Authors

Vasanthan Raghavan is a Principal Engineer at Qualcomm Technologies. He received the B.Tech degree in electrical engineering from the Indian Institute of Technology at Madras, Chennai, India, in 2001, and the M.S. and Ph.D. degrees in electrical and computer engineering, in 2004 and 2006, respectively, and the M.A. degree in mathematics, in 2005, all from the University of Wisconsin, Madison, WI, USA.

Junyi Li is a Vice President, Engineering at Qualcomm Technologies. He is an IEEE Fellow and a Qualcomm Fellow. He received the Ph.D. degree in electrical engineering from Purdue University, West Lafayette, IN, USA, and the MBA degree from the Wharton School, University of Pennsylvania, Philadelphia, PA, USA.

Ashwin Sampath was a Senior Director, Technology at Qualcomm Technologies. He is currently with Blue River Technology, Inc. He received the Ph.D. degree in electrical engineering from Rutgers University, New Brunswick, NJ, USA.

Ozge H. Koymen is a Senior Director, Technology at Qualcomm Technologies. He received the B.S. degree in electrical and computer engineering from Carnegie Mellon University, Pittsburgh, PA, USA, in 1996, and the M.S. and Ph.D. degrees in electrical engineering from Stanford University, Stanford, CA, USA, in 1997 and 2003, respectively.

Tao Luo is a Vice President, Engineering at Qualcomm Technologies. He received the B.S.E.E. degree from Shanghai Jiao Tong University, Shanghai, China, the M.Sc. degree from Queen's University, Kingston, Ontario, Canada, and the Ph.D. degree in electrical engineering and computer science from the University of Toronto, Toronto, Ontario, Canada.

Acknowledgments

This book project started many years ago as a means by which the authors could demystify for themselves the fog around the design of mmWave systems when such systems were at their infancy from a practical implementation perspective. As with any book project, we grossly overestimated our capacity to finish this project on time. In the meanwhile, a number of books [1, 2, 3, 4] have appeared on MIMO, wireless systems as well as on 5G-NR with a vision and outlook toward future evolutionary pathways such as 6G. With this background, this book differs from the available set of textbooks in the market in its philosophical underpinning. The main philosophical difference in our work is the belief that the design and evolution of (commercial) mmWave systems is not a recent phenomenon, but owes its origin to pretty much the initial days of MIMO from the early 1990s. As a careful reader can observe, we have tried to draw parallels and connections of 5G system design in FR2 to the *continuing* evolution of the theory and standardization of MIMO systems at 3GPP, as well as other standardization bodies such as IEEE 802.11.

Given that many years have passed since the inception of this book project, a number of life-altering changes have happened in the lives of the authors. The first author is now a proud father of two kids, Mihir and Adya, who are cherubic, exuberant, imitating of each other, and exasperating in their own ways—none of which he would like to change. He would also like to thank the enormous support and help over the years he has received from his in-laws and his brother, in many ways said and unsaid, and for which he would be eternally grateful. It is often difficult to put words where emotions precede and with Mahee (the first author's spouse), it would be blasé to just describe the moral support and indescribable patience—both of which are true. What he would really like to acknowledge are the effortless ease with which we can understand each other, push ourselves to our capabilities to be our true selves, and to remain rooted and nostalgic in the past as much as be enchanted about the future. And as we get wiser with age, we hope to not let *Father Time* ruin the good memories of life. With that, the first author would like to remember the good times that he has had with his parents over the years, and hopes to cherish and remember their innumerable sacrifices and love as the future unfolds in more unpredictable ways. Finally, the first author would like to thank his good fortune of growing up in the vicinity of the abode of Kapaleeshwarar and Karpagambal.

In addition, all the authors would like to acknowledge the immense support they have received from within Qualcomm Technologies. Baaziz Achour, Durga Malladi, John Smee and Tom Richardson have been massive votaries and pillars of support behind this book project. It goes without saying that while the authors have put pen to paper their understanding of commercial mmWave systems, that understanding itself has been framed and refined over many years of interactions with a number of great colleagues at Qualcomm Technologies. The authors acknowledge the contribution of the Qualcomm Technologies Research mmWave project members, as well as the Qualcomm CDMA Technologies (QCT) modem team, with whom the authors have been working and learning in advancing mmWave technology. In particular, Mohammad Ali Tassoudji, Jeremy Darren Dunworth, Raghu Narayan Challa, and Juergen Cezanne have been encyclopedic sources of knowledge and wisdom on antenna, RF design and system-level issues and the authors owe a lot of gratitude to them in terms of the book's shape and outcome.

Further, the authors have benefited greatly by discussions on

- Antenna designs with Yu-Chin Ou, Mei-Li (Clara) Chi, Lida Akhoondzadehasl, Sinan Adibelli, Jordi Fabrega, and Allen Tran
- RF design with Keith Douglas, Joe Burke, Chuck Wheatley, Jefy Jayamon, and Sherif Shakib
- Measurements, modeling and calibration discussions with Andrzej Partyka, Ricardo Motos, Patrick Connor, Andrew Arnett, Kobi Ravid, and Prasanna Madhusudhanan
- Regulatory discussions with Marco Papaleo, Mustafa Emara, Ebraam Khalifa, and Sumant Iyer
- System level deployment, commercialization and technology development discussions with Raghu Challa, Sony Akkarakaran, Mihir Laghate, Kang Gao, Shrenik Patel, Jung Ryu, Jun Zhu, Derrick Chu, Yongle Wu, Yong Li, Ruhua He, Alex Gorokhov, Mike McCloud, Brian Banister, Alberto Cicalini, Aamod Khandekar, Hari Sankar, Ashok Mantravadi, and Sundar Subramanian and
- Field deployment issues with Kausik Ray Chaudhuri, Atanu Halder, Mukesh Mittal, and Patrick Chan

over the years.

The study on beam coherence time in Chapter 2.3.2 has been due to the help of Tianyang Bai. Chapter 3.3.5 on phase noise has benefitted immensely from Juergen Cezanne's notes. We thank Juan Bucheli, Wenjun Li, Eman Ismail, Joshua Beto, Ajay Gupta, Jane Luo, Navid Abedini, Georg Hampel, Silas Fong, Nazmul Islam, Hashim Shaik, Michael Di Mare, and Ori Shental for several contributions to Chapter 5. In particular, it is fair to say that Chapter 5 as it stands now would not have been possible without the help of our colleagues at Qualcomm Technologies for which we are grateful. Chapter 4.2.4 on adaptive beam weights and Chapter 6.3 on antenna module placement tradeoffs are joint work with Raghu Challa. Chapter 6.6 has benefited immensely from the assistance and feedback of Kapil Gulati. The book has also benefited from the extensive feedback of Sony Akkarakaran that has helped improve the flow and organizational style.

Finally, the team consisting of Danielle Olivotto, Jill Abasto, Melissa DeVita, Liz Bongolan, Andrew Moore, Paul McAdams, Ji-Hyun Helena Kim, Marissa Graniero, and Valeria Quezada have helped us in navigating the book contract, intellectual property and legal clearance, and security clearance processes seamlessly for which we are extremely grateful. Beyond Qualcomm Technologies, the authors would like to acknowledge the active support, encouragement and feedback from Akbar Sayeed, David Love, Angel Lozano, Sundeep Rangan, Ness Shroff, and Srikrishna Bhashyam over many years. We are also extremely grateful for the kind words of support, encouragement and advise on the book from Kaushik Sengupta, Ted Rappaport, Andrea Goldsmith, Jeff Andrews, Sundeep Rangan, and Gabriel Rebeiz, which has helped us improve it significantly. In particular, we are indebted to Prof. Rebeiz for his multiple suggestions that helped in improving the exposition of this book. We would like to acknowledge the use of map data from OpenStreetMap [5] with the license agreement governing this usage described at `https://www.openstreetmap.org/copyright`. This data has been used for generating Figures 5.15 and 5.17.

Abbreviations

- mmWave - Millimeter wave
- 5G - Fifth Generation
- 6G - Sixth Generation
- 3GPP - Third Generation Partnership Project
- FR2 - Frequency Range 2
- FR4 - Frequency Range 4
- FR5 - Frequency Range 5
- FR3 - Frequency Range 3
- RF - Radio frequency
- ISD - Inter-site distance
- 3G - Third Generation
- 4G - Fourth Generation
- Capex - Capital expenses
- Opex - Operational expenses
- RFIC - Radio frequency integrated circuit
- 5G-NR - Fifth Generation-New Radio
- MIMO - Multiple-input multiple-output
- IEEE - Institute of Electrical and Electronics Engineers
- UE - User equipment
- CPE - Customer premises equipment
- QCT - Qualcomm CDMA Technologies
- 1G - First Generation
- AMPS - Advanced Mobile Phone System
- NMT - Nordic Mobile Telephone
- TACS - Total Access Communications System
- FDMA - Frequency division multiple access
- 2G - Second Generation
- GSM - Global System for Mobile communication
- D-AMPS - Digital Advanced Mobile Phone System
- TDMA - Time division multiple access
- CDMA - Code division multiple access
- SMS - Short Message Service
- MMS - Multimedia Messaging Service
- EDGE - Enhanced Data rates for GSM Evolution
- 3G - Third Generation
- WCDMA - Wideband Code Division Multiple Access
- EV-DO - Evolved Data-Only
- 4G - Fourth Generation
- LTE - Long Term Evolution
- OFDMA - Orthogonal frequency division multiple access

- OFDM - Orthogonal Frequency Division Multiplexing
- NR - New Radio
- MBB - Mobile broadband
- eMBB - Enhanced mobile broadband
- URLLC - Ultra-reliable low latency communications
- mMTC - Massive machine type communications
- SCS - Subcarrier spacing
- FR1 - Frequency Range 1
- TDD - Time division duplexing
- LOS - Line-of-sight
- WLAN - Wireless local area network
- WWAN - Wireless wide area network
- QoS - Quality-of-service
- PA - Power amplifier
- MPE - Maximum permissible exposure
- NLOS - Non-line-of-sight
- SINR - Signal-to-interference-and-noise ratio
- MPCs - Multipath components
- COMP - Coordinated multipoint
- RRC - Radio resource control
- PLE - Path loss exponent
- SNR - Signal-to-noise ratio
- SDMA - Spatial division multiple access
- MU-MIMO - Multi-user multiple-input multiple-output
- SFFD - Single-frequency full-duplex
- BWP - Bandwidth part
- CSI - Channel state information
- QAM - Quadrature amplitude modulation
- SU-MIMO - Single-user multiple-input multiple-output
- ADC - Analog-to-digital converter
- DAC - Digital-to-analog converter
- CIR - Channel impulse response
- AoD - Angle of departure
- AoA - Angle of arrival
- GCS - Global coordinate system
- ULA - Uniform linear array
- UPA - Uniform planar array
- XPR - Cross-polarization discrimination ratio
- LTE-A - Long Term Evolution-Advanced
- 3D/FD - Three-dimensional/Full-dimensional
- ITU - International Telecommunications Union
- WINNER - Wireless World Initiative New Radio
- UMi - Urban Micro
- UMa - Urban Macro

- InH - Indoor Hotspot
- RMa - Rural Macro
- CI - Close-in
- ABG - Alpha beta gamma
- CDF - Cumulative distribution function
- IRR - Infra-red reflective
- sub-THz - sub-Terahertz
- LCD - Liquid crystal display
- METIS - Mobile and wireless communications Enablers for the Twenty-twenty Information Society
- 5GCM - Fifth Generation Channel Modeling
- RoI - Region-of-interest
- RMS - Root-mean squared
- RSS - Received signal strength
- RSRP - Reference signal received power
- CDL - Cluster delay line
- PADP - Power-angular-delay profile
- EIRP - Effective (or equivalent) isotropic radiated power
- MAPL - Maximum allowable path loss
- MCS - Modulation and coding scheme
- DFT - Discrete Fourier transform
- SSB - Synchronization signal block
- PSS - Primary synchronization signal
- SSS - Secondary synchronization signal
- RE - Resource element
- RB - Resource block
- PBCH - Physical broadcast channel
- PDCCH - Physical downlink control channel
- PDSCH - Physical downlink shared channel
- HARQ - Hybrid automatic repeat request
- TRP - Transmission reception point
- WSS - Wide-sense stationary
- US - Uncorrelated scattering
- WSSUS - Wide-sense stationary uncorrelated scattering
- PA - Power amplifier
- VSWR - Voltage standing wave ratio
- TRP - Total radiated power
- PCB - Printed circuit board
- GPS - Global positioning system
- PIFA - Planar inverted-F antenna
- RIS - Reconfigurable intelligent surface
- LNA - Low-noise amplifier
- I/Q - In-phase/Quadrature
- DSP - Digital signal processing/processor

- VGA - Variable gain amplifier
- LO - Local oscillator
- IF - Intermediate frequency
- DC - Direct current
- EVM - Error vector magnitude
- ACLR - Adjacent channel leakage ratio
- PAE - Power added efficiency
- IAB - Integrated access and backhaul
- IM - Intermodulation
- ACPR - Adjacent channel power ratio
- IP_n - Intercept points of order n
- AM-to-AM - Amplitude-to-amplitude
- AM-to-PM - Amplitude-to-phase
- QAM - Quadrature amplitude modulation
- FM - Frequency modulation
- IC - Integrated circuit
- QPSK - Quadrature phase shift keying
- MEMS - Microelectromechanical systems
- PLL - Phase locked loop
- TCXO - Temperature compensated crystal oscillator
- DMRS - Demodulation reference signal
- CPE - Common phase error
- ICI - Inter-carrier interference
- PTRS - Phase tracking reference signal
- ENOB - Effective number of bits
- IF VGA - Intermediate frequency variable gain amplifier
- VCO - Voltage-controlled oscillator
- CDRX - Connected mode discontinuous reception
- RRM - Radio resource management
- DPD - Digital pre-distortion
- DPoD - Digital post-distortion
- FTL - Frequency tracking loop
- SVD - Singular value decomposition
- i.i.d. - Independent and identically distributed
- LMMSE - Linear minimum mean squared error
- AAS - Active antenna systems
- CSI-RS - Channel state information reference signal
- TXRU - Transceiver unit
- CRS - Common reference signal
- RDN - Radio distribution network
- KPI - Key performance indicator
- PPS - Progressive phase shift
- CPO - Constant phase offset
- MUSIC - Multiple signal classification

- ESPRIT - Estimation of signal parameters via rotational invariance techniques
- SAGE - Space-alternating generalized expectation maximization
- MDL - Minimum description length
- MRC - Maximum ratio combining
- BIST - Built-in self test
- EGC - Equal gain combining
- TCI - Transmission configuration indicator
- SRS - Sounding reference signal
- MRT - Maximum ratio transmission
- CFO - Carrier frequency offset
- MSE - Mean squared error
- RS - Reference signal
- CQI - Channel quality indicator
- ML - Machine learning
- WAB - Wireless access and backhaul
- ProSe - Proximity-based services
- DCI - Downlink control information
- MAC-CE - Medium access control channel-control element
- FDDF - Full-duplex decode-and-forward mode
- HDDF - Half-duplex decode-and-forward
- SIR - Signal-to-interference ratio
- TX - Transmitter
- RX - Receiver
- GIS - Geographic information system
- GE - Google earth
- GSV - Google streetview
- CNN - Convolutional neural network
- TP - Traffic point
- ILP - Integer linear programming
- MILP - Mixed integer linear programming
- SLA - Service level agreement
- FWA - Fixed wireless access
- CC - Component carrier
- DU - Distributed unit
- RU - Radio unit
- RACH - Random access channel
- TA - Timing advance
- XR - eXtended Reality
- JCAS - Joint communications and sensing
- AI/ML - Artificial intelligence/machine learning
- CMOS - Complementary metal Oxide semiconductor
- RA - Radio altimeter
- OEM - Original equipment manufacturer
- UX - User experience

- BWP - Bandwidth part
- RAT - Radio access technology
- CU - Central unit
- QoE - Quality-of-experience
- UAV - Unmanned/uncrewed aerial vehicle

1 Introduction

1.1 A BACKGROUND ON 5G

5G stands for the Fifth Generation cellular mobile wireless technology. Before 5G, four generations of cellular technologies have been developed and widely adopted over the past 40 years across different geographies. The first generation (1G) was introduced around 1980 and included the Advanced Mobile Phone System (AMPS), Nordic Mobile Telephone (NMT) and Total Access Communications System (TACS) standards. The 1G system established the basic framework of mobile wireless communications. A mobile terminal can be in operation over a wide service area, which is partitioned into a number of cells, each served by a base station. The mobile terminal is connected to a base station over a radio link and may be handed over to another base station as it moves from one cell to another.

While this basic cellular framework holds for all the subsequent generations, the radio link technology and the enabled services have evolved dramatically. Specifically, 1G employed analog transmissions and frequency division multiple access (FDMA) on the radio link and provided only voice services. The second generation (2G) was introduced around 1990 and included Global System for Mobile communication (GSM), Digital AMPS (D-AMPS) or IS-54/136 and IS-95. The 2G system used digital transmissions on the radio link. While GSM and D-AMPS were based on time division multiple access (TDMA) or the FDMA variant evolved from 1G FDMA, IS-95 introduced code division multiple access (CDMA), a radically different spread spectrum technology. The primary service in 2G was still voice, although some preliminary data services such as Short Message Service (SMS) and Multimedia Messaging Service (MMS) were introduced especially with Enhanced Data rates for GSM evolution (EDGE). The third generation (3G) was introduced around 2000 and included three flavors of CDMA technology: Wideband CDMA (WCDMA), CDMA2000 and Evolved Data-Only (EV-DO). The 3G system witnessed a major service transition from voice to high-speed data.

The fourth generation (4G) was introduced around 2008. Unlike its predecessors where multiple technologies competed in different parts of the world, 4G *converged on a single global technology*—Long Term Evolution (LTE), which was based on another multiple access scheme, orthogonal frequency division multiple access (OFDMA). Using orthogonal frequency division multiplexing (OFDM) transmissions, 4G supported wide channel bandwidths and various multiple-input multiple-output (MIMO) technologies. Mobile broadband (MBB) has been the main focus of 4G where the peak downlink data rate increased in low mobility scenarios from 2 Mbps (megabits per second) in 3G to 1 Gbps (gigabits per second) in 4G, and in high mobility scenarios from 384 kbps (kilobits per second) in 3G to 100 Mbps in 4G.

Built upon the success of the previous generations, 5G continues to be a single global technology, known as new radio (NR), based on OFDMA. The scope

DOI: 10.1201/9781032703756-1

of envisioned services of 5G has significantly expanded from those addressed by 1G through 4G. Specifically, MBB of 4G evolves to enhanced mobile broadband (eMBB), where the peak downlink data rate is increased to as high as 20 Gbps and the peak uplink data rate to 10 Gbps. In addition, 5G supports two new classes of use cases, namely Ultra Reliable Low Latency Communications (URLLC) targeting vertical applications such as industrial automation, intelligent transportation and remote health care, and massive Machine Type Communications (mMTC) for a large number of low-cost, low-power devices that transmit a small amount of traffic sporadically.

The main goal of 5G is to be a unified platform or technology that can meet the vastly different requirements of eMBB, URLLC, and mMTC. To this end, 5G is characterized by a high level of design flexibility. For example, the subcarrier spacing (SCS) of OFDM transmissions can be $15 \cdot 2^i$ kHz in 5G, where $i = 0, \ldots, 6$ with $i = 5$ and 6 added in Rel. 17 of the standardization process. In contrast, it is always 15 kHz in LTE. The flexibility of the SCS leads to a variety of symbol and slot durations with which the network can optimally balance the tradeoff between overhead, latency and robustness to fading.

Another important reason for the design flexibility is that 5G utilizes a wide range of frequency bands. The previous generations of cellular mobile wireless technologies (1G–4G) use frequency bands under 7.125 GHz, somewhat informally called as "sub-7 GHz" frequencies or *Frequency Range 1* (FR1) in 3GPP parlance. Millimeter wave (mmWave) bands are used in cellular wireless for the first time in 5G. The term loosely refers to bands above 24.25 GHz with the first release of 5G-NR specifying bands from 24.25 to 52.6 GHz, also known as Frequency Range 2 (FR2). MmWave bands exhibit very different wireless channel characteristics from sub-7 GHz bands. Moreover, the amount of channel bandwidths in these bands can be very different. RF circuitry and antenna design constraints in these bands are also more stringent than at sub-7 GHz. These differences impose quite distinct constraints and requirements on 5G design. As a result, careful attention in design needs to be paid to handle the diverse frequency bands and bandwidths supported by 5G.

Figure 1.1 depicts the evolution of the cellular technologies from 1G to 5G. While 5G-NR provides a number of salient design features distinct from the previous generations, the focus of this book is on mmWave bands.

1.2 OPPORTUNITIES AND CHALLENGES OF MMWAVE COMMUNICATIONS

The mmWave bands considered in 5G-NR are above 24.25 GHz[1]. Specifically, Table 1.1 lists the operating mmWave bands defined in NR so far, all of which are

[1] Strictly speaking, carrier frequencies below 30 GHz correspond to wavelengths over 10 mm. Thus, it is common to treat the 30–300 GHz range as corresponding to mmWave bands. However, due to practical deployment considerations such as licensed operation over a wider geographic coverage region, 24.25 GHz and above are considered as mmWave bands at 3GPP.

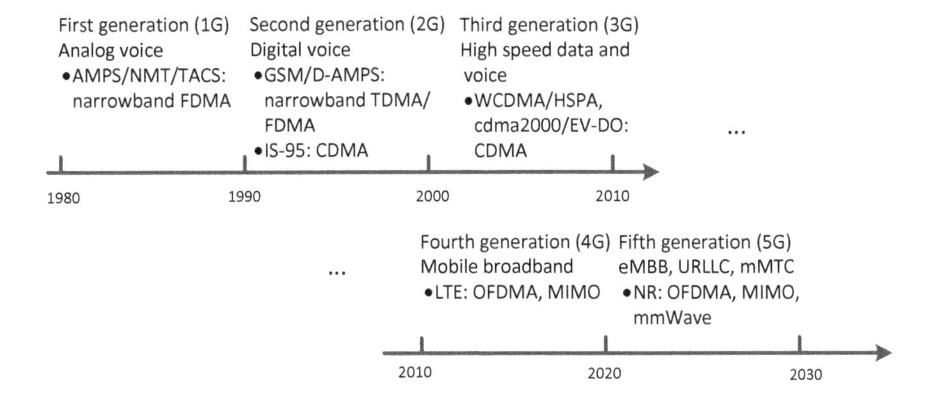

Figure 1.1 Evolution of cellular technologies from 1G to 5G.

unpaired, meaning that the downlink and uplink operate on Time Division Duplexing (TDD). With the advent of Rel. 17 and 18, 3GPP has started addressing upper mmWave bands beyond 52.6 GHz.

The most distinct feature of mmWave bands in NR lies in the vast amount of total bandwidth on the order of a few GHz. In contrast (with only a few exceptions at 2.6, 3.7, 4.7 GHz), a sub-7 GHz 5G band has no more than 100 MHz in total bandwidth. This reflects the fact that the radio spectrum in sub-7 GHz bands for the most part is saturated or crowded for cellular communications. Globally, the total allocation for all 2G, 3G and 4G cellular technologies is less than 800 MHz. Also typically, a cellular operator has no more than 200 MHz in any given geographic market (even less so in highly competitive markets). Moreover, spectrum allocation is often fragmented into disjoint pieces, which makes it difficult to aggregate carriers and achieve very high data rates for a given user. Thus, the newly allocated mmWave bands provide great opportunities to fulfill the demanding needs of very high data rates and

Table 1.1
Operating mmWave bands defined in 5G-NR

Band	Frequency range (GHz)	Total bandwidth (GHz)	Channel bandwidths (MHz)
n257	26.50–29.50	3.00	50, 100, 200, 400
n258	24.25–27.50	3.25	50, 100, 200, 400
n259	39.50–43.50	4.00	50, 100, 200, 400
n260	37.00–40.00	3.00	50, 100, 200, 400
n261	27.50–28.35	0.85	50, 100, 200, 400
n262	47.20–48.20	1.00	50, 100, 200, 400
n263	57.00–71.00	14.00	100, 400, 800, 1600, 2000

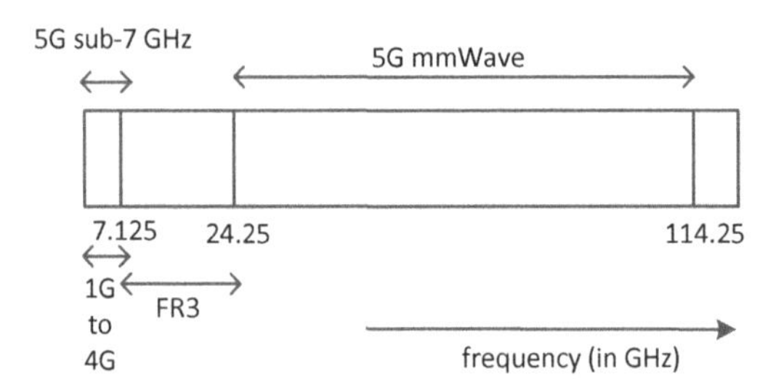

Figure 1.2 Pictorial illustration (not to scale) of available bandwidths in 5G and previous generations.

low latencies of 5G applications. Figure 1.2 pictorially illustrates the extremely large bandwidths available in 5G mmWave bands as compared with the combined allocations in the previous generations.

The use of mmWave bands in cellular systems is itself not new. Proprietary mmWave technologies at 60, 70 and 80 GHz have been used for Gbps backhaul between two base stations in a line-of-sight (LOS) channel condition [6]. The communications range for such systems can be on the order of a few kilometers. High-gain antennas are employed at both ends to form very narrow directional beams. One of the challenges faced by these systems is the misalignment of the narrow beamwidth transmit and receive beams that can occur with vibrations of the antenna mounting structures, for example, due to strong winds.

In contrast with these backhaul applications, the use of mmWave technologies in consumer devices has only emerged recently. The primary reason for this late surge has been due to regulatory policies that opened up this spectrum for commercial licensed use only recently. Beyond this, the major challenges lie in low-cost, low-power hardware implementations—antenna arrays, RFIC chips, as well as high speed baseband to deal with mmWave carrier frequencies and bandwidths up to a few GHz. Advances in hardware implementations have made it possible to commercialize mmWave technologies in the mass market. In particular, wireless local area network (WLAN) has developed the IEEE 802.11ad standard [7], which uses the unlicensed 60 GHz band (specifically, the 57–64 GHz regime) for short-range communications. With a channel bandwidth of 2.16 GHz, the data rate can be as high as 6.76 Gbps. A typical use-case of the IEEE 802.11ad standard is to connect a laptop with a television screen within the same room in an LOS channel condition. The range is about 10–20 meters. The form factor limitation of consumer devices makes it impossible to employ bulky high-gain antennas (e.g., horn antennas) as in wireless backhaul; instead, phased array-based analog or hybrid beamforming transceivers are used to form steerable directional beams. Beam search and tracking are necessary for directional beamforming because the laptop is not fixed, and the indoor channel environment is not stationary.

The key novelty of 5G mmWave communications is that it is designed for *mobile use in wireless wide area networks (WWAN)*. Like the previous generations, 5G services are managed services in the sense that the Quality-of-Service (QoS) has to be managed across the network. In particular, the expectation of coverage and service robustness is higher for 5G than it has been for WLAN. To achieve this goal, a number of challenges need to be addressed:

- The implementation including antenna arrays, RFIC chip and baseband has to fit into the smartphone form factor, yet meeting all the power, thermal and RF exposure requirements. WWAN requires a much longer communication range than WLAN. To meet the link budget requirement, a 5G device needs to transmit higher power, which imposes more stringent demands on power amplifier (PA) capability, power consumption, thermal dissipation and maximum permissible exposure (MPE).
- WWAN is expected to be robust in covering both indoor and outdoor environments. Indoor coverage is more than just coverage within a room and has to include large areas such as office buildings, home environments, shopping malls, stadiums, hospitals, and transportation hubs. Outdoor coverage has to deal with more complex radio channel environments in which a 5G device may be in a non-line-of-sight (NLOS) channel condition relative to a serving base station, as it is shadowed by buildings, trucks, cars, and human bodies. In the outside-to-in or inside-to-out use cases, a device inside a building is served by a base station deployed outdoor, or *vice versa*, and depending on the building materials and their electromagnetic properties, penetration loss can be significant.
- The signal radiation pattern depends on the electromagnetic fields of the device's antenna elements, which are affected by the way a user holds the device. For example, hands/fingers may block the signal from one angle and yet reflect and boost the signal from another angle. A small movement of the hand or body can change the orientation of the device and thus the beam directions. Such beam mobility can happen rapidly in mmWave bands, a phenomenon unseen at sub-7 GHz. Mobility at vehicular speeds, especially in an NLOS environment, makes beam tracking a difficult task.
- The use of directional beams statistically reduces the interference between simultaneous transmissions. However, if interference happens to be directed to the receiver's beam direction, then its detrimental effect can be quite significant. This can happen in a dense deployment where a large number of simultaneous transmissions take place in a small area. Furthermore, because of the form factor limitation, the beam at the device is not very narrow and therefore is susceptible to interference from a wide range of angles of arrival.

1.3 HIGH-LEVEL ASPECTS OF MMWAVE COMMUNICATIONS

The high-level ideas of system solutions to address the above challenges are now introduced. The details, as well as practical physical layer implementation considerations, will be presented in the remaining chapters of the book. The data rate delivered to a device increases with the allocated bandwidth and the signal-to-interference-and-noise ratio (SINR), the ratio between the received power of the desired signal and the total interference and noise power. Let us study the unique features of channel propagation, coverage, interference, and deployment considerations of mmWave communications.

1.3.1 CHANNEL PROPAGATION

Consider an overly simplified scenario of freespace propagation between a transmitter and a receiver. The fundamental relationship between the transmit power P_t and the received power $P_r(d)$ at a distance d is given by the Friis equation (see (2.27) for more details):

$$\frac{P_r(d)}{P_t} = G_t G_r \cdot \left(\frac{\lambda}{4\pi d}\right)^2, \tag{1.1}$$

where λ is the wavelength, and G_t and G_r are the transmit and receive antenna gains, respectively. For a given G_t and G_r, $P_r(d)$ decays as λ decreases, which appears to put the mmWave bands in a disadvantageous position compared to the sub-7 GHz bands. However, rewriting (1.1) in terms of the effective antenna apertures at the transmitter and receiver (A_t and A_r, respectively), we have

$$\frac{P_r(d)}{P_t} = A_t A_r \cdot \left(\frac{1}{\lambda d}\right)^2, \tag{1.2}$$

with more details available in (2.26). From (1.2), we obtain a very different picture. That is, given an effective antenna aperture of A_t and A_r, $P_r(d)$ increases as λ decreases.

As λ decreases, maintaining the effective antenna aperture is equivalent to packing that aperture with multiple antenna elements. Since the inter-antenna element spacing is typically half-wavelength (see more discussion on antenna spacings in Chapters 3.2.2 and 6.2), which reduces as λ decreases, more antenna elements can be packed in the same effective antenna aperture at mmWave bands as at sub-7 GHz frequencies. The increased array gain at either the transmit or receive ends via beamforming over multiple antenna elements can mitigate the freespace propagation losses. When beamforming gains are considered at both ends assuming LOS paths, in fact, mmWave bands show better theoretical performance than sub-7 GHz bands, in terms of received power.

However, the *practical* realizability of these gains is of more importance than the theoretical possibilities. The above analysis applies to LOS paths. If an LOS path is not available and communications have to rely on NLOS propagation, the received signal strength is weaker than that given in (1.1) and relatively speaking, a larger fraction of the power is contributed from reflection than via diffraction and penetration. Significant attenuation can be incurred with blockage by body or hand or even the device itself. On the other hand, NLOS paths offer diversity in urban environments where glass or metallic objects lead to rich multipath scattering. At sub-7 GHz, one can categorize a place as in-coverage or out-of-coverage on the basis of the channel environment, and the orientation of the device is not particularly important. In contrast, the notion of coverage in mmWave is dynamic and spotty, meaning that whether a device is in-coverage (or not) depends heavily on nearby small objects and the orientation of the device at the moment. Further, this can change quickly from one place to another.

In addition to the sensitivity to blockage, another important feature of mmWave channels is their sparsity. The number of multipath components (MPCs) is than what is typically seen in a sub-7 GHz channel because of much fewer significant scatterers and reflectors. The angular spreads are smaller too. As a comparison, the Jakes' model commonly used at sub-7 GHz assumes a uniform distribution of arrival paths in the azimuth plane (360°). This sparse nature of mmWave channels leads to a very different class of MIMO precoding and beamforming strategies that are useful.

1.3.2 COVERAGE

The robustness of signal coverage at mmWave bands can be enhanced by the following approaches.

First, *antenna placement at the device.* To provide good spherical coverage (see Chapter 3.2.3 for more discussions), multiple antenna subarrays are placed at different locations of the device where each subarray covers a part of the sphere. These subarrays can therefore provide diversity against hand blockage and point to base stations in a variety of directions. In practice, antenna placement has to consider form factor and real-estate constraints at the device. Moreover, the added cost of mmWave antenna modules and associated RF front end circuitry (e.g., power amplifiers, low-noise amplifiers, mixers, etc.) and the concomitant power increase put a practical limit on how many antennas can be gainfully employed in the device.

Second, *multi-panel, multi-beam and multi-connectivity.* The base station in a cell is equipped with multiple panels facing different directions and provides coverage over multiple directions simultaneously (or multi-beams). The device can be simultaneously connected with different panels of one base station or with multiple base stations. Coordinated MultiPoint (COMP) transmissions and receptions are supported to realize macro-diversity, interference suppression and spatial multiplexing gains.

Third, *beam management.* This includes a few key components:

- Facilitation of beamformed initial acquisition to bootstrap the connection establishment process
- Beam codebook design that trades off robustness of broad beamwidth beams with increased array gains of narrow beamwidth beams
- Beam search, refinement and tracking procedures to cater to rapid beam mobility
- Beam failure detection and recovery procedures to be error-resilient
- Beam switching procedure that switches between different beams.

The system has to operate efficiently with low overhead, which becomes particularly challenging given a large number of antenna elements and antenna modules.

Fourth, *network densification.* Because of challenging signal propagation, a mmWave cell is naturally a small cell. This is consistent with the trend of network densification [8], [9], an effective way to increase area spectral efficiency or coverage by shrinking cell sizes. In 4G, network densification is achieved by deploying small cells in hotspot areas or places where coverage is known to be problematic. On the other hand, densification can become significant at mmWave frequencies if a very large coverage requirement is imposed. A place may be covered by multiple base stations with large angular separation, not necessarily because of capacity demands but because of robust coverage needs against dynamic shadowing and blockage. To make massive network densification commercially viable, the design goal is to make a majority of base stations low cost with reduced software stack and feature sets (e.g., relays or repeaters), low power (ideally powered by solar panels), and connected via high bandwidth low latency mmWave wireless backhaul.

Fifth, *aggregation of sub-7 GHz and mmWave bands*. When both sub-7 GHz and mmWave bands are available, sub-7 GHz can be used as a fallback option when a mmWave link is temporarily broken because sub-7 GHz is more robust against blockage and shadowing. Alternatively, control signaling[2] can sometimes reside at sub-7 GHz and data traffic can switch between sub-7 GHz and mmWave depending on mmWave link availability. The performance with aggregation depends on whether the sub-7 GHz and mmWave base stations are co-located, and if not, the viability of the backhaul between the two base stations.

1.3.3 INTERFERENCE

The challenging signal propagation of desired signals in mmWave also applies to interference. Fortunately, because the device is usually farther from the interfering base stations than from the serving ones, compared with the useful signal, interference is more likely in an NLOS condition, rather than LOS conditions. This is evident from channel measurements that show that the LOS probability decreases and the path loss exponent (PLE) increases as the distance increases. Another important observation is that because of the use of narrow beamwidth beams in mmWave systems, it is likely that a receiver's beam does not align with an interferer's transmit beam, therefore leading to a high level of interference isolation.

Hence, with an increased likelihood, interference power is not dominant in the case of large inter-site distances and SINR is close to the signal-to-noise ratio (SNR). This is a very different operating regime from 4G sub-7 GHz, where interference is of primary concern and various techniques have been developed to suppress interference. For 5G mmWave, the primary concern is robust coverage instead of interference. However, as inter-site distances become smaller and networks become densified, interference can be a problem even in 5G mmWave.

The above observations hold not only for interference from neighboring cells, but also for interference within the same cell. In the latter case, a base station can use different panels or beams to simultaneously communicate with multiple devices in a

[2]For example, radio resource control (RRC)-based control signaling can be fitted within a dual connectivity framework.

Spatial Division Multiple Access (SDMA) sense without much concern about intra-cell interference, therefore greatly simplifying multi-user MIMO (MU-MIMO).

However, it should be pointed out that in the event that an interferer's transmit beam happens to point toward a receiver's beam, the interference power can be quite strong due to the beamforming effect. While the probability is low, the effect cannot be safely ignored. Moreover, as cell sizes shrink, the number of active devices in a cell drops. The ON/OFF activity cycle and the choice of downlink vs. uplink directions are subject to per-cell random traffic patterns. The "averaging" effect arising from the use of the law of large numbers is less applicable. In effect, the interference distribution is no longer Gaussian, but peaky with heavy tails. This necessitates interference management because traffic patterns and interference beam patterns can be learned *a priori*, and thus if needed, scheduling can be coordinated to deal with peaky interference.

The high level of interference isolation with good beam separation also facilitates the implementation of Single-Frequency Full-Duplex (SFFD) radio. So far, most commercial consumer devices do not transmit and receive at the same time on a single frequency, the so-called half-duplex constraint. To overcome this constraint, an SFFD radio has to deal with self-interference which consists of two contributions:

- Leakage of signal from a transmitter to a co-located receiver
- Impact of clutter from the surrounding environment.

Self-interference suppression is a daunting task because the leakage power is much stronger than the desired power received from a remote transmitter. The techniques studied in the literature are a combination of digital, analog and antenna cancellation. At mmWave bands, a possible solution is to use separate transmit and receive antenna panels. As will be discussed in Chapters 5 and 6, SFFD can be particularly useful when using relay nodes to enhance mmWave coverage.

1.3.4 DEPLOYMENT CONSIDERATIONS

Carrier aggregation has been used in 4G to increase the available bandwidths for transmissions. An interesting use-case in 5G is to aggregate different mmWave bands using the same antenna elements. For example, component carriers or bandwidth parts (BWPs) in the 28 and 39 GHz bands in Table 1.1 can be aggregated assuming that multi-band antenna modules and RFIC chips are made viable for this operation. Another challenge in this domain is to determine the inter-antenna element spacing that robustly works for both bands. The typical rule of thumb of half-wavelength spacings cannot be applied to satisfy operations at both bands. One has to carefully design and manage the beam patterns resulting from non-half wavelength spacings.

One can take complementary approaches to go massive and increase the spatial bandwidth. Network densification with small cells increases the spatial reuse and the total bandwidth in a geographic area. In the interregnum, this small cell approach also creates large cell boundary areas. Inter-cell interference in boundary areas can be significant. One way to deal with inter-cell interference is to employ COMP where neighboring cells cooperate for coordinated or joint signal processing. The performance of COMP depends on the quality of backhaul that connects the cells and the associated channel state information (CSI) exchange.

Alternatively, a massive MIMO approach [10] is to cover the same geographic area with fewer cells and equip the base station in every cell with a massive number of antenna elements and huge processing power. The massive MIMO approach is more attractive from a deployment perspective because of the theoretically fewer number of cells needed for the same performance.

While these two approaches may be comparable at sub-7 GHz, the small cell approach is superior in terms of area throughput and coverage at mmWave. The reason is clear from the previous discussion: mmWave benefits the most when a transmitter is close to a receiver. Further, a dense deployment of small cells provides macro-diversity and inter-cell interference is not dominant. On the other hand, a key challenge in the small cell approach remains to ease massive network deployment. We will discuss these challenges in Chapter 5 where we study practical network deployment lessons.

Figure 1.3 depicts a 5G mmWave deployment scenario that illustrates several features highlighted in the above discussion. In particular, each of the devices A and B are equipped with three antenna subarrays facing three different directions. Each of the base stations 1,2 and 3 are equipped with three panels. Device A is connected with one panel of base station 1 via the LOS and an NLOS path, and one panel of base station 2 with the LOS path. Base station 3 is an interfering base station to device A. However, interference is greatly attenuated. Base station 3 is serving device B

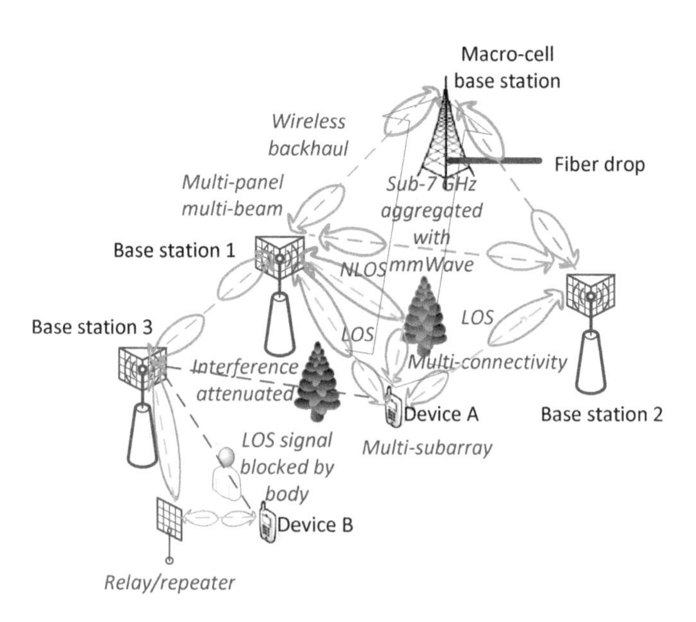

Figure 1.3 Illustration of a 5G mmWave deployment.

Table 1.2
Key differences between sub-7 GHz and mmWave bands

Issue	Sub-7 GHz networks	Millimeter wave networks
Deployment scenarios	Macro cells and small cells with tens of devices per cell	Small cells with many devices
Channel characteristics	Rich multipath propagation. Blockage by small objects is insignificant	Smaller number of paths with small-to-medium angular spreads. Heavily relies on LOS and reflection. Blockage by small objects (e.g., body, hand and finger) can be severe and dynamic
Bandwidth	Small	Large
Objective of bandwidth aggregation	To increase bandwidth by aggregating disjoint frequency pieces	To enhance robustness with sub-7 GHz and mmWave aggregation. To increase bandwidth by aggregating mmWave bands
Higher-order MIMO	Limited to 256-Quadrature Amplitude Modulation (QAM) on downlink. Can go up to 1024-QAM on uplink, but rarely used in practice. Often uses 4 layers in practice	Limited to 256-QAM on downlink. Limited to 64-QAM on uplink. Limited by power consumption to 2 layers in UEs and 4 layers in customer premises equipments (CPEs)
System-level limiting factors	Inter-cell interference and intra-cell interference due to MU-MIMO	Robust coverage. Link budget limited by shadowing and blockage. Interference reduced by directional transmissions and blockage
Use of multiple antennas	For diversity, beamforming and spatial multiplexing with single-user MIMO (SU-MIMO) and MU-MIMO	For directional beamforming. Large arrays needed to overcome increased path loss. Multi-beam for robustness and capacity
Antenna placement	Not critical at device. Maximally spaced to minimize correlation across antennas	Multiple subarrays at device and multi-panels at base station for diversity and spatial multiplexing. Good placements incorporate impact of blockage and typical hand holdings
Spatial densification	Massive MIMO realized by full digital transceivers and large array apertures and used in large cells	Massive MIMO realized by analog or hybrid transceivers and used in *massively* deployed small cells. Low-cost and low-power nodes (e.g., relays or repeaters) connected via mmWave wireless backhaul
Multi-connectivity	COMP for spatial multiplexing and interference suppression	For macro-diversity
Mobility	High user mobility at vehicular speeds	Rapid beam mobility and high user mobility
Additional mmWave physical layer implementation challenges	—	Low PA efficiency, high analog-to-digital and digital-to-analog converter (ADC/DAC) sampling rates, phase noise, MPE constraints

through a relay/repeater because the LOS path is blocked by a human body. The base stations are connected via wireless backhaul to a macro-cell base station, which has a fiber drop. Device *A* is connected with the macro-cell base station in a sub-7 GHz channel. The rest of this book will discuss how these links can be established and sustained over time as environmental disruptions and fading happen. In summary, 5G mmWave communications operates in a very different regime from a conventional sub-7 GHz paradigm, characterized by distinct design constraints and tradeoffs. Table 1.2 summarizes the key differences between mmWave and sub-7 GHz. For further insights, the readers are referred to works such as [11, 12, 13, 14, 15, 16, 17] for a description of the opportunities and challenges of communications at mmWave frequencies.

2 Millimeter Wave Channel Modeling

The scope of this chapter is on the development of the channel model used in performance studies of mmWave systems. We consider the following three aspects in this chapter:

- Relevant modeling framework for 5G mmWave systems that is distinct from classical sub-7 GHz systems
- Large-scale fading component modeling and
- Small-scale fading component modeling.

We begin by extending to the context of multi-antenna wideband systems, the classical single-antenna fading channel model as a superposition of multipath components. We identify different time- and frequency-specific aspects of the developed channel model. Particular attention is also paid to the spatial and polarization domain aspects that are relevant for beamforming design in the subsequent chapters. We then consider the modeling of large-scale components relevant for mmWave systems, including the likelihood of finding an LOS path, path loss modeling for LOS and NLOS paths, and material properties and their impact on propagation at mmWave frequencies. We then focus on a unique mmWave impairment which can lead to significant performance degradation: hand/body blockage. In addition to signal strength loss due to blockage, we also focus on the signal strength enhancement due to the reflection response of the body.

We then consider the modeling of small-scale components such as delay spread, angular spread, and cluster properties and their statistics. The classical version of Doppler spread and coherence duration is generalized to a study on the time-scales at which mmWave signals can be disrupted. Two disruptions are considered: blockage induced loss in signal strength corresponding to physical movements which happen at a slower time-scale, and fading induced changes after beamforming corresponding to beam coherence duration. We finally study how these different mmWave-specific aspects broadly impact and influence physical layer design. We also draw connections between the general modeling framework and the approach adopted by the 3GPP channel modeling document (TR 38.901 [18]) for mmWave carrier frequencies.

2.1 MULTI-ANTENNA FADING

The time-varying channel impulse response (CIR) function between a single-antenna transmitter and a single-antenna receiver can be described as a superposition of contributions from a discrete set of multipath components (MPCs). Each MPC attenuates

DOI: 10.1201/9781032703756-2

and delays the transmitted signal and the CIR at time t and a delay τ is given as:

$$h(\tau,t) = \sum_n \alpha_n(t)\delta(t - \tau_n(t)). \tag{2.1}$$

Here, $\alpha_n(t)$ and $\tau_n(t)$ denote the time-varying amplitude and delay of the n-th MPC and $\delta(\cdot)$ denotes the Dirac delta function. More details on the derivation of (2.1) can be found in [19, 20] and a self-contained derivation is provided in Appendix 2.5. In addition, a background to wireless channel metrics such as delay and Doppler spread and time-frequency partitioning over the delay-Doppler domain are also provided in Appendix 2.5.

We now consider the generalization of the channel modeling framework in (2.1) to a practical scenario with multi-antenna dual-polarized transmission/reception over a wideband setting. For this, we start with the model in (2.1) and assume that the n-th MPC corresponds to contributions from M rays over an angular spread at both the transmitter and receiver ends. The amplitude and the delay of the m-th ray ($m = 1,\ldots,M$) of the n-th MPC ($n = 1,\ldots,N$) are now denoted as $\alpha_{n,m}(t)$ and $\tau_{n,m}(t)$ and this ray corresponds to an angle of departure (AoD) pair (in azimuth and elevation) of $(\phi^{\mathsf{T}}_{n,m}, \theta^{\mathsf{T}}_{n,m})$ and an angle of arrival (AoA) pair (in azimuth and elevation) of $(\phi^{\mathsf{R}}_{n,m}, \theta^{\mathsf{R}}_{n,m})$. Note that the zenith angle θ denotes the angle between the Z axis of a global coordinate system (GCS) and the observation point. On the other hand, the elevation angle is $90^\circ - \theta$ and is the angle between the XY plane of the GCS and the observation point. The azimuth angle ϕ is the angle between the X axis and the observation point in the XY plane. The azimuth and elevation/zenith angles are illustrated in Figure 2.1.

The term *cluster* is often interchangeably used for the MPC in the channel modeling literature with the intention that this MPC corresponds to propagation via a distinct object in the channel environment such as a reflector (e.g., glass window,

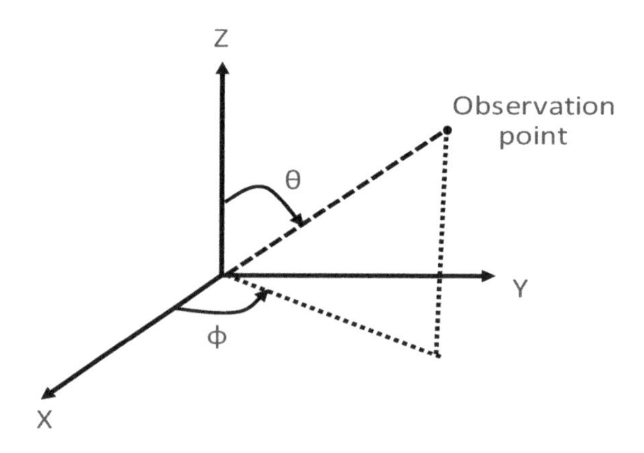

Figure 2.1 Illustration of a global coordinate system, azimuth angle ϕ and zenith angle θ.

metallic object, etc.), or a diffractor (e.g., tree, building corner, etc.), or a diffuse scatterer. With transmission from the s-th transmit antenna (of the available N_t antennas) and reception with the u-th receive antenna (of the available N_r antennas), the CIR $H_{u,s}(\tau,t)$ is given as

$$H_{u,s}(\tau,t) = \sum_{n,m} \alpha_{n,m}(t) \cdot \mathsf{a}_u(\phi^{\mathsf{R}}_{n,m}, \theta^{\mathsf{R}}_{n,m}) \cdot \mathsf{a}^{\star}_s(\phi^{\mathsf{T}}_{n,m}, \theta^{\mathsf{T}}_{n,m}) \cdot \delta(t - \tau_{n,m}(t)) \tag{2.2}$$

where $\mathsf{a}_u(\cdot)$ and $\mathsf{a}_s(\cdot)$ denote the u-th receive and s-th transmit antenna's (complex) gains along the AoA and AoD directions, respectively, and $(\cdot)^{\star}$ denotes complex conjugation operation. With the assumption that the receiver moves along a direction (ϕ_v, θ_v) with a speed of v m/s, we have an additional/explicit Doppler shift term (see [21, p. 49] for details) to lead to

$$\begin{aligned} H_{u,s}(\tau,t) &= \sum_{n,m} \alpha_{n,m}(t) \cdot \mathsf{a}_u(\phi^{\mathsf{R}}_{n,m}, \theta^{\mathsf{R}}_{n,m}) \cdot \mathsf{a}^{\star}_s(\phi^{\mathsf{T}}_{n,m}, \theta^{\mathsf{T}}_{n,m}) \\ &\quad \cdot e^{j2\pi B_d (\overrightarrow{r^{nm}_{\mathsf{R}}} \cdot \overrightarrow{v}) \cdot t} \cdot \delta(t - \tau_{n,m}(t)). \end{aligned} \tag{2.3}$$

In (2.3), $\overrightarrow{r^{nm}_{\mathsf{R}}}$ and $\overrightarrow{v}$ are unit-norm directional vectors

$$\overrightarrow{r^{nm}_{\mathsf{R}}} = \begin{bmatrix} \sin(\theta^{\mathsf{R}}_{n,m})\cos(\phi^{\mathsf{R}}_{n,m}) \\ \sin(\theta^{\mathsf{R}}_{n,m})\sin(\phi^{\mathsf{R}}_{n,m}) \\ \cos(\theta^{\mathsf{R}}_{n,m}) \end{bmatrix}, \quad \overrightarrow{v} = \begin{bmatrix} \sin(\theta_v)\cos(\phi_v) \\ \sin(\theta_v)\sin(\phi_v) \\ \cos(\theta_v) \end{bmatrix} \tag{2.4}$$

with $\overrightarrow{r^{nm}_{\mathsf{R}}} \cdot \overrightarrow{v}$ being the dot/scalar product of the two vectors capturing the Doppler's impact due to the projection of the m-th ray of the n-th cluster's AoA on the receiver's mobility. Further, $B_d = \frac{v}{\lambda}$ is the maximum Doppler frequency that can be seen by the receiver.

It is typical to arrange the multiple antennas at the transmitter and the receiver in a regular array structure with a constant inter-antenna element spacing. A common scenario is one where the transmit/receive antennas are placed uniformly as a linear array with inter-antenna element spacing of d_{T} (or d_{R}), along the X-axis of a GCS. This antenna array structure is commonly called a *uniform linear array* (ULA). In this scenario, the ideal antenna gains are given as

$$\mathsf{a}_s(\phi^{\mathsf{T}}_{n,m}, \theta^{\mathsf{T}}_{n,m}) = \frac{1}{\sqrt{N_t}} \cdot e^{-\frac{j2\pi d_{\mathsf{T}}(s-1)\cdot\sin(\theta^{\mathsf{T}}_{n,m})\cos(\phi^{\mathsf{T}}_{n,m})}{\lambda}}, \quad s = 1,\ldots,N_t \tag{2.5}$$

$$\mathsf{a}_u(\phi^{\mathsf{R}}_{n,m}, \theta^{\mathsf{R}}_{n,m}) = \frac{1}{\sqrt{N_r}} \cdot e^{-\frac{j2\pi d_{\mathsf{R}}(u-1)\cdot\sin(\theta^{\mathsf{R}}_{n,m})\cos(\phi^{\mathsf{R}}_{n,m})}{\lambda}}, \quad u = 1,\ldots,N_r. \tag{2.6}$$

The vector $\left[\mathsf{a}_1(\phi^{\mathsf{T}}_{n,m}, \theta^{\mathsf{T}}_{n,m}), \ldots, \mathsf{a}_{N_t}(\phi^{\mathsf{T}}_{n,m}, \theta^{\mathsf{T}}_{n,m})\right]^T$ is commonly called the *array steering vector* along the direction pair $(\phi^{\mathsf{T}}_{n,m}, \theta^{\mathsf{T}}_{n,m})$. In the scenario where the ULA is placed along the Y or Z axes, the $\sin(\theta)\cos(\phi)$ term in the formulas in (2.5) and (2.6) are replaced with corresponding $\sin(\theta)\sin(\phi)$ and $\cos(\theta)$ terms, respectively. In a more general uniform planar array (UPA) structure with $N_{tx} \times N_{ty}$ antennas placed on the XY-plane where $N_{tx} \cdot N_{ty} = N_t$, the antenna gain of the s-th antenna is given as

$$\begin{aligned} \mathsf{a}_s(\phi^{\mathsf{T}}_{n,m}, \theta^{\mathsf{T}}_{n,m}) &= \frac{1}{\sqrt{N_t}} \cdot e^{-\frac{j2\pi d_{\mathsf{T}} \cdot \left(n_{\mathsf{x}} \cdot \sin(\theta^{\mathsf{T}}_{n,m})\cos(\phi^{\mathsf{T}}_{n,m}) + n_{\mathsf{y}} \cdot \sin(\theta^{\mathsf{T}}_{n,m})\sin(\phi^{\mathsf{T}}_{n,m})\right)}{\lambda}}, \\ & s = 1, \ldots, N_t, \; s - 1 = n_{\mathsf{y}} \cdot N_{tx} + n_{\mathsf{x}}, \\ & n_{\mathsf{x}} = 0, \ldots, N_{tx} - 1, \; n_{\mathsf{y}} = 0, \ldots, N_{ty} - 1. \end{aligned} \tag{2.7}$$

Similar formulas can be written down for UPAs on the XZ- or YZ-planes.

While the formulas in (2.5), (2.6) and (2.7) capture the ideal/theoretical gains, in practice, these gains could be distorted by the material properties of the substrate between the antenna and the ground plane as well as the precise construction of the radiating element. Thus, it is typical to capture the true antenna gains in a practical channel model. Further, while the above formulas correspond to transmission and reception with a single polarization, a practical channel description has to accommodate transmissions and receptions with dual-polarized antennas. See Chapter 3.1.2 for a detailed description of antenna polarization and associated impact on the channel structure.

Such a generalized channel model is given as

$$\begin{aligned} H_{u,s}(\tau, t) &= \sum_{n,m} \alpha_{n,m}(t) \cdot \left[F^{\Theta}_{\mathsf{R},u}(\phi^{\mathsf{R}}_{n,m}, \theta^{\mathsf{R}}_{n,m}) \; F^{\Phi}_{\mathsf{R},u}(\phi^{\mathsf{R}}_{n,m}, \theta^{\mathsf{R}}_{n,m}) \right] \cdot \begin{bmatrix} e^{j\nu^{nm}_{\Theta\Theta}} & \frac{e^{j\nu^{nm}_{\Theta\Phi}}}{\sqrt{\mathsf{XPR}}} \\ \frac{e^{j\nu^{nm}_{\Phi\Theta}}}{\sqrt{\mathsf{XPR}}} & e^{j\nu^{nm}_{\Phi\Phi}} \end{bmatrix} \\ & \cdot \begin{bmatrix} \left(F^{\Theta}_{\mathsf{T},s}(\phi^{\mathsf{T}}_{n,m}, \theta^{\mathsf{T}}_{n,m})\right)^H \\ \left(F^{\Phi}_{\mathsf{T},s}(\phi^{\mathsf{T}}_{n,m}, \theta^{\mathsf{T}}_{n,m})\right)^H \end{bmatrix} \cdot e^{j2\pi B_d(\overrightarrow{r^{nm}_{\mathsf{R}}} \cdot \overrightarrow{v}) \cdot t} \cdot \delta(t - \tau_{n,m}(t)). \end{aligned} \tag{2.8}$$

In the above formula, $(\cdot)^H$ denotes the complex conjugate Hermitian operation, $F^{\Theta}_{\mathsf{R},u}(\cdot,\cdot)$ and $F^{\Phi}_{\mathsf{R},u}(\cdot,\cdot)$, respectively denote the practical antenna gains in the Θ and Φ polarizations of the u-th receive antenna in the AoA direction, whereas $F^{\Theta}_{\mathsf{T},s}(\cdot,\cdot)$ and $F^{\Phi}_{\mathsf{T},s}(\cdot,\cdot)$, respectively denote the gains in the Θ and Φ polarizations of the s-th transmit antenna in the AoD direction.

Note that the dual-polarizations are commonly represented as Vertical (V) and Horizontal (H) polarizations. This classification is dependent on coordinate system and location. To avoid this complexity, it is common to denote the dual polarizations as Θ and Φ, respectively. The 2×2 mixing matrix captures the mixing between the Θ and Φ polarizations at the transmit and receive ends with $\mathsf{XPR} \in [1, \infty)$ standing for the cross-polar discrimination ratio (sometimes also called the cross-polarization ratio) which captures the ability to discriminate between the two polarizations. A larger XPR value says that the transmitted polarization is easily discriminable at the receiver, whereas a smaller XPR value implies that the transmitted polarization is easily confusable as the opposite polarization leading to a mixing of these polarizations. The channel model in (2.8) also assumes that the initial polarization phases $\{\nu_{\Theta\Theta}^{nm}, \nu_{\Theta\Phi}^{nm}, \nu_{\Phi\Theta}^{nm}, \nu_{\Phi\Phi}^{nm}\}$ are cluster/ray dependent allowing us to capture cluster-specific impact on polarized transmissions.

Proceeding as before in the single-antenna fading case (see Appendix 2.5), the frequency domain response of the CIR is given as

$$\begin{aligned} H_{u,s}(f,t) &= \mathscr{F}\left(H_{u,s}(\tau,t)\right) \\ &= \sum_{n,m} \alpha_{n,m}(t) \cdot \left[F_{\mathsf{R},u}^{\Theta}(\phi_{n,m}^{\mathsf{R}}, \theta_{n,m}^{\mathsf{R}}) \; F_{\mathsf{R},u}^{\Phi}(\phi_{n,m}^{\mathsf{R}}, \theta_{n,m}^{\mathsf{R}}) \right] \cdot \begin{bmatrix} e^{j\nu_{\Theta\Theta}^{nm}} & \frac{e^{j\nu_{\Theta\Phi}^{nm}}}{\sqrt{\mathsf{XPR}}} \\ \frac{e^{j\nu_{\Phi\Theta}^{nm}}}{\sqrt{\mathsf{XPR}}} & e^{j\nu_{\Phi\Phi}^{nm}} \end{bmatrix} \\ & \quad \cdot \begin{bmatrix} \left(F_{\mathsf{T},s}^{\Theta}(\phi_{n,m}^{\mathsf{T}}, \theta_{n,m}^{\mathsf{T}})\right)^{H} \\ \left(F_{\mathsf{T},s}^{\Phi}(\phi_{n,m}^{\mathsf{T}}, \theta_{n,m}^{\mathsf{T}})\right)^{H} \end{bmatrix} \cdot e^{j2\pi B_d (\overrightarrow{r_{\mathsf{R}}^{nm}} \cdot \overrightarrow{v}) \cdot t} \cdot e^{-j2\pi f \tau_{n,m}(t)}. \end{aligned} \tag{2.9}$$

In (2.9), $\mathscr{F}(\bullet)$ denotes the Fourier transform operation. We note that the frequency selective impact of the channel is seen with the $e^{-j2\pi f \tau_{n,m}(t)}$ term, whereas the time selective impact of the channel is seen with the $e^{j2\pi B_d (\overrightarrow{r_{\mathsf{R}}^{nm}} \cdot \overrightarrow{v}) \cdot t}$ term.

We now assume that the transmitter and receiver use a multi-carrier scheme such as OFDM with a subcarrier spacing of $\delta = \frac{W}{N_{\mathsf{FFT}}}$ where N_{FFT} denotes the number of subcarriers in the multi-carrier scheme. The CIR over the k-th subcarrier is then given as

$$\begin{aligned} H_{u,s}(k) &\triangleq H_{u,s}(f,t)\Big|_{f=k\delta} \\ &= \sum_{n,m} \alpha_{n,m}(t) \cdot \left[F_{\mathsf{R},u}^{\Theta}(\phi_{n,m}^{\mathsf{R}}, \theta_{n,m}^{\mathsf{R}}) \; F_{\mathsf{R},u}^{\Phi}(\phi_{n,m}^{\mathsf{R}}, \theta_{n,m}^{\mathsf{R}}) \right] \cdot \begin{bmatrix} e^{j\nu_{\Theta\Theta}^{nm}} & \frac{e^{j\nu_{\Theta\Phi}^{nm}}}{\sqrt{\mathsf{XPR}}} \\ \frac{e^{j\nu_{\Phi\Theta}^{nm}}}{\sqrt{\mathsf{XPR}}} & e^{j\nu_{\Phi\Phi}^{nm}} \end{bmatrix} \\ & \quad \cdot \begin{bmatrix} \left(F_{\mathsf{T},s}^{\Theta}(\phi_{n,m}^{\mathsf{T}}, \theta_{n,m}^{\mathsf{T}})\right)^{H} \\ \left(F_{\mathsf{T},s}^{\Phi}(\phi_{n,m}^{\mathsf{T}}, \theta_{n,m}^{\mathsf{T}})\right)^{H} \end{bmatrix} \cdot e^{j2\pi B_d (\overrightarrow{r_{\mathsf{R}}^{nm}} \cdot \overrightarrow{v}) \cdot t} \cdot e^{-j2\pi k\delta \cdot \tau_{n,m}(t)}. \end{aligned} \tag{2.10}$$

In terms of a MIMO channel matrix representation, we have

$$\begin{aligned} H(k) &= \left[\; \cdots \quad H_{u,s}(k) \quad \cdots \; \right] \\ &= \sum_{n,m} \alpha_{n,m}(t) \cdot e^{j2\pi B_d (\overrightarrow{r_{\mathsf{R}}^{nm}} \cdot \overrightarrow{v}) \cdot t} \cdot e^{-j2\pi k\delta \cdot \tau_{n,m}(t)} \\ & \quad \cdot \left[\; F_{\mathsf{R}}^{\Theta}(\phi_{n,m}^{\mathsf{R}}, \theta_{n,m}^{\mathsf{R}}) \quad F_{\mathsf{R}}^{\Phi}(\phi_{n,m}^{\mathsf{R}}, \theta_{n,m}^{\mathsf{R}}) \; \right] \begin{bmatrix} e^{j\nu_{\Theta\Theta}^{nm}} & \frac{e^{j\nu_{\Theta\Phi}^{nm}}}{\sqrt{\mathsf{XPR}}} \\ \frac{e^{j\nu_{\Phi\Theta}^{nm}}}{\sqrt{\mathsf{XPR}}} & e^{j\nu_{\Phi\Phi}^{nm}} \end{bmatrix} \\ & \quad \cdot \begin{bmatrix} \left(F_{\mathsf{T}}^{\Theta}(\phi_{n,m}^{\mathsf{T}}, \theta_{n,m}^{\mathsf{T}})\right)^{H} \\ \left(F_{\mathsf{T}}^{\Phi}(\phi_{n,m}^{\mathsf{T}}, \theta_{n,m}^{\mathsf{T}})\right)^{H} \end{bmatrix} \end{aligned} \tag{2.11}$$

with the generalized array steering vectors given as

$$F_{\mathsf{R}}^{\Theta}(\cdot,\cdot) = \begin{bmatrix} F_{\mathsf{R},1}^{\Theta}(\cdot,\cdot) \\ \vdots \\ F_{\mathsf{R},N_r}^{\Theta}(\cdot,\cdot) \end{bmatrix}, \quad F_{\mathsf{R}}^{\Phi}(\cdot,\cdot) = \begin{bmatrix} F_{\mathsf{R},1}^{\Phi}(\cdot,\cdot) \\ \vdots \\ F_{\mathsf{R},N_r}^{\Phi}(\cdot,\cdot) \end{bmatrix}, \tag{2.12}$$

$$F_{\mathsf{T}}^{\Theta}(\cdot,\cdot) = \begin{bmatrix} F_{\mathsf{T},1}^{\Theta}(\cdot,\cdot) \\ \vdots \\ F_{\mathsf{T},N_t}^{\Theta}(\cdot,\cdot) \end{bmatrix}, \quad F_{\mathsf{T}}^{\Phi}(\cdot,\cdot) = \begin{bmatrix} F_{\mathsf{T},1}^{\Phi}(\cdot,\cdot) \\ \vdots \\ F_{\mathsf{T},N_t}^{\Phi}(\cdot,\cdot) \end{bmatrix}. \tag{2.13}$$

Some important observations can be made from the above representation:

- The time-evolution part of the channel matrix is only in the delays, gains and the Doppler shift term with the other terms in (2.11) remaining time-invariant.
- There are MN impulses in the frequency domain representation of the channel matrix, each at $B_d(\overrightarrow{r_{\mathsf{R}}^{nm}} \cdot \vec{v})$ Hz, which is the Doppler shift of the m-th ray in the n-th cluster.
- In the simpler case of transmission and reception with a single (say, Θ) polarization, $H(k)$ reduces to

$$\begin{aligned} H(k) &= \sum_{n,m} \alpha_{n,m}(t) \cdot e^{j2\pi B_d(\overrightarrow{r_{\mathsf{R}}^{nm}} \cdot \vec{v}) \cdot t} \cdot e^{-j2\pi k\delta \cdot \tau_{n,m}(t)} \\ & \quad \cdot F_{\mathsf{R}}^{\Theta}(\phi_{n,m}^{\mathsf{R}}, \theta_{n,m}^{\mathsf{R}}) \cdot \left(F_{\mathsf{T}}^{\Theta}(\phi_{n,m}^{\mathsf{T}}, \theta_{n,m}^{\mathsf{T}}) \right)^H \cdot e^{j\nu_{\Theta\Theta}^{nm}} \end{aligned} \tag{2.14}$$

  which is sometimes called the Saleh-Valenzuela model [22] and is studied widely in the academic literature. The typical representation of this model is to consider the different rays per MPC, each as an individual path ($\ell = 1, \ldots, L'$ with $L' = MN$) and write

$$H \triangleq H(k) = \sum_{\ell=1}^{L'} \alpha_\ell \cdot u_\ell v_\ell^H \tag{2.15}$$

  where

$$\alpha_\ell = \alpha_{n,m}(t) \cdot e^{j2\pi B_d(\overrightarrow{r_{\mathsf{R}}^{nm}} \cdot \vec{v}) \cdot t} \cdot e^{-j2\pi k\delta \cdot \tau_{n,m}(t)} \tag{2.16}$$

$$u_\ell = F_{\mathsf{R}}^{\Theta}(\phi_{n,m}^{\mathsf{R}}, \theta_{n,m}^{\mathsf{R}}) \text{ and} \tag{2.17}$$

$$v_\ell = F_{\mathsf{T}}^{\Theta}(\phi_{n,m}^{\mathsf{T}}, \theta_{n,m}^{\mathsf{T}}). \tag{2.18}$$

  Here, ℓ is mapped to an appropriate cluster/ray index (n,m).

We can rearrange the last term in (2.10) as $\exp\left(-j\frac{2\pi}{N_{\mathsf{FFT}}} \cdot k \cdot \frac{\tau_{n,m}(t)}{\frac{1}{\delta \cdot N_{\mathsf{FFT}}}}\right)$ and define the discrete delay as

$$\ell_{n,m} = \left\lfloor \frac{\tau_{n,m}(t)}{\frac{1}{\delta \cdot N_{\mathsf{FFT}}}} \right\rfloor. \tag{2.19}$$

With $L'' = \max_{n,m} \ell_{n,m}$ and under the assumption that $L'' \gg 1$ (or, equivalently that $W = \delta \cdot N_{\mathsf{FFT}} \gg 1$), we can approximate $H_{u,s}(k)$ as follows:

$$H_{u,s}(k) \approx \sum_{\ell=0}^{L} e^{-j\frac{2\pi}{N_{\mathsf{FFT}}}\cdot k\ell} \cdot \left[\sum_{n,m} \alpha_{n,m}(t) \cdot \left[F^{\Theta}_{\mathsf{R},u}(\phi^{\mathsf{R}}_{n,m}, \theta^{\mathsf{R}}_{n,m}) \; F^{\Phi}_{\mathsf{R},u}(\phi^{\mathsf{R}}_{n,m}, \theta^{\mathsf{R}}_{n,m}) \right] \right.$$
$$\left. \cdot \begin{bmatrix} e^{j\nu^{nm}_{\Theta\Theta}} & \frac{e^{j\nu^{nm}_{\Theta\Phi}}}{\sqrt{\mathsf{XPR}}} \\ \frac{e^{j\nu^{nm}_{\Phi\Theta}}}{\sqrt{\mathsf{XPR}}} & e^{j\nu^{nm}_{\Phi\Phi}} \end{bmatrix} \cdot \begin{bmatrix} \left(F^{\Theta}_{T,s}\left(\phi^{\mathsf{T}}_{n,m}, \theta^{\mathsf{T}}_{n,m}\right)\right)^H \\ \left(F^{\Phi}_{T,s}\left(\phi^{\mathsf{T}}_{n,m}, \theta^{\mathsf{T}}_{n,m}\right)\right)^H \end{bmatrix} \cdot e^{j2\pi B_d (\overrightarrow{r^{nm}_{\mathsf{R}}} \cdot \overrightarrow{v}) \cdot t} \cdot \delta\left(\ell - \ell_{n,m}\right) \right] \tag{2.20}$$

where $L = \max(L', L'')$. The above approximation gets accurate as L increases and is the typical channel model used in standardization efforts. In particular, the models in (2.10) and (2.20) are used by 3GPP for LTE, LTE-Advanced as well as three-/full-dimensional (3D/FD) beamforming based systems [23, 24] and in 5G-NR [18].

2.2 LARGE-SCALE PARAMETERS OF MMWAVE CHANNELS

2.2.1 PROBABILITY OF AN LOS PATH

The existence of an LOS path between the transmit and receive nodes is preferable for high data rate communications (relative to an NLOS path) since the LOS path encounters a smaller propagation loss through the medium. Thus, it is imperative to understand the likelihood of observing an LOS path, $P_{\mathsf{LOS}}(d)$, at a distance of d m between the transmit and receive nodes. In general, this probability decreases as d increases due to the increased possibility of some obstruction between the transmit and receive nodes.

A common model used for $P_{\mathsf{LOS}}(d)$ is the International Telecommunications Union (ITU) model where

$$P_{\mathsf{LOS}}(d) = \begin{cases} 1 & \text{if } d \leq d_1 \\ \frac{d_1}{d} + \left(1 - \frac{d_1}{d}\right) \cdot e^{-\frac{d-d_1}{d_2}} & \text{if } d > d_1. \end{cases} \tag{2.21}$$

This model can be rearranged as

$$P_{\mathsf{LOS}}(d) = \min\left(\frac{d_1}{d}, 1\right) + \left[1 - \min\left(\frac{d_1}{d}, 1\right)\right] \cdot e^{-\frac{d-d_1}{d_2}} \tag{2.22}$$

where the first part of (2.22) captures the probability of an LOS path when the receive node is in the same main street or thoroughfare that contains the transmit node, and the second part captures the probability of an LOS path when the receive node is not in the main street that contains the transmit node. In the latter case, as d increases, it is assumed that the effect of the intervening buildings that block the signal follows a regular spatial point process and this effect grows homogenously with distance. These two effects are assumed to be dominant in two distinct distance regimes

($d \leq d_1$ for the former effect and $d > d_1$ for the latter effect). The exponentially decaying term with a "distance constant" of d_2 captures the effect of the intervening buildings. Thus, this model can be intuitively viewed as capturing two distinct and independent effects that contribute to the existence or lack of an LOS path with the two effects being:

- The LOS probability getting smaller even on the main street as d increases
- The LOS path getting blocked by buildings beyond the main street.

A simple extension of the model in (2.22) is denoted as the extended ITU model, where

$$P_{\mathrm{LOS}}(d) = \begin{cases} 1 & \text{if } d \leq d_1 \\ e^{-\frac{d-d_1}{d_2}} & \text{if } d_1 < d \leq d_3 \\ \gamma & \text{if } d > d_3 \end{cases} \tag{2.23}$$

designed for appropriate choices of d_1, d_2, d_3 and γ. This model extends the ITU model in (2.22) by assuming that the two independent effects that lead to the ITU model occur independently *and* disjointly over distinct distance ranges (exponentially decaying LOS probability over $d_1 < d \leq d_3$ and main thoroughfare effect over $d > d_3$). Naturally, the intuition from ITU model is lost here due to the separation of coupling between the two effects. Another extension of the ITU model is proposed by the Wireless World Initiative New Radio (WINNER) project, which is given as

$$P_{\mathrm{LOS}}(d) = \begin{cases} 1 & \text{if } d \leq d_1 \\ e^{-\frac{d-d_1}{d_2}} & \text{if } d > d_1 \end{cases} \tag{2.24}$$

for appropriate choices of d_1 and d_2. The WINNER model is a capped version of the extended ITU model where the main thoroughfare effect is completely removed from this model.

With some minor modifications and correction factors, the ITU model is used for capturing LOS probability in Urban Micro (UMi) and Urban Macro (UMa) settings in TR 38.901 [18]. For Indoor Hotspot (InH) and Rural Macro (RMa) settings, the extended ITU and WINNER models are seen to offer better fits, respectively. The precise parameter values used for these models are based on accumulation of simulated and measured data from multiple companies participating in the 3GPP modeling/standard setting process.

2.2.2 PATH LOSS EXPONENTS

Once the existence (or lack thereof) of an LOS path is determined, we are interested in determining the strengths of the paths between the transmit and receive nodes. For this, we consider an ideal (omni-directional and lossless) antenna with a transmit power of P_t. The power $P_r(d)$ received by an antenna in the far field of the transmit antenna over freespace at a distance of d meters is given as

$$P_r(d) = \frac{P_t}{4\pi d^2} \cdot G_t \cdot A_r \tag{2.25}$$

with G_t being the antenna gain of the transmit antenna along the direction of the receive antenna and A_r being the effective aperture of the receive antenna. Note that $\frac{P_t}{4\pi d^2}$ captures the power density seen over the surface area of a sphere of radius d meters. With G_r denoting the antenna gain of the receive antenna along the direction of the transmit antenna, the effective aperture[1] satisfies the relationship:

$$A_r = \frac{\lambda^2}{4\pi} \cdot G_r \tag{2.26}$$

leading to

$$P_r(d) = P_t \cdot \left(\frac{\lambda}{4\pi d} \right)^2 \cdot G_t G_r, \tag{2.27}$$

where λ is the wavelength. This equation is called the Friis transmission equation. With a unit-gain omni-directional antenna at both ends (that is, $G_t = G_r = 1$), the path loss (in dB) incurred over freespace transmissions at a distance of d meters is thus given as

$$\mathsf{PL}(d)\,[\text{in dB}] = 10\log_{10}\left(\frac{P_t}{P_r(d)} \right) = 20\log_{10}\left(\frac{4\pi d}{\lambda} \right). \tag{2.28}$$

In practice, transmissions at mmWave carrier frequencies encounter reflection, diffraction or scattering from material objects in the propagation environment. Stated simply, these are all phenomena that are dependent on λ and its relation to the size or nature of the encountered object. In scattering, energy bounces off in different directions. It occurs when the object is not smooth and its dimensions are smaller than λ. In reflection, incident energy is bounced off in a precise reflected direction. It occurs when the object is smooth and its dimensions are larger than λ. In diffraction, energy bends and spreads around the object. It is typically observed when the object's dimensions are comparable with λ [26]. At mmWave frequencies, the relative smoothness of the objects in the environment implies that reflection and diffraction are common propagation modes. However, the bending of waves over corners in diffraction can lead to significantly impaired link margins relative to reflection [27].

To capture these effects, we consider unobstructed or freespace propagation till a distance of d_0 meters (it is typical to assume a small nominal value for d_0 such as $d_0 = 1$ meter) and propagation that is harsher than freespace beyond d_0 meters corresponding to a path loss exponent, denoted by PLE. That is, the received power $P_r(d)$ at a distance of d meters is given as

$$\begin{aligned} P_r(d) &= \left. \frac{P_r(d_0)}{A_r} \right|_{\mathsf{freespace}} \cdot \int_{\theta,\phi} \frac{d\theta d\phi \sin(\theta)}{4\pi (d/d_0)^{\mathsf{PLE}}} \cdot A_r & (2.29) \\ &= \frac{P_t G_t}{4\pi d_0^2} \cdot \frac{1}{(d/d_0)^{\mathsf{PLE}}} \cdot \frac{\lambda^2 G_r}{4\pi}. & (2.30) \end{aligned}$$

[1] See (3.12) and the discussion therein. Also, see [25, Chapter 2] for a derivation.

This approach is known as the *close-in* (CI) reference model corresponding to a reference distance of d_0 meters [16, 20]. With $G_t = G_r = 1$, the path loss (in dB) is given as

$$\begin{aligned} \mathsf{PL}(d)\,[\text{in dB}] &= 10\log_{10}\left(\frac{P_t}{P_r(d)}\right) && (2.31) \\ &= 20\log_{10}\left(\frac{4\pi d_0}{\lambda}\right) + 10\cdot\mathsf{PLE}\cdot\log_{10}\left(\frac{d}{d_0}\right). && (2.32) \end{aligned}$$

While (2.32) captures the path loss with propagation over a single (point) dominant reflector or diffractor, contributions from substantial obstacles partly or fully blocking the Fresnel zone can lead to randomness around a best PLE fit for measurements. This random component, denoted as X_{CI}, corresponds to the phenomenon known as *shadow fading*. Following central limit theorem assuming a large number of paths that exist between the transmitter and receiver nodes, this component is typically modeled as a zero mean Gaussian with a certain standard deviation parameter σ_{CI} (in the dB domain). In other words, the shadow fading is modeled as a log-normal random variable with $X_{\mathsf{CI}} \sim \mathcal{N}(0, \sigma_{\mathsf{CI}}^2)$ and we have

$$\mathsf{PL}(d) = \mathsf{PL}(d_0) + \mathsf{PLE}\cdot 10\log_{10}(d/d_0) + X_{\mathsf{CI}} \quad [\text{in dB}]. \tag{2.33}$$

An alternate approach forgoes the physical interpretation of reference distance d_0 possible with the CI model, and takes a general mathematical modeling viewpoint. It considers general pre-log parameters for distance and frequency dependence as well as an offset parameter. This approach leads to the Alpha-Beta-Gamma (ABG) model [28] and is given as

$$\mathsf{PL}(d) = 10\alpha\cdot\log_{10}\left(\frac{d}{d_0}\right) + \underbrace{\beta + 10\gamma\cdot\log_{10}\left(\frac{f_c}{1\,\mathsf{GHz}}\right)}_{=\beta'} + X_{\mathsf{ABG}} \;\; [\text{in dB}], \tag{2.34}$$

where α and γ capture how the path loss changes with distance and frequency, respectively, β is an optimized offset parameter and $X_{\mathsf{ABG}} \sim \mathcal{N}(0, \sigma_{\mathsf{ABG}}^2)$ models log-normal shadowing. In scenarios where the ABG model is fitted across a single frequency, β and γ can be combined to lead to a simplified parameter β' as indicated in (2.34). The CI and ABG models trade off explanatory power at the cost of more model parameters. In particular, a better fit can be expected with the ABG model since the two parameter CI framework can be subsumed within the four parameter ABG framework. Whether the increase in number of parameters results in a substantially better model fit is a question of interest in channel modeling and a number of works explore this tradeoff [29, 30, 31].

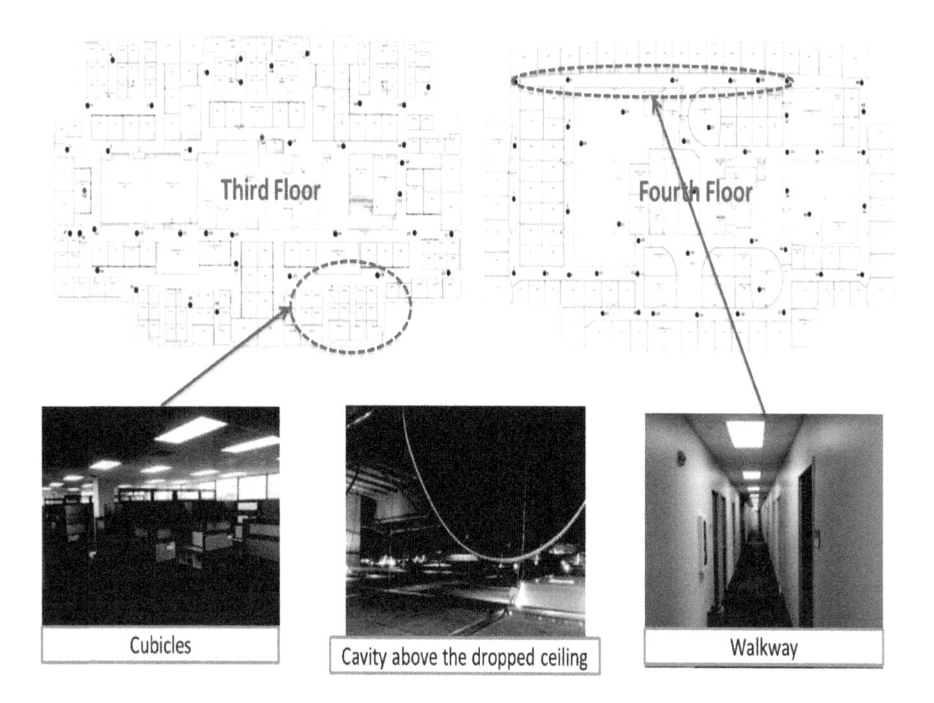

Figure 2.2 Pictorial view of the floor plan and some prominent features from inside the Qualcomm Technologies building.

PLE Estimates via Measurements: To understand path loss exponents and how they compare across different frequencies, *simultaneous*[2] measurements at 2.9, 29 and 61 GHz are performed across different indoor and outdoor environments. A more detailed account of the channel sounder used in these studies, as well as the environment description are available in [32].

In terms of the studied environments, indoor office measurements are made across two floors of the Qualcomm Technologies building in Bridgewater, NJ. These studies encompass two floors of a typical North American indoor office building with cubicles, walled offices, conference rooms and long walkways, as well as a large indoor shopping mall with multiple business locations, long spacious walkways and high ceilings. Figure 2.2 presents a floor plan of the third and fourth fooors of the Qualcomm Technologies building with a pictorial view of the cubicles, walkway and cavity above the dropped ceiling. The first set of outdoor measurements are obtained

[2]Measurements across multiple carriers but at simultaneous locations are important because they offer insights into the different propagation mechanisms at different frequencies by controlling the nature of the propagation environment. A large number of works in channel modeling consider independent studies across different locations and different carriers. While such works offer some insights into propagation, they do not control the environment and hence comparison across frequencies becomes difficult.

Table 2.1
Path loss coefficients for the different measurement settings. f_c is in GHz, $\sigma_{CI}, \sigma_{ABG}, \beta'$ are in dB

Indoor office										
	LOS					NLOS				
f_c	PLE	σ_{CI}	α	β'	σ_{ABG}	PLE	σ_{CI}	α	β'	σ_{ABG}
2.9	1.62	5.49	2.11	35.47	5.28	3.08	6.60	4.36	23.08	5.81
29	1.46	4.25	1.48	61.36	4.25	3.46	8.31	4.96	39.79	7.45
61	1.59	4.81	1.03	75.47	4.50	4.17	13.83	4.23	67.18	13.83
Shopping mall										
	LOS					NLOS				
f_c	PLE	σ_{CI}	α	β'	σ_{ABG}	PLE	σ_{CI}	α	β'	σ_{ABG}
2.9	1.93	5.32	1.74	45.09	5.29	2.61	9.08	2.81	37.61	9.07
29	1.98	3.56	1.62	68.43	3.45	2.76	9.47	2.96	57.57	9.45
61	2.05	4.29	1.90	70.86	4.27	2.98	12.86	2.27	82.05	12.70
UMi, street canyon										
	LOS					NLOS				
f_c	PLE	σ_{CI}	α	β'	σ_{ABG}	PLE	σ_{CI}	α	β'	σ_{ABG}
2.9	2.18	4.41	3.23	18.87	3.35	2.95	7.82	4.32	12.50	7.60
29	2.19	4.37	3.11	42.31	3.47	3.07	8.16	4.40	33.39	7.97
61	2.22	4.84	3.12	48.91	4.19	3.27	10.70	5.18	27.56	10.41
Outdoor, open areas										
	LOS					NLOS				
f_c	PLE	σ_{CI}	α	β'	σ_{ABG}	PLE	σ_{CI}	α	β'	σ_{ABG}
2.9	2.41	4.60	3.03	28.54	4.56	3.01	4.00	5.91	−21.29	3.07
29	2.73	5.73	2.46	67.31	5.72	3.39	8.03	8.70	−53.36	6.53
61	2.83	6.78	5.40	13.38	6.24	3.42	1.97	0.08	137.81	0.83
Outdoor, parking structures										
	LOS					NLOS				
f_c	PLE	σ_{CI}	α	β'	σ_{ABG}	PLE	σ_{CI}	α	β'	σ_{ABG}
2.9	2.82	13.54	0.82	83.95	8.26	3.23	8.54	2.85	49.94	8.44
29	2.94	21.02	−0.49	132.71	9.57	3.44	10.50	2.21	88.41	9.63

in downtown New Brunswick, NJ, corresponding to an UMi-type environment. The second set of outdoor measurements are obtained outside the Qualcomm Technologies building and represent a tree-lined open square-type setting with some parts of a street canyon-type environment. Specific points of interest in this scenario include parking lots with bordering buildings, vegetation which is a mix of pine and spruce trees, and highways and a large shopping mall in close vicinity.

The total received power from omni-directional antenna measurements is used to estimate the path loss model for 2.9, 29 and 61 GHz. For the CI model, with $d_0 = 1$ m, the reference path loss $\mathrm{PL}(d_0)$ is removed from the measurement data to normalize the path loss to 0 dB at d_0 for all the three frequencies thus allowing a direct

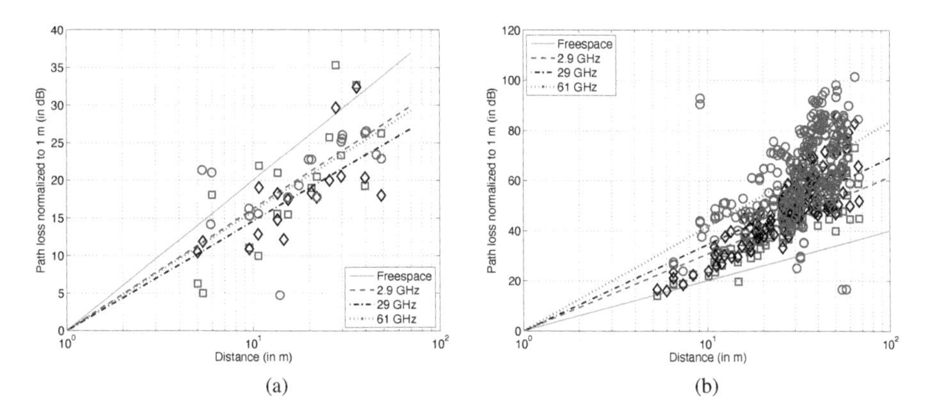

Figure 2.3 Path loss fits with the CI model for (a) LOS and (b) NLOS links in the indoor office setting.

comparison across them. Estimates of the model parameters are obtained through a least-squares fit of the parameters to the measurement data. Table 2.1 presents the parameters for the CI and ABG models in both LOS and NLOS settings for these scenarios. In the indoor office setting, data from both floors are combined together in obtaining a global estimate of PLE and shadowing factors with both models across the building. The best fit PLEs and shadowing factors for NLOS links at 2.9, 29 and 61 GHz are 3.1, 3.5 and 4.2, and 6.6, 8.3 and 13.8 dB, respectively. Figures 2.3(a, b) present the path loss fits with the CI model in the indoor office setting. PLEs for LOS links are considerably lower: 1.6, 1.5 and 1.6 at 2.9, 29 and 61 GHz, respectively. In general, a lower PLE can be expected at 29 GHz (than at 2.9 GHz) due to waveguide effects of long walkways and false ceilings in the indoor office setting and/or changes in material properties at higher frequencies. The β' parameter estimated with the ABG model shows wide variations, as also documented in other works such as [33].

In contrast with the above approach, path loss fits conditioned on locations across a single floor suggest a better fit with a dual-slope model corresponding to a break-point distance of d_{BP} than a single-slope model:

$$\mathsf{PL}(d) - \mathsf{PL}(d_0) = \begin{cases} \mathsf{PLE}_1 \cdot 10\log_{10}(d/d_0) + X_{\mathsf{CI}}^1 \ \text{if } d < d_{\mathsf{BP}} \\ \mathsf{PLE}_2 \cdot 10\log_{10}(d/d_{\mathsf{BP}}) + \mathsf{PLE}_1 \cdot 10\log_{10}(d_{\mathsf{BP}}) + X_{\mathsf{CI}}^2 \\ \qquad\qquad\qquad\qquad\qquad\qquad \text{if } d \geq d_{\mathsf{BP}}. \end{cases} \tag{2.35}$$

For example, at 2.9 GHz, we obtain $d_{\mathsf{BP}} = 11.5$ m, $\mathsf{PLE}_1 = 2.35$, $\mathsf{PLE}_2 = 5.12$, $\sigma_{\mathsf{CI}}^1 = 2.03$ dB, $\sigma_{\mathsf{CI}}^2 = 5.98$ dB leading to a net shadowing factor of 5.68 dB. On the other hand, the single slope model results in $\mathsf{PLE} = 3.13$ and $\sigma_{\mathsf{CI}} = 6.69$ dB. These observations suggest that two distinct modes of communications are likely in indoor settings (long walkways and office rooms in one floor vs. primarily cubicles and conference rooms in another floor): predominantly LOS or reflected paths at $d < d_{\mathsf{BP}}$ and diffracted paths at $d \geq d_{\mathsf{BP}}$, respectively.

Table 2.1 also reports the PLEs and shadow fading parameters for the indoor shopping mall as well as outdoor settings. The main conclusions from these studies are:

- A consistent increase in the PLE in both LOS and NLOS cases in all the scenarios
- While the shadow fading parameters generally increase with frequency, inconsistent trends are occasionally seen at higher carrier frequencies due to waveguide effect (in indoor settings) and radar cross-section effect[3] of certain reflectors (typically in outdoor settings).

From a performance comparison study between the CI and ABG models, in all the settings considered here, we observe that σ_{CI} is comparable with σ_{ABG} provided that there are enough measurements to ensure parameter consistency. Thus, the CI model appears to provide a comparable fit relative to the ABG model with a smaller number of parameters and is hence preferable. Similar conclusions have also been made in [33, 34] from more general parameter stability considerations. Nevertheless, both the CI and ABG models have been used in TR 38.901 for different scenarios.

2.2.3 MATERIAL PROPERTIES AT MMWAVE FREQUENCIES

Given that mmWave signals are mostly LOS or are reflected or diffracted if NLOS, it is of interest in understanding how different inanimate materials (e.g., glass, metal, concrete, sheetrock, wood, etc.) behave in terms of reflection and penetration response. In particular, out-to-in coverage (an important use–case of mmWave systems such as with fixed wireless access and CPEs) is critically dependent on such an understanding of materials in residential and office buildings.

2.2.3.1 Reflection Response

Towards understanding the reflection response, measurements with different materials are performed over the 22–43 GHz range. The antenna is placed at about 1.5–2.5 foot distance (or 0.45–0.76 m) from the tested sample and incidence angles are varied in the studies. Absorber panels are used to contain reflections from the background objects surrounding the test site. A reference curve is obtained by placing a perfect reflecting plate (a 2×2 sq ft or 0.61×0.61 sq meters aluminum plate) and sweeping over the same frequency range to obtain the reflected energy.

Reflection tests are conducted with different materials across a large range of incidence angles and for both parallel and perpendicular polarizations. Figure 2.4(a) illustrates the reflection response with a $5/8$ inch or 15.875 mm sheetrock material over the 22–43 GHz range at a perpendicular polarization with an incidence angle of 18.5°. The main observation here is that periodic notches that are several GHz wide and often with more than 35 dB variations in gain are seen. These variations

[3]Radar cross-section tells us how much more reflected energy is received when compared to reflection from a sphere having a cross-section of 1 sq m, or equivalently how much bigger a sphere is needed to have the same effect.

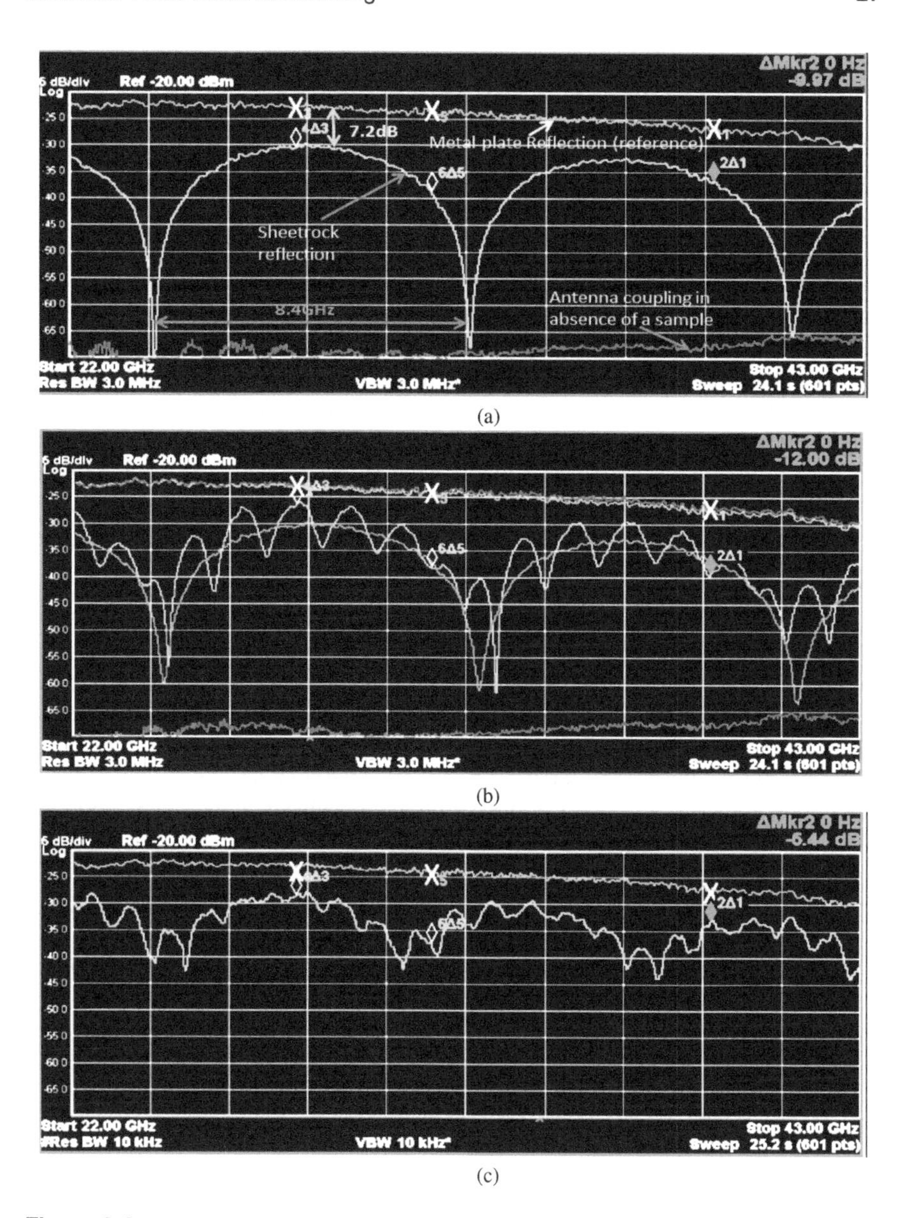

(a)

(b)

(c)

Figure 2.4 Material measurements illustrating reflection response over (a) sheetrock using perpendicular polarization. Response with (b) a structured partition wall and (c) an external wall in the Qualcomm Technologies building.

are attributed to widely changing material properties with frequency due to which signals from different surfaces that make the material constructively or destructively interfere. While a similar trend is observed across these experiments for both polarizations and different choices of incidence angles, the precise response at a frequency and the depth of the notches depend on the material, incidence angle and polarization.

Figure 2.4(b) shows the more realistic response of a structured partition wall with multiple layers of materials (two sheetrock plates separated by a 4 inch or 0.1 m air gap) at 18.5° incidence angle and perpendicular polarization. The superposition of the response from the individual layers leads to periodic patterns across the frequency range (yellow curve), whereas the response of the single sheetrock alone is presented in the green curve. Figure 2.4(c) illustrates the reflection response with a typical external wall material in the Qualcomm Technologies building, which is similar in behavior to Figure 2.4(b). Given the wide notches, these studies motivate the need for system designs that support *both* frequency and spatial diversity.

2.2.3.2 Penetration Loss

From a penetration loss perspective, a broadband sweep over the 22–43 GHz and 50–67 GHz range is done with common residential wall materials. Note that a broadband sweep is necessary to mitigate multi-surface reflections encountered in the wall structure of residential materials. For reference, omni-directional antenna measurements over the 2.5–3.5 GHz range are also obtained.

These studies, illustrated in Figure 2.5, show that the cumulative distribution function (CDF) of penetration loss over three different broadband regimes (2.5–3.5 GHz,

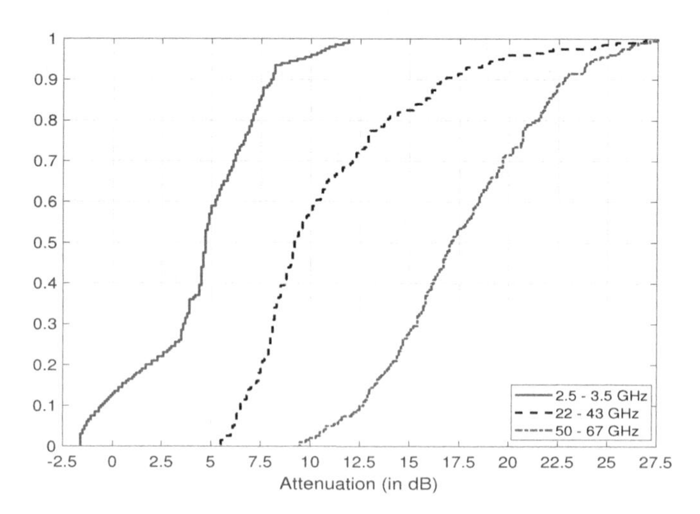

Figure 2.5 CDF of penetration loss/attenuation with exterior walls in typical residential buildings.

22–43 GHz and 50–67 GHz). The median values of loss are 4.7, 9.2 and 17.1 dB, which shows that a 4.5 and 12.4 dB worse median performance over the mmWave bands. These excess losses are associated with sheathing material in walls made of strand boards (wood chips) which often involve the heavy use of glue that has increased attenuation at higher frequencies. For walls made of plywood material, median values of 2.2 and 3.0 dB are seen in the 22–43 and 50–67 GHz regimes, respectively. Since strand board is typically lower cost than plywood, it is likely that newer or urban constructions (in North America and elsewhere) as well as exterior residential walls are more likely to use strand board material than plywood material. Interior residential walls are more likely to be made of plywood material.

Regarding measurement comparisons, TR 38.901 indicates that at 30 GHz, wood and standard multi-pane glass have a small loss of 8 and 8.5 dB, respectively. On the other hand, materials such as infra-red reflective (IRR) glass and concrete have a large loss of 32 and 125 dB, respectively. A "low-loss" model of penetration loss which has a 30–70% mix of glass and concrete has a mean loss of $\approx$ 13.7dB, whereas a "high-loss" model of penetration loss which has a 70–30% mix of IRR glass and concrete has a mean loss of $\approx$ 33.5dB, at 30GHz. These numbers appear to be consistent with the trends reported from the broadband sweep earlier.

2.2.3.3 Additional Losses

In addition to these impairments, various frequency-specific absorptions or attenuations are observed at mmWave carrier frequencies. For example, the 60 GHz regime (53–67 GHz) is well-understood to suffer from oxygen and water vapor-related losses [18]. Similar losses are also seen at sub-Terahertz frequencies (also called as sub-THz) beyond 114.25 GHz where molecular absorption by different materials could dominate the link budget calculations [35]. Other impairments could include attenuation due to rainfall. This loss peaks at 15 dB/km at 60 GHz, which is still only (a relatively small) 1.5 dB of loss at a $d = 100$ m small cell coverage distance, rendering these components minor relative to other material-related impairments.

2.2.4 IMPACT OF HUMAN HAND AND BODY

2.2.4.1 Blockage Loss

In general, objects that are electrically small at microwave frequencies become electrically large at mmWave frequencies, and small objects (which have the size of a few millimeters) located in the proximity of the antennas affect the antenna performance and deteriorate both their efficiencies and their radiation patterns. For example, antennas placed on the display side can be affected by the liquid crystal display (LCD) shielding, LCD glass, component shields, as well as other objects such as camera(s), speaker, microphone, sensors, etc.

Further, blockage by the human hand/body is an important impairment in realizing practical mmWave UEs. Regarding modeling of blockage, TR 38.901 builds on prior work that uses horn antennas (instead of phased arrays) for studying loss models and proposes a flat 30 dB loss over a defined blockage region for the UE in either the portrait or landscape modes. The loss region is in itself modeled using data from studies with a form factor accurate experimental mmWave UE prototype at 28 and 60 GHz. Similar models have also been proposed for IEEE 802.11(ad) systems using ray tracing studies [7], Mobile and wireless communications Enablers for the Twenty-twenty Information Society (METIS) project using a double-knife edge diffraction modeling framework [21], and by the 5G Channel Modeling (5GCM) alliance using horn antenna studies [29]. A number of other works such as [31, 36, 37, 38], etc. have also proposed blockage models.

In this direction, we consider five controlled experiments with a commercial form factor UE operating at 28 GHz. The UE is equipped with a commercial grade mmWave modem (Qualcomm Snapdragon®[4] X50), and antenna module solution and is driven by a beam management software solution that adheres to the 3GPP system level protocol specifications in Rel. 15. The antenna modules incorporated here use a 4×1 dual-polarized patch array and two 2×1 dipole arrays across two polarizations or layers similar to the description in Figure 3.9.

These controlled experiments are performed in an anechoic chamber by studying beam patterns over a sphere with freespace/no blockage and a human holding the phone with the hand and body of the human blocking the signals. By studying the beam patterns over a sphere, the impact of the channel used to establish a beamformed link is removed and we can showcase the true impact of blockage in different directions. The reported studies correspond to different targeted antenna arrays of different dimensions (4×1 patch array vs. 2×1 dipole array), different UE orientations (portrait vs. landscape), and different hand holdings or grips. The grips studied here include a "hard" hand holding grip where the hand completely engulfs all the antenna elements in the array with minimal air gaps between the fingers, a "loose" hand holding grip where only a few fingers engulf some of the antenna elements in the array with the remaining antenna elements seeing unobstructed signals, and an "intermediate" hand holding grip where a few fingers engulf some antenna elements with a big air gap between the palm of the hand and the remaining antenna elements.

Typical beam pattern responses with freespace/no blockage and with blockage in the hard and loose hand holding grip modes are illustrated in Figures 2.6 and 2.7(a, b), respectively. These beam patterns correspond to the use of 4×1 microstrip patch array. From these plots, we observe that the hard hand grip can significantly distort the beam pattern in the targeted directions, while the loose hand grip only moderately distorts the beam pattern. More specifically, depending on the antenna type (dipole vs. patch), array size (4×1 vs. 2×1), type of beam used (scan angle and beamwidth), material property of UE, and the user's hand properties (such as hand grip, hand size, skin properties), etc., we can observe different responses:

- The hand can attenuate signals in a certain set of directions
- The hand can reflect energy in some set of directions and

---

[4]Snapdragon and Qualcomm branded products are products of Qualcomm Technologies, Inc. and/or its subsidiaries.

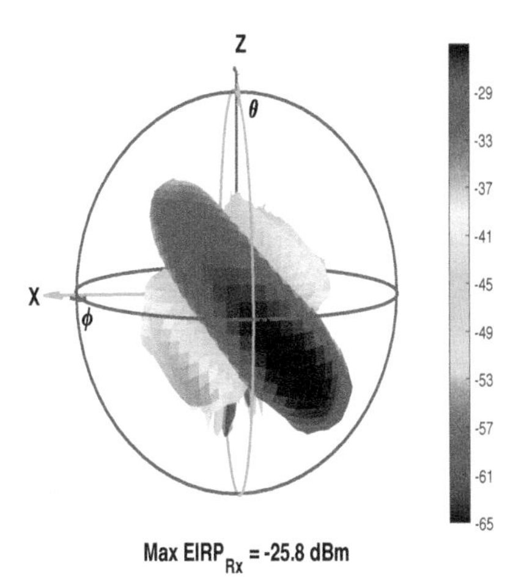

Figure 2.6 Beam patterns of the boresight beam with a 4×1 microstrip patch array in freespace.

- The hand can leave signal energy essentially unchanged in the remaining set of directions.

We can make a gross estimate of blockage losses by comparing the CDFs of the radiated signal power with freespace/no blockage and with blockage. These losses vary from 8.5–17 dB in the hard hand holding grip mode to 3.5–11 dB in the loose hand holding grip mode for the 4×1 subarray. These loss estimates are *significantly* lower than loss estimates in TR 38.901. To capture the impact of hand reflections,

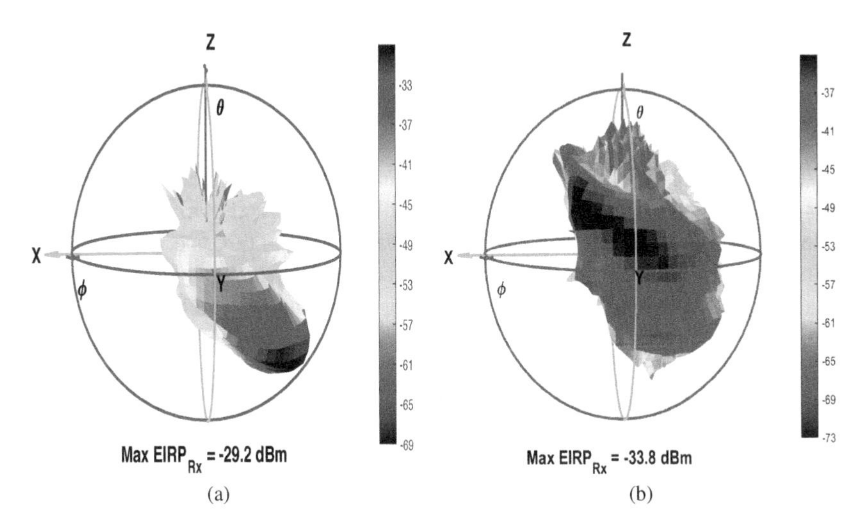

Figure 2.7 Beam patterns of the boresight beam with a 4×1 microstrip patch array in (a) hard hand holding grip and (b) with loose hand holding grip.

Table 2.2
Summary statistics from the five blockage studies

Study	Subarray type	UE orientation	Hand grip	Gross loss estimate (in dB)	Relative RoI improvement (in %)
1	4×1 patch	Portrait	Hard	8.6 to 17.2	2.1% to 4.6%
2	4×1 patch	Portrait	Loose	3.6 to 10.6	6.7% to 8.0%
3	2×1 dipole	Portrait	Hard	15.9 to 19.7	0% to 1.7%
4	2×1 dipole	Portrait	Loose	0.4 to 10.8	8.5% to 12.9%
5	4×1 patch	Landscape	Intermediate	9.5 to 12.7	8.9% to 15.7%

we define a "region of interest" (RoI) where blockage is observed. With the RoI obtained from the freespace/no blockage beam pattern as the benchmark, we then augment it with the region where signal strength in the blockage mode is also above a signal strength threshold. A comparative analysis of these two RoIs shows that hand reflections can improve performance over naïve estimates of blockage loss in certain scenarios. Specifically, gains are observed in scenarios with loose or intermediate hand holding grip, where a few antenna elements are unobstructed, or where a significant air gap can be seen between some fingers and the antenna elements. The summary statistics on blockage performance for these five studies are briefly described in Table 2.2. These summary statistics illustrate the gross loss estimate (in dB) with blockage and how the augmentation in RoI due to hand reflection relatively improves spherical coverage at different effective (sometimes, also called equivalent) isotropic radiated power (EIRP) levels.

2.2.4.2 Reflection Response

While the previous section focused on individual hand or body response, in a practical use-case such as a stadium deployment, we are more interested in the collective response/behavior of multiple human bodies on an individual's effective observation. For this, a controlled experiment is illustrated in Figure 2.8 with two rows of improvised seats where four people are seated in the back row and three people are seated in the front row emulating a stadium setting. The chairs on which people sit are made of metal covered with vinyl cushion. Absorbing panels are placed behind the back row and ground bounces are mitigated by absorbing panels on the floor.

In general, the human body scatters energy on to nearby geographic locations thereby aiding in stadium deployments by providing secondary bounces (alternate paths) for signaling when the LOS path is blocked. To understand this observation, four controlled experiments are performed:

- No persons and no chairs (for baseline/reference)
- One person with no surrounding chairs

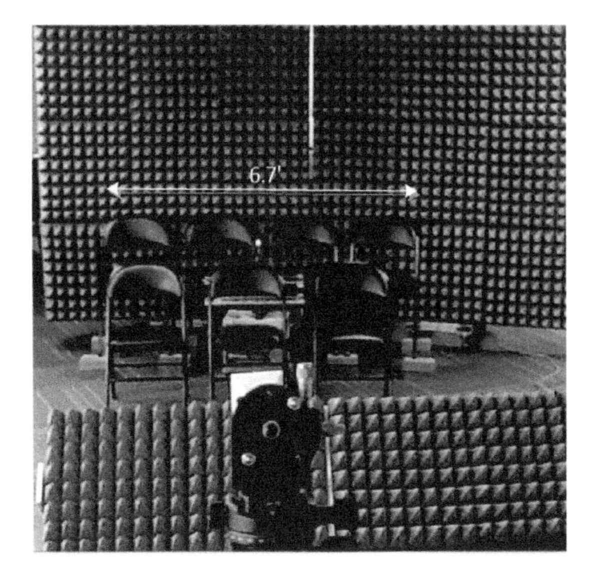

Figure 2.8 Illustration of multiple chairs set up to emulate a stadium setting.

- One person with all the surrounding chairs and
- Seven persons in their respective chairs.

Blockage loss is computed for the three latter scenarios relative to the baseline scenario of no persons and no chairs. Figure 2.9 illustrates the CDF of blockage loss in these three scenarios. In particular, the effect of adjacent chairs reduces the blockage loss significantly from a median of 19.2–14.9 dB, and the presence of nearby humans improves the median further to 10.8 dB. Thus, this study illustrates that these additional reflections and energy accumulation assist with communications in stadium use cases.

2.3 SMALL SCALE PARAMETERS OF MMWAVE CHANNELS

2.3.1 DELAY SPREAD

As noted in Chapter 2.5.1, the excess delay denoted as τ_{excess} and root-mean squared (RMS) delay spread denoted as τ_{rms} with omni-directional scans across different environments are important figures-of-merit in understanding the frequency selectivity, and thus, the optimal subcarrier spacing necessary in an OFDM design. In particular, if τ_i and p_i denote the delay and power (in linear domain) corresponding to the i-th tap in a certain omni-directional scan, the excess delay and the RMS delay spread

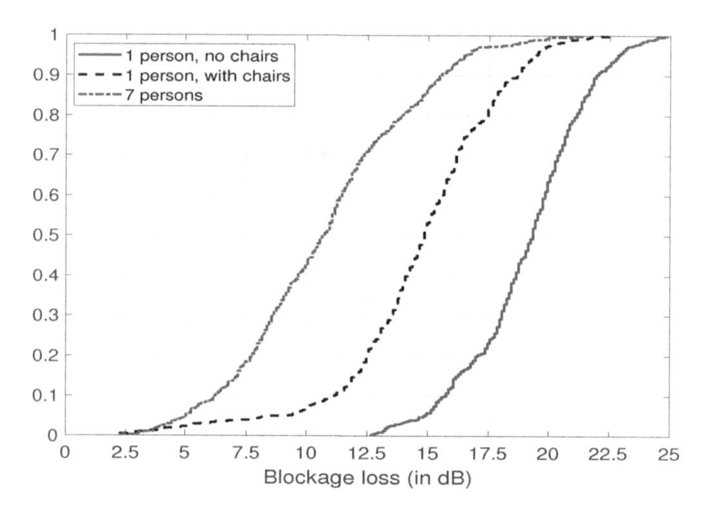

Figure 2.9 CDF of blockage loss in a stadium emulation.

are defined as:

$$\tau_{\text{excess}} = \frac{\sum_i \tau_i p_i}{\sum_i p_i} \tag{2.36}$$

$$\tau_{\text{rms}} = \sqrt{\frac{\sum_i \tau_i^2 p_i}{\sum_i p_i} - \left(\frac{\sum_i \tau_i p_i}{\sum_i p_i}\right)^2}. \tag{2.37}$$

Delay Spread Estimates via Measurements: We now report some measurements and estimates of delay spread. In the indoor office setting in the Qualcomm Technologies building, the longest end-to-end delay is 250 ns. Any delay beyond this value is a result of reflections. For excess delay, an exponential distribution is fitted to the data for each link type (LOS or NLOS) and frequency band. The means of the exponentials at 2.9, 29 and 61 GHz[5] for the post-beamformed excess delay of the combined third and fourth floor measurements with NLOS links are given by $\mu^{-1} = 93.4, 82.3$ and 52.2 ns, respectively. In contrast, a representative value for the RMS delay spread at 1 GHz carrier frequency is 1000 ns [39, p. 34]. This trend is as expected given the difference in propagation characteristics at higher carrier frequencies. The CDF of RMS delay spreads and the parameters associated with an exponential fit for NLOS links are provided in Figure 2.10 and Table 2.3, respectively. The corresponding numbers for the exponential fit to the excess delay at 2.9, 29 and 61 GHz in the LOS case are $\mu^{-1} = 65.8$, 71.9 and 33.3 ns. For LOS links, the mean of the excess delay is actually higher at 29 GHz. The RMS delay spread for LOS links illustrates this difference through a heavier tail at larger delay values.

[5]Data in some measurement settings are missing for 61 GHz.

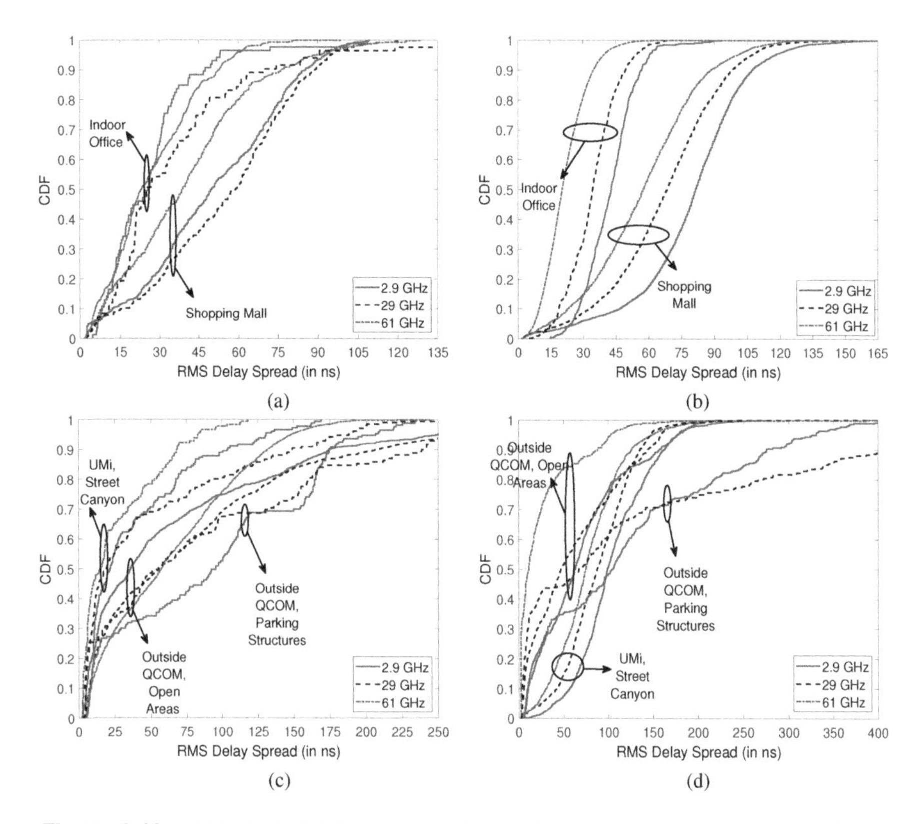

Figure 2.10 CDF of RMS delay spread in (a)–(b) indoor LOS and NLOS and (c)–(d) outdoor LOS and NLOS settings.

The parameters associated with the RMS delay spread across all transmitter and receiver locations with omni-directional antenna scans at 2.9, 29 and 61 GHz in the shopping mall are also presented in Table 2.3 for NLOS and LOS links. As in the indoor office setting, an increase in frequency reduces the RMS delay spread for NLOS links. Similar behavior is seen for the excess delay distributions. Table 2.3 also presents the RMS delay spread parameters for different scenarios in the outdoor case. In general, the delay spread in the outdoor setting is larger than in the indoor setting with some tail values corresponding to strong but significantly delayed subdominant clusters/paths.

The main conclusions from our studies are:

- Delay spread for NLOS links generally decreases with increase in frequency
- Delay spread for LOS links decreases with frequency in dense environments
- Delay spread for LOS links in non-dense environments shows no regular behavior as frequency increases.

Table 2.3
Statistics of RMS delay spread

Metric	Param.	$f_c = 2.9$ GHz		$f_c = 29$ GHz		$f_c = 61$ GHz	
Indoor office		LOS	NLOS	LOS	NLOS	LOS	NLOS
Delay spread (in ns)	Median	25.72	42.89	25.39	34.34	23.10	20.36
$\log_{10}$(Delay spread)	Mean	−7.67	−7.39	−7.56	−7.49	−7.68	−7.72
	Std.	0.28	0.13	0.39	0.17	0.35	0.23
Shopping mall		LOS	NLOS	LOS	NLOS	LOS	NLOS
Delay spread (in ns)	Median	50.0	81.5	59.0	68.5	39.0	57.5
$\log_{10}$(Delay spread)	Mean	−7.40	−7.15	−7.35	−7.23	−7.52	−7.31
	Std.	0.38	0.25	0.35	0.26	0.38	0.27
UMi, Street canyon		LOS	NLOS	LOS	NLOS	LOS	NLOS
Delay spread (in ns)	Median	21.75	99.0	18.75	87.25	14.75	74.5
$\log_{10}$(Delay spread)	Mean	−7.65	−7.02	−7.67	−7.11	−7.85	−7.18
	Std.	0.48	0.20	0.59	0.28	0.51	0.30
Open areas		LOS	NLOS	LOS	NLOS	LOS	NLOS
Delay spread (in ns)	Median	35.5	105.0	55.5	67.0	57.0	11.0
$\log_{10}$(Delay spread)	Mean	−7.45	−7.15	−7.36	−7.36	−7.38	−7.95
	Std.	0.52	0.52	0.54	0.75	0.47	0.57
Parking struct.		LOS	NLOS	LOS	NLOS	LOS	NLOS
Delay spread (in ns)	Median	95.5	62.5	55.0	46.5	–	–
$\log_{10}$(Delay spread)	Mean	−7.26	−7.31	−7.38	−7.44	–	–
	Std.	0.52	0.43	0.62	0.51	–	–

The plausible explanations for these behaviors are:

- *Waveguide effect* where long enclosures such as walkways/corridors, dropped/false ceilings, etc., tend to propagate electromagnetic energy via alternate modes/more reflective paths decreasing the PLE (often even below the freespace PLE of 2.0) and increasing the delay spread as frequency increases.
- *Radar cross-section effect* where seemingly small objects that do not participate in electromagnetic propagation at lower frequencies show up at higher frequencies. Such behavior happens as the wavelength approaches the roughness of surfaces (e.g., walls, light poles, etc.) and is typical when reflection and diffraction dominate propagation.

Since mmWave systems are likely to be used with beamforming, it is of interest in understanding the beamformed delay spread of the channel relative to that with an omni-directional scan. In this context, we note that in general, the beamformed delay spread is smaller than the omni/non-beamformed delay spread. However, for most scenarios of interest in the indoor setting, this reduction is only by a small amount. A simple explanation for this observation is that indoor mmWave channels are sparse with few dominant clusters/paths. From (2.36) and (2.37), we observe that beamformed delay spread is dominated by the delays associated with the dominant clusters/paths. On the other hand, in the outdoor setting, the beamformed delay spread for the tail values can be significantly smaller than the omni-directional delay spread. Thus, the effect of the significantly delayed sub-dominant clusters/paths get mitigated with beamforming.

2.3.2 TIME-SCALES OF MMWAVE LINK DISRUPTION

The time-scales at which a mmWave link can get disrupted are of two types. The first time-scale is one at which a blockage phenomenon (either materials in the environment or the human hand/body) can disrupt an established link. This time-scale is specific to mmWave carrier frequencies and thus requires a careful exploration. The second time-scale is one at which Doppler shifts of rays within a cluster and across clusters can make a link incoherent. This time-scale is similar to the classical coherence time framework at sub-7 GHz frequencies described in Chapter 2.5.3 (a representative value for the coherence time at 1 GHz carrier frequency is 2.5 ms [39, p. 34]). Further, given the focus on beamformed systems at mmWave frequencies, the notion of coherence time should be appropriately extended to the notion of beam coherence time.

2.3.2.1 Time-Scales of Blockage

We first focus on the time-scales at which blockage events disrupt mmWave signals. We assume that an established mmWave link with a steady-state received signal strength (RSS) gets disrupted by a controlled introduction of the hand or body. We define a *link degradation time* as the time required for the RSS of the unblocked link to drop from its steady-state value to its minima in the case of good-to-moderate channel condition, or the time required for the RSS to drop from the steady-state value to a complete link loss in the case of poor channel condition. The link degradation time serves as the worst-case time by which a beam switching or link adaptation procedure must be enabled to ensure that mmWave coverage remains robust, reliable and seamless.

We report on six experiments studying link degradation time here. These experiments are performed with a 28 GHz experimental prototype [40] with directional beamforming at both the base station and user terminal. A beam scanning periodicity/latency of 40 ms is needed for scanning all the beam pairs at both ends thus resulting in the identification of any link degradation/loss to within an accuracy of ± 20 ms. The first two experiments (denoted as Experiments 1–2) concern hand blockage with a medium-to-poor and a good link condition, respectively. Note that these classifications are relative. Experiments 3–6 concern body blockage with a good, good-to-medium, medium and poor link condition, respectively. The precise connections between the six link conditions and the associated SNR values are provided in Table 2.4. Multiple independent tests (ranging from 32–44 tests) are performed in these settings and the link degradation times are recorded.

Table 2.4
Description of link degradation experiments

Experiment	Blockage type	Channel condition	Number of tests
1	Hand	Medium-to-poor (SNR $\approx$ 14.5 dB)	38
2	Hand	Good (SNR $\approx$ 29.5 dB)	34
3	Body	Good (SNR $\approx$ 28.5 dB)	36
4	Body	Good-to-medium (SNR $\approx$ 26 dB)	32
5	Body	Medium (SNR $\approx$ 20 dB)	44
6	Body	Poor (SNR $\approx$ 7.5 dB)	39

Figures 2.11(a, b) capture the CDF of link degradation time across different channel conditions with hand and body blockage, respectively. From these plots, we note that the link degradation times generally decrease as the channel condition deteriorates with similar behaviors across hand and body blockage dynamics. It is important to note that these time-scales are determined by the dynamics of blockage, and in particular, the speed at which the hand grabs the UE and blocks the link, or the speed at which humans walk (or other blockers emerge and depart) to block a link. These time-scales are on the order of a few 100s of ms or slower, depending on the link condition. Thus, from Figure 2.11, it is not entirely surprising to see that the median value of link degradation time being on the order of 240 ms for hand blockage and 200–480 ms for body blockage. Note that even the worst-case link degradation times are better than 100 ms.

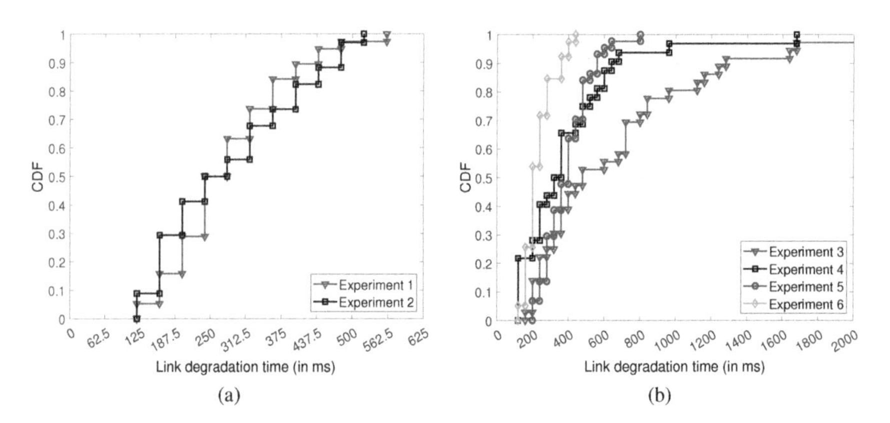

Figure 2.11 CDF of link degradation time for (a) hand blockage and (b) body blockage studies.

2.3.2.2 Beam Coherence Time

We now complement the previous study with another study (albeit done via simulations) on the time-scales at which Doppler impairs beamformed mmWave channels. To understand this, we define the notion of a *cluster stay time*. Given a user's mobility setting (direction of movement and speed), let $P_n(t)$ denote the reference signal received power (RSRP) measured in dB as a function of time t with beamforming along the direction of the n-th cluster in the channel. Let $B(t_0)$ denote the index of the cluster with the best RSRP at time t_0, defined as,

$$B(t_0) \triangleq \arg\max_n P_n(t_0). \tag{2.38}$$

We define the *cluster stay time*, $T_{\rm cst}(X)$, of the channel at time t_0 as the smallest time beyond which the RSRP of beamforming along the strongest cluster $B(t_0)$ falls X dB below the RSRP of the instantaneous best cluster for the first time. That is,

$$T_{\rm cst}(X, t_0) = \arg\min_\tau \left\{ P_{B(t_0)}(t_0+\tau) \leq P_{B(t_0+\tau)}(t_0+\tau) - X \right\}, \tag{2.39}$$

where X is the hysteresis associated with the beam switching process. The statistics of $T_{\rm cst}(X)$ captures the beam dynamics in a beamformed setting due to Doppler spread in the channel. In particular, the smaller the cluster stay time, the more dynamic the channel and consequently, the shorter the required periodicity of reference signals for beam management (and *vice versa*).

We investigate the trends of $T_{\rm cst}(X)$ using the 5G-NR cluster delay line (CDL) model profiles defined in TR 38.901 for typical mmWave deployments via numerical studies. For example, the CDL-A and CDL-B profiles correspond to indoor NLOS and UMi NLOS settings, respectively. Other examples of CDL model profiles include CDL-C, CDL-D and CDL-E settings, which cover other environments or LOS conditions. In addition to the spatial/angular information in azimuth and elevation, relative average channel gain and relative delay spread values (that can be scaled appropriately) are specified for each cluster in the channel. Each cluster is assumed to have 20 rays with equal gain, random phase, angular deviation from the center direction of the cluster (within a defined angular spread) and an intra-cluster delay on top of the cluster-level delay (see [18, Section 7.7.1] for details).

In a mmWave system, when directional beamforming is applied toward the direction of a single cluster, the contributions to delay and Doppler spread from other clusters can be significantly reduced. Even if all the clusters except the beamformed cluster are ignored, signal reception for a moving user can still suffer from a non-negligible Doppler spread due to the angle difference between the rays within a cluster which leads to a time-varying RSRP. Similarly, the fading can be frequency selective even with a single cluster due to the intra-cluster delay spread between rays.

We study the temporal dynamics with CDL-A and CDL-B model profiles assuming a pedestrian speed of $v = 3$ kmph. The CDL-A model has two dominant cluster directions that are separated by ≈ 6 dB in terms of average path gain. The CDL-B model has four dominant cluster directions whose average path gains are within ≈ 3 dB of each other. Further, the azimuth angular spreads of CDL-B are 22° at

Table 2.5
Cluster stay times (in ms) with bandwidths of $W = 50$ and 200 MHz

	W = 50 MHz				W = 200 MHz			
Hysteresis (X in dB)	0	1	3	5	0	1	3	5
Median, CDL-A	37.1	44.9	62.2	82.0	103.2	146.3	294.3	736.1
Median, CDL-B	15.0	19.8	29.0	40.0	18.9	26.8	45.0	70.2
10-th percentile, CDL-A	5.0	7.3	11.3	15.1	9.5	15.0	33.8	66.4
10-th percentile, CDL-B	2.4	4.6	9.1	12.6	2.5	6.9	13.7	20.5

the user/receiver side and 10° at the base station/transmitter side. These numbers are approximately twice as large as those for the CDL-A model. We consider $W = 50$ and 200 MHz bandwidths as illustrations of narrowband and wideband operations, respectively. We also assume 8×4 and 4×2 uniform planar arrays at the base station and user ends, respectively.

Table 2.5 presents the statistics of cluster stay times for the CDL-A and CDL-B models with $W = 50$ and 200 MHz bandwidths corresponding to different hysteresis levels. From this table, we observe that the cluster stay time depends significantly on the propagation environment with consistently smaller cluster stay time values in CDL-B over CDL-A. That is, the richer the mmWave channel, the smaller the cluster stay time, and *vice versa*. A larger number of clusters/rays with comparable gains can lead to a higher level of RSRP dynamics. Further, addition of a hysteresis level in beam switching can substantially increase the cluster stay time. For example, a 3 dB hysteresis can increase the cluster stay time approximately by 1.5 to 3 times for both models (relative to the case of 0 dB hysteresis). Also, leveraging frequency selectivity with a wideband channel can stabilize the RSRP dynamics and increase the cluster stay time significantly. These studies motivate the use of wideband reference signals for beam management in 5G-NR.

2.3.3 NUMBER OF CLUSTERS AND ANGULAR SPREAD

Clustering of multipath [31, 41, 42] is an important channel property that needs to be understood since the dominant clusters/angles capture the modes of propagation and are hence useful for higher-rank beamforming. In order to determine the number of clusters in (2.10), a robust clustering methodology is applied to power-angular-delay profile (PADP) measurements obtained at mmWave frequencies. Critical to the use of PADPs is the notion of an absolute delay measurement relative to the first tap in the channel that requires high-precision time synchronization between the transmitter and the receiver and directional scans obtained typically with horn antennas [43]. In the academic literature, methodologies such as K-means, K-power means, K-moments algorithms, etc. are applied to the PADP to classify the main clusters [44, 45, 46]. Due to the specific construction of the first-generation Qualcomm Technologies sounder, the disambiguation of absolute delays is not possible with our measurements. The sounder only allows a measure of relative delays across angles. This is because multiple angular measurements (over multiple channels) cannot be obtained with our measurement apparatus.

Thus, in lieu of clustering with the PADP, we propose the following clustering methodology based on azimuthal scans at 29 and 61 GHz. We assume that most rays from a cluster lead to similar angle of arrival/departure profiles. This is a reasonable assumption to make in practice since a cluster often embodies a geographically distinct object made of multiple reflecting, diffracting or scattering points on a certain surface and can thus be assumed to produce rays within a certain small (and reasonable) angular spread around a main departure and arrival angle. Leveraging this assumption, we propose to collect the power from all the taps in a certain azimuthal angle spread (for azimuthal scans) and within a certain set of azimuth and elevation angles (for spherical scans) as corresponding to that cluster. Angles that are within a certain appropriately chosen power level P_{cutoff} of the dominant cluster/angle are determined to be dominant clusters capturing the modes of propagation and hence useful for higher-rank beamforming.

The specific choice of P_{cutoff} used in the classification methodology depends on the relevance of a cluster. To understand the scope of this claim, the role of multiple clusters from a mmWave system level perspective is in providing diversity in blockage conditions and in higher-rank signaling. Thus, significant multiplexing gains can be reaped only if the clusters are well-separated directionally (and thus easily disambiguated in signaling) and are of high enough power to result in significant performance improvement. We choose P_{cutoff} values of 5, 7 or 10 dB in our study, but other choices can also be considered.

Cluster Estimates via Measurements: Two transmit locations are used for measurements in the shopping mall. The first transmit location is located on the second floor in a central foyer-type location (and is marked in red in Figure 2.12(a)) with a number of retail outlets with glass windows and long walkways allowing strong reflections and the LOS path to propagate to the different receiver locations. The second transmit location (see Figure 2.12(b)) is placed on the opposite side of the foyer in the third floor with a nearby food court and multiple retail outlets allow strong reflections off the glass and metal enclosures.

Regarding the number of azimuthal scan measurements at distinct locations, we report on 50 unique locations for 29 GHz and 64 unique locations for 61 GHz. Table 2.6 presents the macroscopic cluster statistics such as the mean number of clusters (and the corresponding inter-cluster angular spread) within a P_{cutoff} of 5, 7 or 10 dB of the dominant cluster/angle at the two transmitter locations based on azimuthal scans at 29 and 61 GHz. From this table, we note that (on an average) 4–5 clusters are within a power differential of 5 dB across both transmit locations suggesting diversity for blockage and higher-rank schemes. These observations also provide some evidence for the relevance of CDL-A to CDL-E channel profiles in TR 38.901 for mmWave studies. Also, while the cluster statistics appear to be broadly similar at both frequencies, the mean number of clusters is smaller at 61 GHz relative to 29 GHz. Given that an indoor office setting is similar (in terms of material properties) to a shopping mall setting, the mean number of useful clusters is expected to be similar.

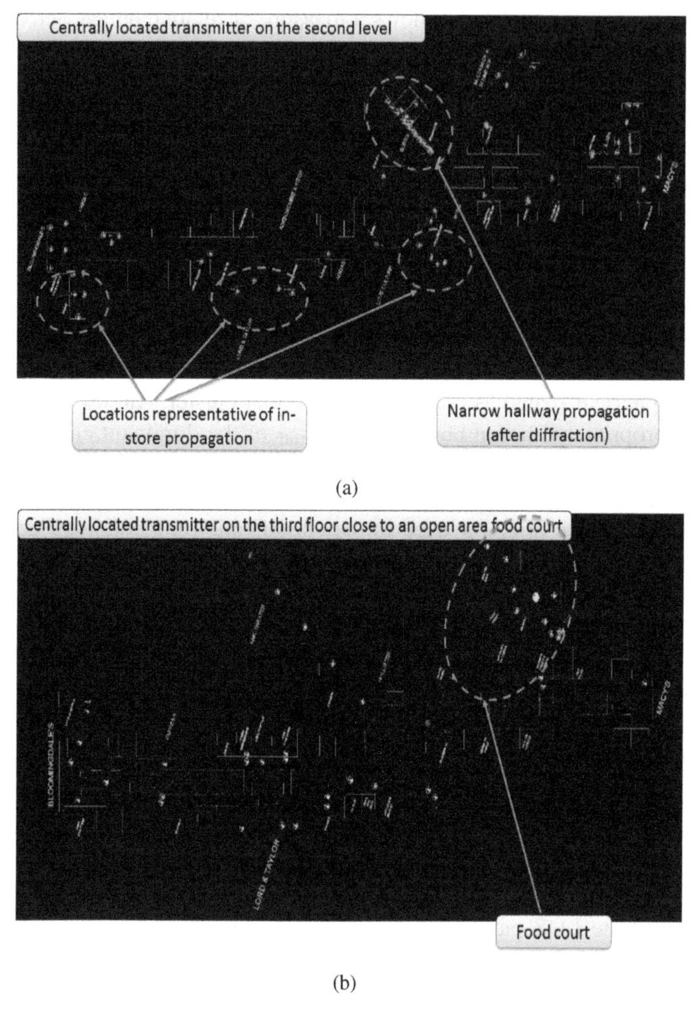

Figure 2.12 Location of (a) transmitter 1 and (b) transmitter 2 (marked in red) relative to different shops in a shopping mall setting.

Figures 2.13(a,b) plot the CDF of the number of clusters as classified by the above clustering algorithm for different choices of P_{cutoff} at 29 and 61 GHz across all transmit and receive locations. Clearly, from these figures, we see that the median number of clusters is less than 5 at both frequencies and at the 80-th percentile level, the number of clusters is approximately 10 and 7, respectively suggesting a moderate number of available directions for diversity at these carrier frequencies. Figures 2.13(c,d) also plot the CDF of the inter-cluster angular spreads or the angle between adjacent clusters at 29 and 61 GHz. From these figures, we note that the median inter-cluster angular spread is on the order of 20° at 29 GHz and 30° at

Table 2.6
Cluster statistics at 29 and 61 GHz

Metric	$f_c = 29$ GHz			$f_c = 61$ GHz		
P_{cutoff}	10 dB	7 dB	5 dB	10 dB	7 dB	5 dB
Mean no. clusters (TX Location 1)	12.6	7.9	5.4	10.5	6.9	4.9
Inter-cluster ang. spread (TX Location 1)	34.1°	36.2°	46.7°	22.1°	25.8°	38.8°
Mean no. clusters (TX Location 2)	13.1	8.0	5.1	10.1	6.1	4.2
Inter-cluster ang. spread (Tx Location 2)	20.5°	29.2°	55.4°	24.9°	27.5°	53.7°
Mean no. clusters (Both TX)	12.7	7.9	5.3	10.4	6.7	4.7
Inter-cluster ang. spread (Both TX)	29.8°	34.0°	49.5°	22.8°	26.3°	42.5°

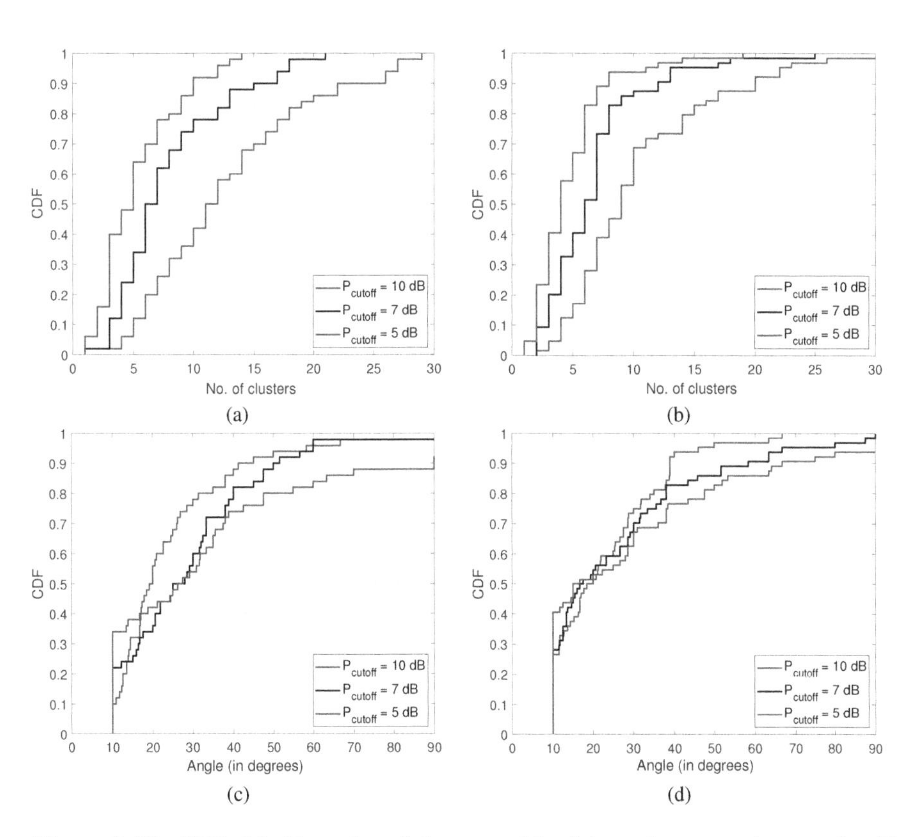

Figure 2.13 CDF of (a, b) number of clusters and (c, d) inter-cluster angular spread at 29 and 61 GHz, respectively.

61 GHz. Such a wide separation allows easy beamforming with low-complexity RF beamformers with minimal loss and low power amplifier backoff due to beam combining.

The readers are also referred to other works such as [45, 47, 48] that provide measurements-based evidence of a moderate number of clusters in a channel at mmWave frequencies.

2.4 IMPLICATIONS OF MMWAVE CHANNEL STRUCTURE ON SYSTEM DESIGN

The EIRP at the transmitter with a multi-antenna array is given as (all loss/gain values are in dB/dBm)

$$\mathsf{EIRP} = P + 10 \cdot \log_{10}(N_t) + P_{\mathsf{elem}} + G_{\mathsf{pol}} + G_{\mathsf{tx}} - \mathsf{Loss}, \tag{2.40}$$

where P is the power from a single transmitter[6] or power amplifier, N_t is the number of antenna elements in the array, P_{elem} denotes the peak elemental gain of each antenna element, G_{pol} captures the polarization-based gains, G_{tx} denotes the transmit side beamforming array gain and Loss captures the losses in the transmit circuitry including feedline losses, antenna matching losses, etc. The factor $10 \cdot \log_{10}(N_t)$ corresponds to an assumption of an independent power amplifier for each antenna element. Note that the total power scales with N_t with this assumption. In a different (extremal) beamforming architecture where a high rating PA is shared across all the antenna elements with a power divider network, this factor reduces to 0 dB.

The noise power over a bandwidth of W Hz is given as

$$\mathsf{Noise\ power} = N_0 + 10 \cdot \log_{10}(W) + \mathsf{NF} \tag{2.41}$$

where N_0 is the noise power spectral density (defined as ≈ -174 dBm/Hz) and NF is the noise figure of the receiver. With this background, the maximum allowable path loss (MAPL) corresponding to a desired/target SNR of $\mathsf{SNR}_{\mathsf{target}}$ is then given as

$$\mathsf{MAPL} = \mathsf{EIRP} - G_{\mathsf{pol}} + G_{\mathsf{rx}} - \mathsf{Noise\ power} - \mathsf{SNR}_{\mathsf{target}} \tag{2.42}$$

with G_{rx} denoting the receive side beamforming gain. The target SNR is a function of the modulation and coding scheme (MCS) used as well as the link conditions (e.g., additive white Gaussian noise channel, fading channel following a certain PADP, etc.).

We use practically motivated representative values for some parameters of interest in the MAPL computation: $P = 7$ dBm, $P_{\mathsf{elem}} = 3$ dBi, $G_{\mathsf{pol}} = 3$ dB, $\mathsf{Loss} = 1$ dB, $\mathsf{NF} = 13$ dB. We also assume that a narrow beamwidth discrete Fourier transform (DFT) beam (see Chapter 4 for how these beams are designed) is steered at both the transmit and receive sides to lead to an array gain of $10 \cdot \log_{10}(N_t)$ and $10 \cdot \log_{10}(N_r)$

[6] In RFIC chip terminology, P is called the power available at the bump assuming the use of a ball grid array-type packaging.

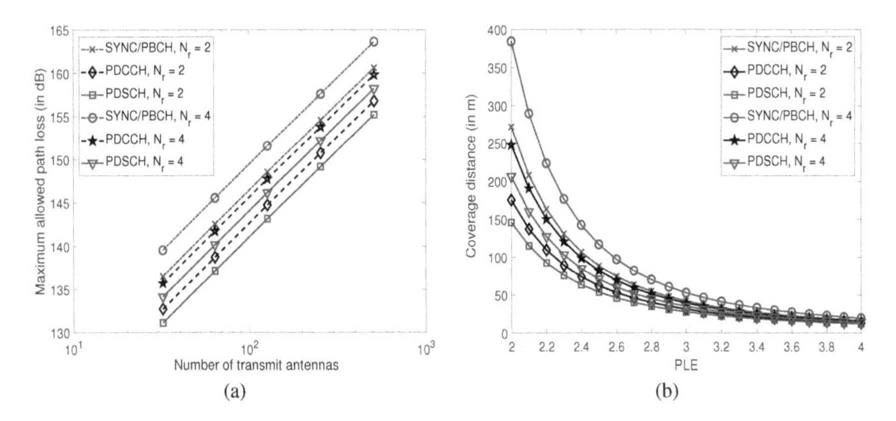

Figure 2.14 (a) MAPL as a function of N_t for different channels and N_r values. (b) Coverage distance for the MAPL in different scenarios.

dB, respectively. With these assumptions, Figure 2.14(a) plots the MAPL as a function of N_t (for the range of 32 to 512 antennas) for two scenarios of $N_r = 2$ and $N_r = 4$. In this study, we also assume that $\mathsf{SNR}_{\mathsf{target}} = -5$ dB, which is a reasonable assumption from a practical deployment perspective for the lowest-order MCS to be used in cell edge operations. We consider three channels from a 5G-NR perspective:

- The synchronization channel over the synchronization signal block (SSB) also known as the "SYNC" channel with a bandwidth of $W = 28.8$ MHz and subcarrier spacing of 120 kHz. The primary and secondary synchronization signals (PSS/SSS) use a sequence length of 127 which is ≈ 11 resource blocks (RBs) whereas the physical broadcast channel (PBCH) uses 20 RBs.
- The physical downlink control channel (PDCCH) is used for exchanging control signaling between the base station and a user with a bandwidth of $W = 69.1$ MHz in typical practical deployments.
- The physical downlink shared channel (PDSCH) is used for data transmissions from the base station to the user and in a typical scenario uses a bandwidth of $W = 100$ MHz.

This plot shows that with practical $N_t = 256$ or $N_t = 128$ element antenna arrays at the base station, we can sustain a path loss over 145 and 140 dB, respectively for the three physical layer channels with $N_r = 2$.

Further, for the MAPL in different scenarios[7], we compute the coverage distance (in meters) for different values of PLE assuming a low-loss material penetration value of 13.7 dB and a blockage loss of 12.75 dB (estimated as the average of the

[7] In practice, the UE can combine SYNC over time to get a few dB of non-coherent combining gains and the PDSCH can also obtain hybrid automatic repeat request (HARQ) gain. We will not simulate these aspects here.

8.5–17 dB spread in the hard hand holding grip). This coverage distance is plotted as a function of PLE in Figure 2.14(b) for $N_t = 32$ and $N_r = 2$ or 4 for the different channels. This fits in well with the theme of network densification (ISD < 500 meters) further explored in Chapter 5. As the PLE increases (LOS to strong NLOS scenario), the coverage distance significantly degrades to a value even less than 50 meters. This suggests the need for an intermediate class of devices that can regeneratively repeat and amplify mmWave signals (see Chapter 5 for more details). Also, a doubling of receive antenna elements leads to an increase in coverage distance of around 41% and 19% for small and large PLEs, respectively. On the other hand, a doubling of transmit antenna elements leads to a doubling of the coverage distance at small PLEs that reduces to $\approx 41\%$ improvement for large PLEs.

We now perform a different computation assuming $N_t = 32$, $N_r = 2$, $W = 100$ MHz and the same set of nominal values for material penetration and blockage losses. Here, we note that the received SNR observed with a path loss of 100, 120 and 140 dB are 23.75, 3.75 and –16.25 dB, respectively. The considered path loss numbers approximate a coverage distance of $d = 100$ m with a PLE of 2.0, 3.0 and 4.0, respectively. This computation shows that low post- and pre-beamforming SNRs are the norm in mmWave systems if path losses seen are inordinately high (from Table 2.1, note that PLEs above 2.5–3 are typically observed in most scenarios of interest). Thus, a viable mmWave system design critically relies on beamforming gains for good performance and an insufficient gain can substantially deteriorate the network coverage and performance. Given the energy and complexity tradeoffs associated with large arrays, typical antenna geometries at the base station are 64×4, 32×8, etc., with 2 to 16 layer transmissions.

To be more specific, from (2.32), we observe that if the PLEs and shadow fading parameters across two frequencies f_1 and f_2 remain the same, then the path losses across these frequencies differ as:

$$\mathsf{PL}(d)\Big|_{f=f_1} - \mathsf{PL}(d)\Big|_{f=f_2} = 20 \cdot \log_{10}\left(\frac{f_1}{f_2}\right). \tag{2.43}$$

Thus, the path loss seen at 29 GHz is 20 dB worse than the path loss seen at 2.9 GHz under the assumption that the PLEs and shadow fading parameters remain the same. On the other hand, more antennas can be deployed at the higher carrier frequency than the lower carrier frequency under the assumption that the same physical aperture is used for an antenna array. This array gain is $10\log_{10}\left(\frac{f_2}{f_1}\right)$ dB at *both* the base station and user ends, thus effectively negating the path loss depreciation. In addition, given that independent power amplifiers are used at the transmit side of this link, increase in EIRP translates to an additional gain of $10\log_{10}\left(\frac{f_2}{f_1}\right)$ dB which suggests that improved performance can be seen at higher carrier frequencies. However, this increased gain has to compensate for increased penetration and blockage losses that dominate propagation at these frequencies effectively leading to reduced ISDs and small cell operations.

At the user end, subarray diversity is critical to overcome near field obstructions such as those due to different parts of the human body that can significantly impair the received signal quality. This is also important to ensure coverage at the user side over the entire angular space or sphere (also called as spherical coverage and studied more carefully in Chapter 3). Due to the smaller λ at 60 GHz (relative to 28 GHz), more subarrays can be packed in the same area and such capabilities should be leveraged for better performance to overcome the higher PLEs observed at 60 GHz (relative to 28 GHz). While a large number of subarrays can be theoretically envisioned in a UE design, cost and complexity considerations suggest the use of 2–4 layers with each layer independently controlling a subarray of 2–8 antenna elements.

Practical beamforming algorithms should simultaneously optimize multiple criteria such as:

- Good beamforming gain
- Less unintended interference
- A link margin-dependent hierarchical solution for beam weight learning allowing a smooth tradeoff between beamforming gain and number of training samples
- Robustness to channel dynamics
- Ability to work with different beamforming architectures
- A simpler network architecture that allows for a broadcast solution in initial UE discovery
- Scalability of the beamforming algorithm from a single-user to a multi-user and multiple transmission reception points (multi-TRP) perspective.

The scope of good beamforming algorithms and their scalability will be studied from a link and system level in Chapters 4 and 5, respectively.

Small cell operations mean that in addition to the likelihood of multiple viable paths to a certain base station, there are also likely to be viable paths to multiple base stations. This observation suggests the criticality of a dense deployment of base stations for robust mmWave operations and inter-base station handover to leverage these paths. One aspect that is critical for such an operation is the efficient and intelligent management of such dense deployments, which is the scope of Chapter 5.

2.5 APPENDIX

We now provide a self-contained background into the fading channel model development in the single antenna case. The readers are referred to a number of textbooks such as [19, 49, 20, 39, 50, 51, 16, 44, 52] for details on many of these aspects.

2.5.1 FADING IN A SINGLE ANTENNA SCENARIO

Let $s_\ell(t)$ be a baseband signal that is frequency modulated to a carrier frequency of f_c (leading to a bandpass signal of $s(t) = \mathscr{R}\left(s_\ell(t)e^{j2\pi f_c t}\right)$ where $\mathscr{R}(\cdot)$ denotes the real part) and transmitted over a single antenna fading channel described by a discrete set of MPCs. The noise-free passband received signal $r(t)$ is a linear combination of delayed and attenuated copies of the transmitted signal and is given as

$$r(t) = \sum_n \alpha_n(t) s(t-\tau_n(t)) \tag{2.44}$$

$$= \sum_n \alpha_n(t) \cdot \mathscr{R}\left(s_\ell(t-\tau_n(t)) e^{j2\pi f_c(t-\tau_n(t))}\right) \tag{2.45}$$

$$= \mathscr{R}\left(\sum_n \alpha_n(t) \cdot e^{-j2\pi f_c \tau_n(t)} \cdot s_\ell(t-\tau_n(t)) e^{j2\pi f_c t}\right) \tag{2.46}$$

with $\alpha_n(t)$ and $\tau_n(t)$ denoting the real amplitude and delay of the n-th MPC. The equivalent baseband received signal

$$r_\ell(t) = s_\ell(t) \star h_\ell(t) \tag{2.47}$$

(where $\star$ denotes the convolution operation) is given as

$$r_\ell(t) = \sum_n \alpha_n(t) \cdot e^{-j2\pi f_c \tau_n(t)} \cdot s_\ell(t-\tau_n(t)). \tag{2.48}$$

From this, the CIR in baseband can be written as

$$h_\ell(\tau,t) \triangleq h_\ell(t) = \sum_n \alpha_n(t) e^{-j2\pi f_c \tau_n(t)} \cdot \delta\left(t-\tau_n(t)\right) \tag{2.49}$$

with the CIR in passband given as

$$h(\tau,t) = \sum_n \alpha_n(t) \delta\left(t-\tau_n(t)\right), \tag{2.50}$$

where $\delta(\cdot)$ denotes the Dirac delta function. Note that the difference between (2.49) and (2.50) is that of CIR in the baseband and passband domains, respectively.

We will now attempt to understand the statistics of the CIR $h_\ell(\tau,t)$ since it is a random process. To simplify our understanding, we will make the reasonable assumption that $h_\ell(\tau,t)$ is a wide-sense stationary (WSS) process. The WSS assumption implies that the first and second moments of $h_\ell(\tau,t)$ are independent of t. If $h_\ell(\tau,t)$ is also a Gaussian, then all of its relevant statistics are captured by the first and second moments alone and are also independent of t. The WSS assumption is reasonable to make for small time-scales over which we are interested in understanding the performance of fading systems.

2.5.2 DELAY SPREAD

We define the correlation function $\Phi_h(\tau_1,\tau_2,\Delta t)$ between the CIR at a delay of τ_1 due to an impulse transmitted at time t and the CIR at a delay of τ_2 due to an impulse transmitted at time $t+\Delta t$:

$$\Phi_h(\tau_1,\tau_2,\Delta t) = \frac{1}{2} \cdot \mathsf{E}\left[h_\ell(\tau_1,t)^\star h_\ell(\tau_2,t+\Delta t)\right]. \tag{2.51}$$

Typically, we expect this correlation to decay as $|\tau_1 - \tau_2|$ increases. In addition to the WSS assumption, we make the further simplifying uncorrelated scattering (US) assumption (together commonly called the WSSUS assumption), wherein any attenuation and phase at a delay of τ_1 is uncorrelated with the attenuation and phase at a *distinct* delay of τ_2 resulting in the simplification:

$$\Phi_h(\tau_1, \tau_2, \Delta t) = \Phi_h(\tau_1, \Delta t) \cdot \delta(\tau_1 - \tau_2). \tag{2.52}$$

The special case of $\Delta t = 0$ leads to

$$\Phi_h(\tau) \triangleq \Phi_h(\tau, 0) = \frac{1}{2} \cdot \mathsf{E}\left[|h_\ell(\tau,t)|^2\right], \tag{2.53}$$

which captures the average power of the channel at a delay of τ and is hence called the *delay power spectrum* (and sometimes also as the *multipath intensity profile*) of the channel. Given $\Phi_h(\tau)$, we can define the *delay spread* T_m as the spread in τ over which $\Phi_h(\tau)$ is "essentially non-zero." The "essential non-zero" content can be precisely quantified as corresponding to the spread capturing either the 90-th, 95-th percentiles or median of $\Phi_h(\tau)$ (depending on the application of interest). Intuitively, the delay spread captures the notion that $h_\ell(\tau,t)$ and $h_\ell(\tau+\delta,t)$ are correlated only for $|\delta| \leq T_m$. For $|\delta| > T_m$, $h_\ell(\tau,t)$ and $h_\ell(\tau+\delta,t)$ are uncorrelated with a high probability.

We now consider the Fourier transform of $h_\ell(\tau,t)$ where the "time" variable is τ:

$$H_\ell(f,t) = \mathscr{F}(h_\ell(\tau,t)) = \int_{-\infty}^{\infty} h_\ell(\tau,t) e^{-j2\pi f \tau} d\tau. \tag{2.54}$$

Since $H_\ell(f,t)$ is WSS when $h_\ell(\tau,t)$ is, we can analogously define the second-order correlation function:

$$\Phi_H(f_1, f_2, \Delta t) = \frac{1}{2} \cdot \mathsf{E}\left[H_\ell(f_1,t)^\star H_\ell(f_2, t+\Delta t)\right]. \tag{2.55}$$

We can simplify the expression in (2.55) as follows:

$$\Phi_H(f_1, f_2, \Delta t)$$

$$= \frac{1}{2} \cdot \mathsf{E}\left[\left[\int_{-\infty}^{\infty} h_\ell(\tau_1,t) e^{-j2\pi f_1 \tau_1} d\tau_1\right]^\star \int_{-\infty}^{\infty} h_\ell(\tau_2, t+\Delta t) e^{-j2\pi f_2 \tau_2} d\tau_2\right] \tag{2.56}$$

$$= \int_{-\infty}^{\infty}\int_{-\infty}^{\infty} \frac{1}{2} \cdot \mathsf{E}\left[h_\ell(\tau_1,t)^\star h_\ell(\tau_2, t+\Delta t)\right] \cdot e^{j2\pi(f_1\tau_1 - f_2\tau_2)} d\tau_1 d\tau_2 \tag{2.57}$$

$$= \int_{-\infty}^{\infty}\int_{-\infty}^{\infty} \Phi_h(\tau_1, \tau_2, \Delta t) \cdot e^{j2\pi(f_1\tau_1 - f_2\tau_2)} d\tau_1 d\tau_2 \tag{2.58}$$

$$= \int_{-\infty}^{\infty}\int_{-\infty}^{\infty} \Phi_h(\tau_1, \Delta t) \cdot \delta(\tau_1 - \tau_2) \cdot e^{j2\pi(f_1\tau_1 - f_2\tau_2)} d\tau_1 d\tau_2 \tag{2.59}$$

$$= \int_{-\infty}^{\infty} \Phi_h(\tau_1, \Delta t) \cdot e^{-j2\pi(f_2 - f_1)\tau_1} d\tau_1 \tag{2.60}$$

where (2.57) follows from rewriting the product of integrals as a double integral and changing the order of expectation and integral, and (2.59) follows from the US assumption. Thus, under the WSSUS assumption $\Phi_H(f_1, f_2, \Delta t)$ only depends on $\Delta f = f_2 - f_1$ and hence, can be denoted as $\Phi_H(\Delta f, \Delta t)$. Further, $\Phi_H(\Delta f, \Delta t)$ is the Fourier transform of $\Phi_h(\tau, \Delta t)$ over the τ variable. An important subtlety to note here is that Δf is the frequency counterpart[8] of delay τ and not the time difference Δt (the frequency counterpart of Δt is the Doppler frequency ν), which we will focus on next.

The function $\Phi_H(\Delta f, \Delta t)$ is called the *spaced-frequency spaced-time correlation function* of the channel. In the special case of $\Delta t = 0$, this function denoted for simplicity as

$$\Phi_H(\Delta f) \triangleq \Phi_H(\Delta f, 0) \tag{2.61}$$

is called the *spaced-frequency correlation function*. $\Phi_H(\Delta f)$ is the Fourier transform of the delay power spectrum $\Phi_h(\tau)$. Due to the inverse relationship between the essential non-zero content of a signal and its Fourier transform, $\Phi_H(\Delta f)$ is essentially non-zero over a frequency spread of $1/T_m$. This non-zero frequency spread $1/T_m$ is called the *coherence bandwidth* of the channel and is denoted as $(\Delta f)_c$. Often, $(\Delta f)_c$ is represented as κ/T_m for some constant κ corresponding to some cutoff point (e.g., 90-th percentile).

2.5.3 DOPPLER SPREAD

We now consider time variations in the channel corresponding to the Δt domain. For this, we define the Fourier transform of $\Phi_H(\Delta f, \Delta t)$ over the variable Δt as

$$S_H(\Delta f, \nu) = \int_{-\infty}^{\infty} \Phi_H(\Delta f, \Delta t) e^{-j2\pi\nu\Delta t} d\Delta t. \tag{2.62}$$

In the above equation, ν stands for the Doppler frequency. In the special case of $\Delta f = 0$, we write

$$S_H(\nu) \triangleq S_H(0, \nu), \tag{2.63}$$

and we call it the *Doppler power spectrum* of the channel, which captures the signal intensity as a function of ν. Consider

$$\Phi_H(0, \Delta t) = \frac{1}{2} \cdot \mathsf{E}\left[|H_\ell(f,t)^\star H_\ell(f, t+\Delta t)| \right], \tag{2.64}$$

which is called the *spaced-time correlation* function. Note that $\Phi_H(0, \Delta t)$ captures the correlation in the channel in the time domain. Thus, a slowly changing channel

---

[8] The connection $\tau \longleftrightarrow \Delta f$ is not surprising since variations in the τ domain lead to distortions in the frequency response of the channel attributed to the $e^{-j2\pi f_c \tau_n(t)}$ factor.

corresponds to a slow decay of $|\Phi_H(0,\Delta t)|$ in Δt and a quickly changing channel corresponds to a quick decay of $|\Phi_H(0,\Delta t)|$ in Δt.

Consider the extreme case where there are no time variations in the channel (that is, $\Phi_H(\Delta f,\Delta t) = 1$). From the definition of $S_H(\Delta f,\nu)$, we note that this reduces to $\delta(\nu)$. That is, if the signal is transmitted at a frequency f_c, it arrives at f_c and there is no spectral broadening of the transmitted signal in the Doppler domain. In general, if there are time variations in the channel, this is reflected in the spectral broadening captured by $S_H(\Delta f,\nu)$ and in particular, by

$$S_H(\nu) \triangleq S_H(0,\nu) \tag{2.65}$$

which is defined in (2.62). We define the Doppler spread B_d as the range of values in ν over which $S_H(\nu)$ is essentially non-zero. Note that $S_H(\nu)$ and $\Phi_H(0,\Delta t)$ are related to each other via a Fourier transform relationship:

$$S_H(\nu) = S_H(0,\nu) = \int_{-\infty}^{\infty} \Phi_H(0,\Delta t) e^{-j2\pi\nu\Delta t} d\Delta t. \tag{2.66}$$

Thus, $\Phi_H(0,\Delta t)$ is essentially non-zero over a spread in Δt that is $1/B_d$. This spread is called the *coherence time* of the channel and is denoted as $(\Delta t)_c$.

2.5.4 TIME-FREQUENCY PARTITIONING

In addition to the three functions $\Phi_h(t,\Delta t)$, $\Phi_H(\Delta f,\Delta t)$ and $S_H(\Delta f,\nu)$, we define a fourth function $S_h(\tau,\nu)$ called the *scattering function*. This function is the Fourier transform of $\Phi_h(\tau,\Delta t)$ in the Δt domain or the inverse Fourier transform of $S_H(\Delta f,\nu)$ in the Δf domain. With these four functions, we have the picture as described in Figure 2.15 on the interconnections between the various quantities that we have defined so far.

Note that $(\Delta f)_c$ introduced in Chapter 2.5.2 measures the frequency coherence of the channel and has the engineering intuition that two sinusoids with a frequency separation of $(\Delta f)_c$ are affected differently by the channel. Based on how $(\Delta f)_c$ compares with the bandwidth W of the transmitted signal, we can classify the channel in two ways:

- *Frequency selective* channel where $W > (\Delta f)_c$ which implies that different frequency components of the transmitted signal encounter uncorrelated fading
- *Frequency non-selective* channel where $W \leq (\Delta f)_c$ which implies that all the frequency components of the transmitted signal encounter the same correlated fading.

When $W \gg (\Delta f)_c$ as is typically the case in 5G systems, the above classification allows us to partition W into smaller frequency chunks of $(\Delta f)_c$ as illustrated in the left-hand side of Figure 2.16, where within each frequency chunk, the CIR is heavily correlated with independent fades across frequency chunks. Note that in the figure on the left, each box represents a time-frequency resource unit where the channel is approximately constant.

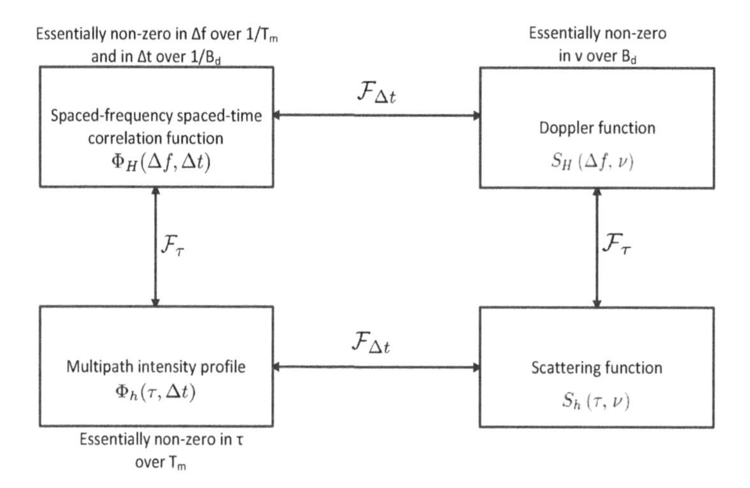

Figure 2.15 Interconnections between the different second-order statistics functions in time and frequency.

Similarly, $(\Delta t)_c$ measures the spread over which the CIRs at times t and $t + \Delta t$ are correlated and allows classification of the channel as follows:

- *Fast fading* channel where $T > (\Delta t)_c$ which implies that different sub-symbol periods of the transmitted signal encounter uncorrelated fading. Since the symbol duration of the transmitted signal is approximately $1/W$, the above condition is equivalent to $W < B_d$ or $TB_d > 1$

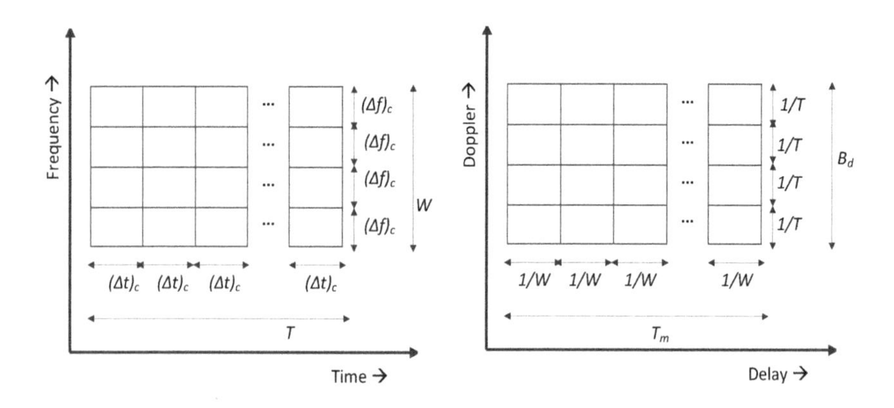

Figure 2.16 Partitioning of the signal space in the classical time and frequency domains vs. the delay-Doppler domains.

- *Slow fading* channel (sometimes called a *flat fading* channel) where $T \leq (\Delta t)_c$ which implies that the entire symbol period of the transmitted signal encounters the same correlated fading. This condition is equivalent to $W > B_d$ or $TB_d < 1$.

As before, we have a similar picture as illustrated in the left-hand side of Figure 2.16 in the time-domain. In the figure on the right, each box represents a resolvable combination of delay-Doppler. That is, if two paths whose delay and Doppler are not within the same box, they are resolvable or distinguishable.

In the case of a frequency selective and fast fading channel, we can partition the time-frequency RBs of the transmitted signal (of duration T and bandwidth W) as described in Figure 2.16 where each chunk in the time-frequency resource grid of size $(\Delta t)_c \times (\Delta f)_c$ encounters independent fading in time and frequency. Thus, given a transmit signal resource of dimensionality $T \times W$, we have

$$\frac{T}{(\Delta t)_c} \cdot \frac{W}{(\Delta f)_c} = \lfloor TWT_m B_d \rfloor \tag{2.67}$$

independent degrees of freedom (or chunks) due to the channel. An alternative interpretation of the time-frequency dependencies, illustrated in the right-hand side of Figure 2.16, is now provided. In the case of frequency selective and fast fading channel, we can *resolve* distinct delays in the delay-Doppler domain in the following ways:

- Ability to distinguish between different delays (this ability is inversely proportional to the signal bandwidth W)
- Ability to parse between different Doppler frequencies (this ability is inversely proportional to the signal duration T)
- The larger the W or T of the transmitted signal, the better our resolvability in the delay and Doppler domains, respectively.

Thus, large bandwidth systems (such as 5G) can easily resolve closely spaced taps in the delay domain and this can pose a problem in terms of equalization. The readers are referred to works such as [53, 54, 55, 56, 57] and [58] for more time-frequency partitioning insights. In particular, the design of new waveforms taking advantage of the delay-Doppler partitioning [59, 60] is to be noted.

2.5.5 RECEIVED SIGNAL MODEL

We will now consider the baseband received signal $r_\ell(t)$ which can be written as:

$$r_\ell(t) = \int_{-\infty}^{\infty} h_\ell(\tau,t) s_\ell(t-\tau) d\tau \tag{2.68}$$

$$= \int_{\tau=-\infty}^{\infty} h_\ell(\tau,t) \left[\int_{f=-\infty}^{\infty} S_\ell(-f) e^{-j2\pi f t} e^{j2\pi f \tau} df \right] d\tau \tag{2.69}$$

where in (2.69), we have used the fact that

$$\mathscr{F}(s_\ell(-\tau+t)) = S_\ell(-f) e^{-j2\pi f t}. \tag{2.70}$$

The above equation can be rearranged as

$$r_\ell(t) = \int_{f=-\infty}^{\infty} S_\ell(-f)e^{-j2\pi ft} \cdot \left[\int_{\tau=-\infty}^{\infty} h_\ell(\tau,t)e^{j2\pi f\tau}d\tau\right] df \quad (2.71)$$

$$= \int_{f=-\infty}^{\infty} S_\ell(-f)e^{-j2\pi ft} \cdot H_\ell(-f,t)df \quad (2.72)$$

$$= \int_{f=-\infty}^{\infty} S_\ell(f)e^{j2\pi ft} \cdot H_\ell(f,t)df. \quad (2.73)$$

From (2.73), we observe that $H_\ell(f,t)$ distorts $S_\ell(f)$ (in general). By restricting attention to transmissions over a single time-frequency RB where the channel behavior is frequency non-selective and slow fading, we have

$$H_\ell(f,t) = H_\ell(\overline{f},t) \quad (2.74)$$

where we have used $\overline{f}$ to denote a sample carrier frequency in the $(\Delta t)_c \times (\Delta f)_c$ region. We thus have

$$r_\ell(t) = H_\ell(\overline{f},t) \cdot \int_{f=-\infty}^{\infty} S_\ell(f)e^{j2\pi ft}df = H_\ell(\overline{f},t) \cdot s_\ell(t). \quad (2.75)$$

In other words, the effect of the channel is as a *multiplicative effect* to the transmitted signal, which allows simplified reception as the impact of the channel is that of a simple constant (but unknown or random) gain.

3 Antenna and RF Constraints in mmWave Systems

The focus of this chapter is on understanding the tradeoffs at the antenna and RF chain levels. Towards this goal, we begin with a development of antenna theory metrics such as gain, bandwidth, efficiency, etc. We then focus on aspects such as polarization, antenna types, and array geometry and array sizes. Antenna module placement is a mmWave-specific issue and we study the face and edge designs (two commercially interesting designs) from a spherical coverage performance and practical design constraints. We then study more advanced antenna module designs such as those that steer energy in multiple directions.

At the RF chain level, we begin with a discussion of the beamforming architectures (analog, hybrid, or digital) as well as their tradeoffs. Fixed passive beamforming networks such as the Butler matrix are studied next. From an RF component perspective, we then focus on power amplifiers (PAs), phase shifters, and phase noise. This then leads to the study of beamforming architecture and down-/up-conversion architecture selection. System level solutions to mitigate RF constraints such as beamforming rank adaptation, increased power levels, PA non-linearity, phase noise are studied next.

3.1 BASIC ANTENNA THEORY

We start with basic antenna theory facts that are applicable for both sub-7 GHz as well as mmWave frequencies. While a number of textbooks such as [61, 25, 62, 63] can provide all this information (and more), a self-contained introduction is really helpful in appropriately understanding the antenna-level impact on mmWave system design.

An antenna is a passive device that transmits and receives signals over the air by converting a guided electromagnetic wave enclosed inside a transmission line into a propagating wave radiating in freespace, and *vice versa*. This is made possible by varying the current distribution in the antenna element over time and by the impact of this distribution due to the antenna's structure and its surrounding environment.

3.1.1 METRICS OF POWER TRANSFER AND DIRECTIVITY

To understand the power and directivity metrics possible with an antenna element, we first develop some basics. From a transmissions perspective, a power generator with an available maximum power that can be delivered to the load $P_{\text{available}}$ is connected

DOI: 10.1201/9781032703756-3

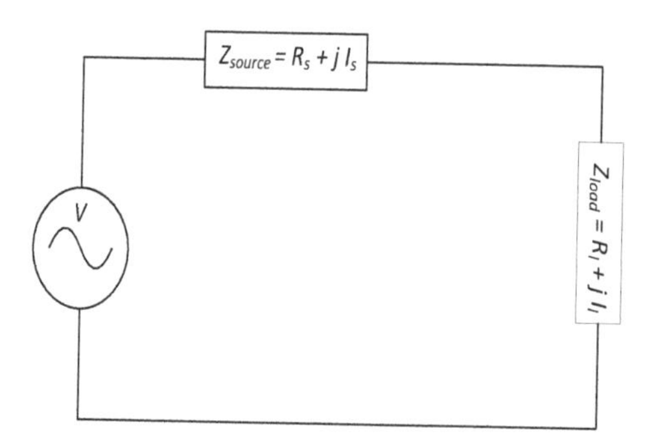

Figure 3.1 Circuit representation of an antenna with a source and a load.

to an antenna via a transmission line (e.g., a coaxial cable) and the process of interconnection leads to loss in power transfer from the power generator to the antenna (let P_{load} be the power transferred to/accepted by the antenna). This loss can be minimized by matching the impedances of the two networks, which reduces the reflection of the power from the transmission line back to the power generator [61, 25, 62]. The degree of reflection of voltage (or current) is captured by the reflection coefficient Γ, which measures the ratio of the voltage of the reflected wave to the incident wave with $0 \leq |\Gamma| \leq 1$. Here, $\Gamma = 0$ corresponds to no reflection and full power transfer from the transmission line to the antenna. Specifically, the voltage standing wave ratio (VSWR), associated with a frequency f, captures the peak-to-trough of voltage on the transmission line due to reflection and is connected to Γ as

$$\text{VSWR}(f) = \frac{|V_{\text{max}}|}{|V_{\text{min}}|} = \frac{1+|\Gamma|}{1-|\Gamma|}. \tag{3.1}$$

As Figure 3.1 illustrates the circuit representation of an antenna with a source/cable impedance of Z_{source} and an antenna/load impedance of Z_{load}, the reflection coefficient Γ can be given as

$$\Gamma = \frac{Z_{\text{load}} - Z_{\text{source}}}{Z_{\text{load}} + Z_{\text{source}}}. \tag{3.2}$$

While the "bandwidth" of an antenna is an antenna property, it is typically associated with its matching network. The bandwidth is defined as the set of all frequencies for which the matching is acceptable in the sense that the VSWR is below a certain appropriately determined threshold γ. That is,

$$\text{BW}(\gamma) = \{f \ : \ \text{VSWR}(f) < \gamma\}. \tag{3.3}$$

The efficiency of power transfer from the power generator to the antenna element, η_{transfer}, is given as

$$\eta_{\text{transfer}} = \frac{P_{\text{load}}}{P_{\text{available}}} = 1 - |\Gamma|^2 = \frac{4 \cdot \text{VSWR}}{(1+\text{VSWR})^2}. \tag{3.4}$$

Note that $|\Gamma|^2$ captures the fraction of power that is reflected and it is sometimes called the "mismatch loss" or "return loss." It can also be represented in dB scale. Typically a γ of 2 is assumed for the bandwidth definition in (3.3). Note that this corresponds to $|\Gamma| = \frac{1}{3}$ meaning that $|\Gamma|^2 \approx 11\%$ of the power is reflected back due to impedance mismatch. Or, equivalently, this corresponds to a return loss of $-10 \cdot \log_{10}(|\Gamma|^2) \approx 9.5$ dB.

The antenna's freespace total radiated power (sometimes denoted as TRP, but not to be confused with a transmission reception point as used in subsequent chapters), denoted as P_{rad}, can be usually much smaller due to resistive losses within the antenna and inductive or capacitive effects that lead to energy stored in the reactive near field. Other sources of loss could include spillover loss, dielectric loss, conduction loss, blockage from antenna supporting structures, radome, surface deviations, reflection losses, polarization mismatches, etc. The effective power radiated, P_{rad}, is determined by the antenna's radiation efficiency η_{rad}, defined as,

$$\eta_{\mathsf{rad}} = \frac{P_{\mathsf{rad}}}{P_{\mathsf{load}}}. \tag{3.5}$$

Typical antenna efficiency numbers range from 50–75% for antennas designed with planar printed circuit board (PCB) design processes, which are commonly used in UE design at mmWave frequencies. Note that higher efficiencies are possible for narrowband antennas. The antenna's total efficiency η is defined as

$$\eta = \frac{P_{\mathsf{rad}}}{P_{\mathsf{available}}} = \eta_{\mathsf{rad}} \cdot (1 - |\Gamma|^2). \tag{3.6}$$

Thus, a good antenna requires *both* a good matching network as well as good radiation efficiency.

The radiation intensity in a direction (θ, ϕ) is the relative field strength of the antenna per unit solid angle in that direction in the far field. Here, θ and ϕ denote the zenith and azimuth directions as illustrated in Figure 2.1. The total radiated power of the antenna is the accumulation of the radiation intensity over the entire sphere and is given as

$$P_{\mathsf{rad}} = \int_{\theta=0}^{\pi} \int_{\phi=0}^{2\pi} I(\theta, \phi) \sin(\theta) d\theta d\phi. \tag{3.7}$$

In (3.7), the $\sin(\theta)$ weighting factor is used because not all directions (θ, ϕ) are made the same. For example, $\theta = 0°$ and $\theta = 180°$ correspond to the polar regions (with infinitesimally small radii) and thus, a uniform sampling of points in ϕ for these θ values would wrongly estimate the total radiated power. The correct weighting factor to use is the Jacobian of the transformation from a rectangular coordinate system to a spherical coordinate system (see Appendix 3.6.1 for details), which turns out to be $\sin(\theta)$ for a given choice of (θ, ϕ). To be precise, the weightage of a set of uniformly sampled points is progressively reduced by the $\sin(\theta)$ factor as we approach the polar regions.

Sometimes, an antenna is better defined in terms of its effective isotropic radiated power (EIRP) along a certain direction (θ, ϕ), which satisfies the relationship: $\mathsf{EIRP}(\theta, \phi) = 4\pi \cdot I(\theta, \phi)$. The antenna is said to be *isotropic* if $\mathsf{EIRP}(\theta, \phi) = \mathsf{EIRP}_{\mathsf{iso}}$ for all (θ, ϕ). While isotropicity is mostly a theoretical concept, it allows us to benchmark the performance of practical or non-isotropical antennas. Since an antenna is a passive device, it does not amplify signals. However, it can spatially concentrate energy in certain directions. To capture this spatial concentration, the gain of an antenna $G(\theta, \phi)$ is defined as the radiation intensity in the direction (θ, ϕ) relative to an isotropic antenna that radiates the entire power accepted by the antenna (that is, $\eta_{\mathsf{rad}} = 100\%$). For an isotropic antenna with $\eta_{\mathsf{rad}} = 100\%$, we have

$$P_{\mathsf{load}} = P_{\mathsf{rad}} = \mathsf{EIRP}_{\mathsf{iso}} \tag{3.8}$$

and thus

$$G(\theta, \phi) = \frac{\mathsf{EIRP}(\theta, \phi)}{\mathsf{EIRP}_{\mathsf{iso}}} = \frac{\mathsf{EIRP}(\theta, \phi)}{P_{\mathsf{load}}}. \tag{3.9}$$

In contrast to the gain, directivity $D(\theta, \phi)$ is the radiation intensity in (θ, ϕ) relative to the radiation intensity of an isotropic antenna radiating P_{rad} (η_{rad} need not be 100%). Thus, we have

$$D(\theta, \phi) = \frac{\mathsf{EIRP}(\theta, \phi)}{P_{\mathsf{rad}}} \Longleftrightarrow G(\theta, \phi) = D(\theta, \phi) \cdot \frac{P_{\mathsf{rad}}}{P_{\mathsf{load}}} = D(\theta, \phi) \cdot \eta_{\mathsf{rad}}. \tag{3.10}$$

Directivity captures the notion of how directive (or non-isotropic) an antenna is relative to the radiated power of the antenna, whereas gain captures the same notion with respect to the accepted power of the antenna. Directive antennas are important in the design of cellular systems as they can be used to focus or steer energy in specific directions and avoid specific directions and thus mitigate interference. Since the total radiated power is conserved, a highly directive antenna produces an energy peak in a specific direction by re-assigning the energy radiated in other directions (with minimal energy radiated in these directions).

In contrast to the gain and directivity, the realized gain of an antenna is defined with respect to $P_{\mathsf{available}}$ and is given as

$$G_{\mathsf{realized}}(\theta, \phi) = \frac{\mathsf{EIRP}(\theta, \phi)}{P_{\mathsf{available}}} = D(\theta, \phi) \cdot \eta_{\mathsf{rad}} \cdot \eta_{\mathsf{transfer}}. \tag{3.11}$$

The need for notions such as gain and directivity (in contrast to a single notion of realized gain) is both from a historical development of antenna theory and also to isolate the performance of an antenna with respect to different power levels that are relevant in its functioning. For terms such as directivity, EIRP and gain, if a specific direction is not mentioned, then it is implicitly assumed that we are talking about the directions corresponding to their peak values.

The effective aperture (or area) A_e of a unit-gain isotropic antenna captures how much power is captured by the antenna from an incoming plane wave. This is given

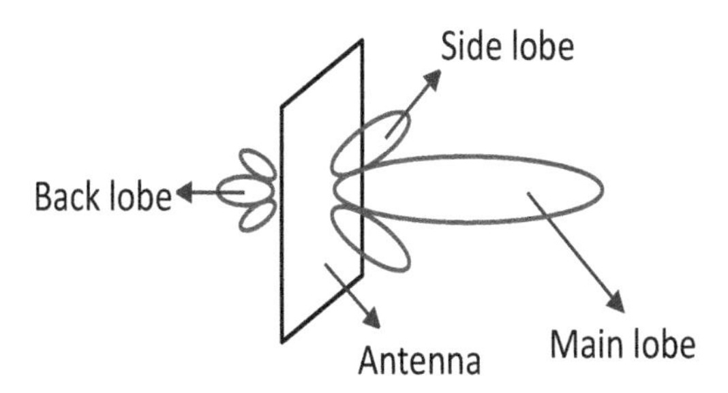

Figure 3.2 Typical antenna radiation pattern with main lobe, side lobes and back lobes highlighted.

by the well-known formula[1]:

$$A_e = \frac{\lambda^2}{4\pi} \cdot G(\theta, \phi). \tag{3.12}$$

Equation (3.12) plays a critical role in the derivation of the Friis transmission and radar equations.

An antenna's radiation properties (e.g., EIRP(θ, ϕ), directivity or gain) can be graphically illustrated or mathematically plotted as a function of the spatial coordinates (θ, ϕ). The antenna's gain as a function of (θ, ϕ) is sometimes referred to as radiation pattern or antenna pattern. The direction or orientation of peak/highest radiation pattern is called the *main lobe* (often also as the *boresight* direction), with other secondary peaks in the radiation pattern called as *side lobes*. Peaks generally opposite to the main lobe are called as *back lobes*. In this setting, the front-to-back ratio is defined as the difference (in dB) between the main lobe and back lobe strengths. A typical antenna's radiation pattern is pictorially illustrated as a three-dimensional polar plot over the sphere as described in Figure 3.2. Alternately, various two-dimensional plots are typically used to illustrate the projection of the three-dimensional view on different planes. For example, the radiation pattern is illustrated over the sphere presented in rectangular (θ, ϕ) coordinates in Figure 3.3(a) and over XYZ coordinates in Figure 3.3(b). In these plots, the radiation pattern corresponds to a typical microstrip patch antenna designed and operating at 28 GHz in a form factor UE with boresight

---

[1]Note that the smallest distance at which an antenna can radiate or capture power from an incoming electromagnetic wave is given by the Fresnel zone characterization, which corresponds to a near field distance of at least $r = \lambda/(2\pi)$. At this distance r, the plane wave intersects the received spherical wavefront leading to a circular aperture of area $\pi r^2 = \pi \cdot \left(\frac{\lambda}{2\pi}\right)^2 = \frac{\lambda^2}{4\pi}$. Thus, the effective aperture of the unit-gain isotropic antenna is given as $A_e = \frac{\lambda^2}{4\pi}$. In the case of an antenna with a gain $G(\theta, \phi)$, the received power scales by this factor $G(\theta, \phi)$. There are other more involved and rigorous ways by which (3.12) can be derived; see [25], for example.

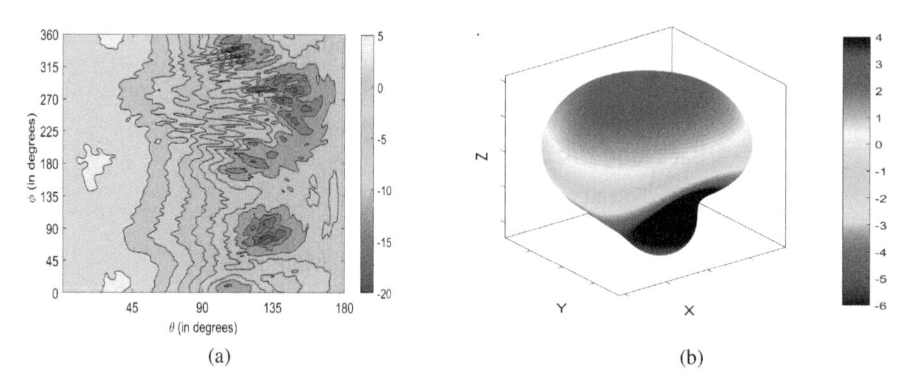

Figure 3.3 Radiation pattern over a sphere illustrated in (a) rectangular θ-ϕ coordinates and (b) over XYZ coordinates.

along the Z axis ($\theta = 0°$). We observe peak gains of $\approx$ 4.6 dBi along $\theta = 0°$ in Figure 3.3.

For an antenna, the term "E plane" is sometimes used to indicate the plane containing the electric field vector and the direction of maximum radiation. The perpendicular plane containing the magnetic field vector and the direction of maximum radiation is then indicated as the "H plane." For a V-pol and a H-pol antenna (See Chapter 3.1.2 for definitions), the E plane coincides with the vertical or elevation plane and horizontal or azimuth plane, respectively. Complementing the plots of the radiation pattern in Figures 3.2 and 3.3, Figure 3.4 illustrates the radiation pattern in the E and H planes for the patch antenna element considered in Figure 3.3. Here, the E and H planes correspond to the YZ and XZ planes, respectively. Given the boresight of this antenna is along Z axis, the radiation patterns in the E and H planes coincide at $\theta = 0°$ (the Z-axis) with relatively minor variations elsewhere.

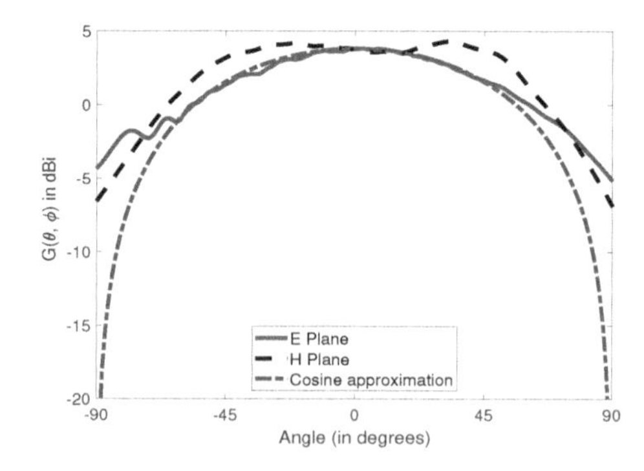

Figure 3.4 $G(\theta, \phi)$ over the E and H planes for a microstrip patch antenna on a finite ground plane designed at 28 GHz.

Also, the variation of the radiation pattern over θ is often approximated as $|\cos(\theta)|^{1.5}$, which is also plotted in Figure 3.4 and is shown to be a good match with the true radiation pattern over $180° \pm 60°$ coverage area. Note that at $\pm 60°$ off the boresight direction, the radiation pattern droops by

$$\left|10 \cdot \log_{10}(1/2^{1.5})\right| \approx 4.5 \text{ dB}. \tag{3.13}$$

Since the use of the $|\cos(\theta)|^{1.5}$ formula implicitly assumes a very good antenna matching or calibration at $\pm 60°$, $|\cos(\theta)|^{1.5}$ is sometimes used for $\pm 45°$ scanning and $|\cos(\theta)|^2$ is used for 45°-to-60° scanning. Note that the $|\cos(\theta)|^{1.5}$ and $|\cos(\theta)|^2$ approximations are good for microstrip patch antenna elements. For other antenna types (e.g., dielectric antennas), different approximations such as $|\cos(\theta)|^n$ (e.g., $n = 1.2, 1.3, 3.5$ or 10) may be used [64]. That said, the correct choice of exponent of $\cos(\theta)$ is a function of the antenna design and the scan range of interest. This is a subject of considerable debate and the readers are referred to advanced antenna theory textbooks for such a discussion.

Beamwidth of an antenna is a metric that captures the spatial extent of gain or the spatial coverage vs. directional gain tradeoff. In radar signal processing, beamwidth captures the antenna's resolvability or its ability to distinguish between adjacent targets. The most common beamwidth metrics are half-power beamwidth and first null-to-null beamwidth. The half-power beamwidth is defined as the spread of angles over which the radiation pattern is within 3 dB of the peak gain. Destructive interference of the radiated power creates nulls in the radiation pattern. The first null-to-null beamwidth is the angle space covering the first null around the boresight direction. Figure 3.5 pictorially illustrates the half-power and null-to-null beamwidths for a typical antenna's radiation pattern.

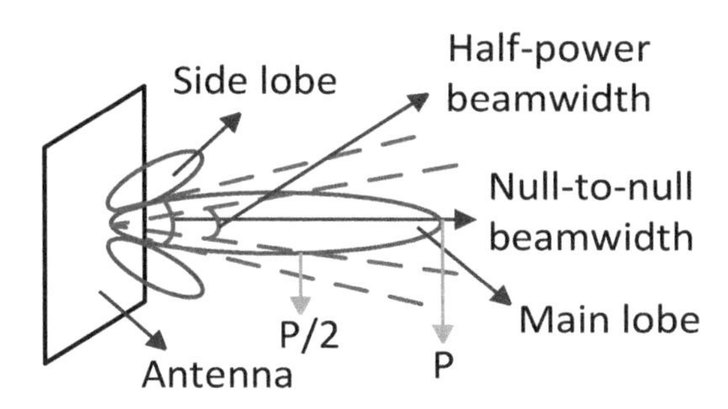

Figure 3.5 Pictorial illustration of half-power and null-to-null beamwidths for a typical antenna with P being the peak power associated with the antenna's radiation pattern.

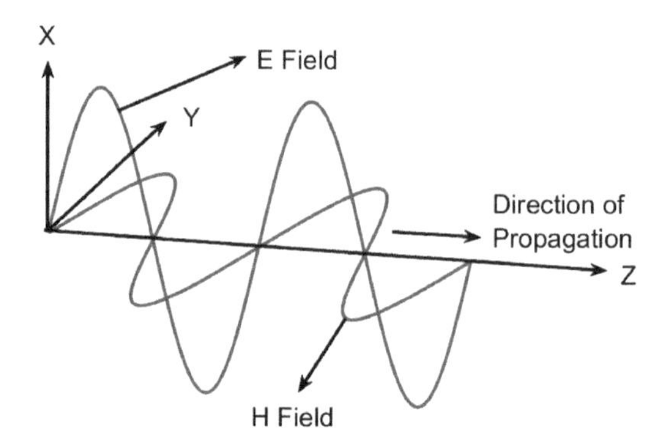

Figure 3.6 Electric and magnetic fields and direction of propagation of electromagnetic wave.

3.1.2 ANTENNA POLARIZATION

Electromagnetic waves are vector fields that consist of electric as well as magnetic fields. The electric field at a point in space is a measure of the force experienced by a unit positive charge, whereas the magnetic field measures the force exerted on a moving charged particle. In the far field typical of traditional communications, the electric and magnetic fields associated with an electromagnetic wave are perpendicular to each other and to the direction of propagation of the wavefront (the direction in which the energy flux of the wave moves). Figure 3.6 pictorially illustrates the behavior of the electric and magnetic fields along with their connection to the direction of propagation of the electromagnetic wave.

Since the electric and magnetic field vectors are contained within the plane perpendicular to the direction of propagation, they can be decomposed (in infinitely many ways) along two basis vectors of this orthogonal plane. Let $\widehat{x}$ and $\widehat{y}$ denote two basis vectors in this plane with $\widehat{z}$ denoting the direction of wave propagation. The general form of the electric and magnetic field vectors are then given as

$$E(z,t) \triangleq \begin{bmatrix} E_x(t) \\ E_y(t) \\ E_z(t) \end{bmatrix} = \begin{bmatrix} E_x e^{j\phi_x} \\ E_y e^{j\phi_y} \\ 0 \end{bmatrix} \cdot e^{j\left(2\pi f t - \frac{2\pi z}{\lambda}\right)} \tag{3.14}$$

$$H(z,t) \triangleq \begin{bmatrix} H_x(t) \\ H_y(t) \\ H_z(t) \end{bmatrix} = \begin{bmatrix} H_x e^{j\gamma_x} \\ H_y e^{j\gamma_y} \\ 0 \end{bmatrix} \cdot e^{j\left(2\pi f t - \frac{2\pi z}{\lambda}\right)} \tag{3.15}$$

with $\{E_x, E_y, H_x, H_y\}$ denoting the non-negative amplitudes and $\{\phi_x, \phi_y, \gamma_x, \gamma_y\}$ denoting the phases. When a single frequency is of interest, for convenience, the $e^{j2\pi f t}$ term is dropped from (3.14) and (3.15). Since the wave travels along the Z direction, there is only a Z component in the exponentials of (3.14) and (3.15). The angular frequency and wave number are denoted as $\omega = 2\pi f$ and $k = 2\pi/\lambda$, respectively.

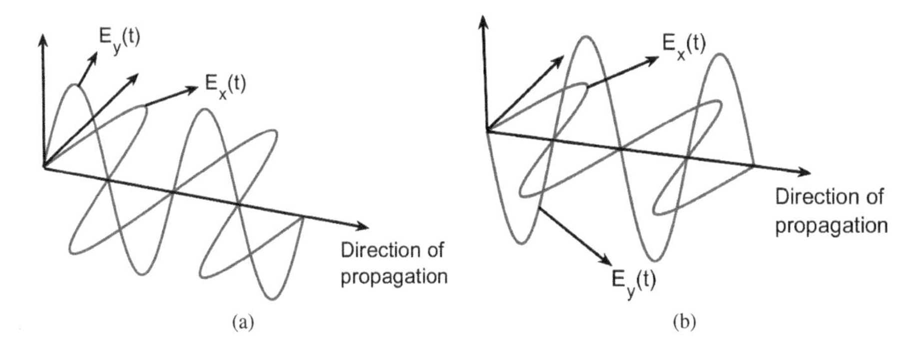

Figure 3.7 Electric field behavior for a linear polarized system with (a) positive and (b) negative phases.

Note that the physically realized components of the electric field are the real parts of (3.14) and are given as

$$E_x(t) = E_x \cdot \cos(\omega t - kz + \phi_x) \tag{3.16}$$
$$E_y(t) = E_y \cdot \cos(\omega t - kz + \phi_y). \tag{3.17}$$

Further, since the magnetic field components are related to the electric field components in the far field as

$$H_y(t) = \frac{E_x(t)}{\eta}, \; H_x(t) = \frac{-E_y(t)}{\eta}, \tag{3.18}$$

where η denotes the wave impedance in freespace[2]. That is, there can be independent relationships only for E_x, E_y, ϕ_x and ϕ_y. Thus, we can ignore the magnetic field components and define polarization in terms of the direction along which the electric field components oscillate. Depending on the relationship between E_x, E_y, ϕ_x and ϕ_y. The shape of the path of the electric field over time (its locus) can be different and we have the following categorizations:

- If $\phi_x = \phi_y$, then the locus of the electric field in the $\widehat{x}$-$\widehat{y}$ plane is a straight line with a non-negative slope satisfying

  $$\frac{E_y(t)}{E_x(t)} = \frac{E_y}{E_x}. \tag{3.19}$$

  Similarly, if $\phi_x = \phi_y + 180°$, the locus of the electric field has a non-positive slope satisfying

  $$\frac{E_y(t)}{E_x(t)} = -\frac{E_y}{E_x}. \tag{3.20}$$

[2]Note that the wave impedance in freespace, η, is given as $\eta = \sqrt{\frac{\mu_0}{\varepsilon_0}} = \mu_0 \cdot c$ where μ_0 and ε_0 are the permeability and permittivity in freespace and c is the speed of light. η is typically approximated as 376.7 Ω.

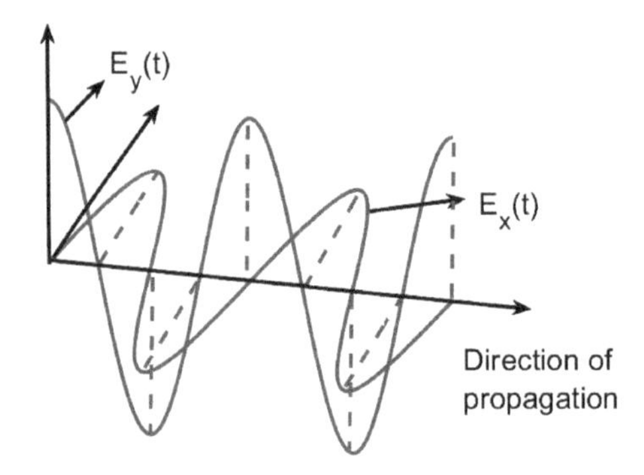

Figure 3.8 Electric field behavior for a circular polarized system.

These two cases are generally called as *linear* polarization since the locus of the electric field is a straight line. In a specific extreme example, the wave is said to be horizontally polarized[3] (H-pol) if $E_y = 0$. Similarly, the wave is said to be vertically polarized (V-pol) if $E_x = 0$. If $E_x = E_y$ and $\phi_x = \phi_y$, the straight line traversed by the electric field vector has a 45° slope, which is often called as *slant* 45° polarization. In many instances, a dual-polarized system based on linear polarization can support H- and V-pol, or slant 45° and slant −45°.

- If $E_x = E_y = E$, but $\phi_x = \phi_y \pm 90°$, the locus of the electric field is a circle

$$E_x(t) = \mp E_x \cdot \sin(\omega t - kz + \phi_y), E_y(t) = E_x \cdot \cos(\omega t - kz + \phi_y) \tag{3.21}$$

$$\Longrightarrow E_x(t)^2 + E_y(t)^2 = E^2. \tag{3.22}$$

Both these scenarios are generally called as *circular* polarization. Note that the case of $\phi_x = \phi_y + 90°$ leads to a clockwise motion of the electric field vector (as seen by a bystander watching the wave moving toward him/herself), whereas the case of $\phi_x = \phi_y - 90°$ leads to a counterclockwise motion of the electric field vector. More classically, the clockwise motion is seen as the curving of the remaining fingers of the left hand as the left hand thumb points toward the direction of propagation, or left-handed circular polarization. The counterclockwise motion is seen as the curving of the right hand as the right hand thumb points toward the direction of propagation, or right-handed circular polarization. If the bystander were to face the wave along the propagating direction, clockwise and counterclockwise motion of the electric field vector would correspond to right-handed and left-handed circular polarization, respectively.

- In the general case where $E_x \neq E_y$ and $\phi_x = \phi_y + \phi$, the locus of the electric field is an ellipse, which is denoted as *elliptical* polarization. The equation of the locus

[3]The notion of H-pol and V-pol depends on a fixed antenna orientation in a global coordinate system (GCS).

satisfies

$$\frac{E_x(t)}{E_x} = \cos(\omega t - kz + \phi_y)\cos(\phi) - \sin(\omega t - kz + \phi_y)\sin(\phi) \quad (3.23)$$

$$\frac{E_y(t)}{E_y} = \cos(\omega t - kz + \phi_y) \quad (3.24)$$

$$\Longrightarrow \left(\frac{E_x(t)}{E_x}\right)^2 + \left(\frac{E_y(t)}{E_y}\right)^2 - 2\left(\frac{E_x(t)}{E_x}\right)\left(\frac{E_y(t)}{E_y}\right)\cos(\phi) = \sin^2(\phi). \quad (3.25)$$

As in the circular polarization case, elliptical polarization could be left-handed or right-handed. Special cases of elliptical polarization are linear and circular polarization.

Note that the polarization of the wave is associated with the polarization of the electric field generated by the wave (by convention and historical terminology). The polarization of an antenna is associated with the polarization of the electric field generated by the antenna in the far field. Thus, a V-pol antenna has its electric field vector perpendicular to the ground, whereas a H-pol antenna has its electric field vector parallel to the ground. Note that the terminology H-pol and V-pol are also arbitrary unless the antenna has a fixed orientation with respect to the earth.

Ignoring the impact of the channel, in theory, there is no fundamental difference between a dual-polarized system based on linear polarization and a dual-polarized system based on circular polarization. In practice, however, circular polarization is difficult to obtain in form factor UE designs at sub-7 GHz bands due to the antennas being imbalanced. While circular polarization is realizable more easily at mmWave frequencies, maintaining a $\pm 90^\circ$ offset across a wideband requires careful circuit design. Circular polarization is commonly used in satellite communication systems since linear polarization can be unequally rotated by the impact of ionosphere (Faraday rotation) and can also suffer a loss due to misalignment between a geostationary satellite and earth station. A circular polarization wave can also flip polarization orientation (left-to-right or right-to-left) upon reflection by a metallic plate or object.

In general, the polarization of the wave generated by an antenna and propagating in freespace remains unchanged until the wave encounters an object in the environment and there is interaction between the field and the object. In particular, when the electromagnetic wave hits an object, the properties of the two orthogonal polarization components get transformed due to the impact on the object. The terms co-polarization (co-pol) and cross-polarization (cross-pol) responses capture how the energy in a certain polarization of the incident wave is retained in the same or intended polarization, and how it is flipped into the orthogonal or unintended polarization, respectively. A 2×2 scattering/mixing matrix as in (2.8) is often used to capture the transformation of energy across the two polarization components:

$$\begin{bmatrix} e^{j\nu_{\Theta\Theta}} & \frac{e^{j\nu_{\Theta\Phi}}}{\sqrt{\mathsf{XPR}}} \\ \frac{e^{j\nu_{\Phi\Theta}}}{\sqrt{\mathsf{XPR}}} & e^{j\nu_{\Phi\Phi}} \end{bmatrix}. \quad (3.26)$$

Table 3.1
Theoretical attenuation due to polarization mismatches

Rx. Ant. Pol.		Incident wave polarization			
		Vertical	Horizontal	Right-hand circular	Left-hand circular
	Vertical	0 dB	∞	3 dB	3 dB
	Horizontal	∞	0 dB	3 dB	3 dB
	Right-hand circular	3 dB	3 dB	0 dB	∞
	Left hand circular	3 dB	3 dB	∞	0 dB

The diagonal terms of this matrix are of unit amplitude and specify how the object keeps/retains the energy in the original polarization components with the off-diagonal terms specifying how the object mixes the energy into the opposite polarization component. Here, cross-polar discrimination ratio (or the XPR) introduced in (2.8) and reproduced in (3.26) captures this aspect from a channel model perspective. In TR 38.901 [18], XPR in the range of 7–12 dB is assumed which is a reasonable assumption for practical UE designs at mmWave frequencies. With an XPR of 10 dB, 10% of the energy in one polarization mixes into the orthogonal polarization. Note that higher the XPR, the less the mixing between the polarization components (and *vice versa*).

Axial ratio is defined as the ratio of the major and minor axes:

$$\text{Axial ratio} = \frac{\max(E_x, E_y)}{\min(E_x, E_y)}. \tag{3.27}$$

For maximal power reception at a receiver antenna, the incident wave and the receiving antenna should have the same axial ratio, same sense of polarization (as measured in a GCS with the direction of propagations matched), and the same spatial orientation. If these quantities are mismatched, loss can be seen in signal reception. A loss factor as described in Table 3.1 captures the theoretical attenuation due to polarization mismatches. From this table, we note that a horizontally polarized signal cannot be received by a vertically polarized antenna (and *vice versa*), and a circularly polarized antenna with both polarization components can capture the signal with a 50% polarization efficiency. In some single polarization transmission and reception scenarios, a linear polarization at one end and circular polarization at the other end can lead to robust performance independent of channel effects. The same property can be leveraged to transmit two independently coded signals (over orthogonal polarizations) in a given time-frequency resource. By adjusting the receiving antenna for either polarization, either signal can be accurately received without interference from the other signal. Thus, polarization offers a simple mechanism[4] to *double* the

[4]Note that a dual-polarized antenna can replicate all possible polarizations by mixing the two polarizations appropriately since two linearly independent two-dimensional vectors can generate any two-dimensional vector.

capacity of the channel without commensurately increasing the chip area, especially at mmWave frequencies.

3.2 PRACTICAL CONSIDERATIONS IN ANTENNA ARRAY DESIGN FOR MMWAVE FREQUENCIES

A single antenna element may not provide the radiation intensity or the flexibility needed to satisfy performance requirements in a mmWave network. In this context, recall from (2.43) that freespace path loss differentials between frequencies can be bridged with the use of more antenna elements at the higher carrier frequency over the same physical aperture. Thus, it is of broad interest to understand the capabilities of an antenna array, which is a collection of multiple antenna elements that can be jointly controlled.

Given form factor (or real-estate) constraints at both the base station and UE and relatively smaller wavelengths at mmWave carrier frequencies, a number of individual antenna elements can be placed or mounted within the same form factor allowing increased array gains that are hitherto not possible at sub-7 GHz. More specific beamforming capabilities of multi-element antenna arrays are studied in Chapter 4. While such a possibility makes a theoretical case for packing as many antennas as possible in mmWave systems (contingent on real-estate constraints at the UE side), the added cost of mmWave antenna modules and associated RF front-end components (e.g., power, low-noise and variable gain amplifiers, phase shifters, mixers, analog-to-digital and digital-to-analog converters, switches/connectors, etc.) and the concomitant power and thermal increase puts a practical limit on how many antennas can be gainfully employed in a mmWave system.

More importantly, while the use of a large number of antennas (and antenna modules) can *theoretically* lead to increased beamforming gains, if these capabilities are not *practically* exercisable with a low beam management overhead, the capabilities can quickly turn out to be onerous and become a curse rather than serve as a blessing. Thus, in translating the ideas described in Chapters 3.1.1–3.1.2 to practical mmWave antenna array designs, a number of optimizations need to be considered. These include:

- Antenna types
- Array geometry and array size
- Antenna module placement tradeoffs (primarily at the UE side).

We will focus mostly on the impact of these considerations at thc UE side with key differences at the base station highlighted.

3.2.1 ANTENNA TYPES

An antenna can be one of many different types: wire antenna (e.g., dipole, helix, loop or Yagi), aperture antenna (e.g., horn or dish), or printed antenna (e.g., patch, printed dipole, spiral, slot or Vivaldi/tapered slot). It can have different gains: high gain which is >20 dBi[5] (e.g., dish), medium gain which is between 10 and 20 dBi (e.g., horn, helix or Yagi), or low gain which is <10 dBi (e.g., dipole, loop, patch, slot or whip). An antenna can have different bandwidths: wide bandwidth (e.g., biconical, conical spiral or log-periodic), moderate bandwidth (e.g., horn or dish), or narrow bandwidth (e.g., dipole, helix, loop, patch, slot, whip or Yagi). Thus, the best type of antenna to use is dependent on how it is intended to be used and the application setting. The design tradeoffs in antenna type selection include frequency of operation and bandwidth desired, radiation pattern and angular coverage (such as half-power beamwidth, front-to-back ratio, null patterns), directional gain, polarization properties including XPR, power delivered, size and weight, cost, complexity of design and fabrication, etc.

Reflection of the incident wave from certain objects (e.g., ground bounces, snow accumulated on the ground, etc.) can reverse the phase of a certain polarization at certain carrier frequencies. Thus, a superposition of the incident and reflected wave at these carrier frequencies can lead to poor signal strength in that polarization resulting in a preference for the use of orthogonal polarizations. This behavior is seen with horizontal polarization transmissions in amplitude and frequency modulation[6] at lower carrier frequencies thereby making transmissions with vertical polarization typical. Thus, antennas in legacy wireless systems such as 3G and 4G have mostly been V-pol designs at the UE side with dual-polarized designs at the base station side (where there are limited real-estate constraints). However, an interest in potential doubling of the spectral efficiency has led to increased interest in dual-polarization (V-pol and H-pol) designs. The possible imbalance in signals across the two layers due to H-pol attenuation has meant that these dual-polarization designs are of the form slant 45° and slant −45°. In some practical implementations, a slant 45° and slant −45° design can be more compact than a V-pol and H-pol design adding to its commercial popularity in UEs.

At sub-7 GHz frequencies, planar inverted-F antennas (PIFAs) or monopoles are typically used for transmissions and receptions at the UE side. PIFAs are similar to patches, but with an additional shorting pin to ground to add inductance for impedance matching and compactness reasons. Most of these antennas rely on the ground plane or chassis of the UE for polarization stability and thus do not allow good polarization control/separation. Thus, they can only support one layer as the polarizations are mixed/coupled. If dual-polarization transmissions/receptions are required, a doubling of the aperture is necessary making it feasible more typically at the base station side. In some practical implementations, an unbalanced $\lambda/4$ length dipole that relies on the UE's ground plane for counterpoise is used. Its performance is comparable with that of a monopole.

---

[5]Note that dBi is a unit of measurement capturing decibels relative to isotropic radiating conditions.

[6]Similar to the above, ghosting in television broadcast transmissions with vertical polarized signals makes horizontal polarization preferable. Other examples where certain polarizations are preferred include satellite communications systems such as Global Positioning System (GPS) where right-handed circular polarization transmissions are preferable for their robustness.

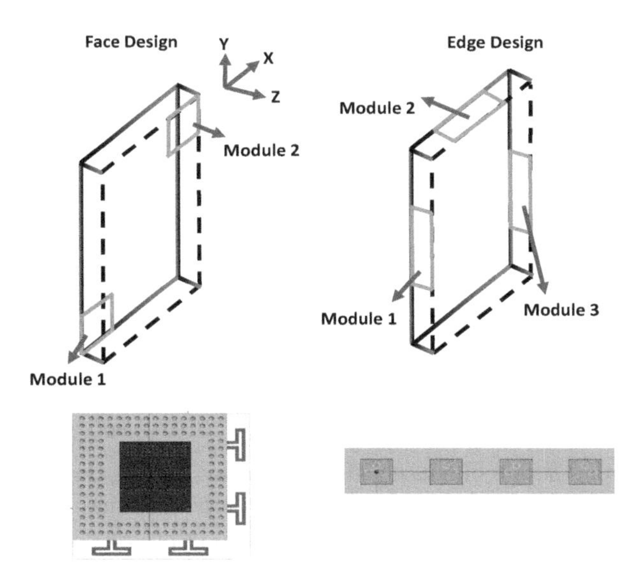

Figure 3.9 Pictorial illustration of the *face* and *edge* designs along with the individual antenna module structure in these designs. The arrows in the figures are just pointers to the antenna modules and have no specific correlation to the boresight directions of the subarrays.

In the sub-7 GHz channel model specifications (e.g., TR 36.873 [23]), a dual-polarized channel as with the mmWave specifications is assumed, but the practical consequences in terms of chip area doubling are not considered in the channel modeling specifications. In practice, due to these consequences, most sub-7 GHz systems are single polarized with spatial MIMO (as opposed to polarization MIMO) being the common mode of transmissions and receptions at the UE side.

In contrast to PIFAs, microstrip patch antennas (without aperture doubling) and dipole antennas designed via PCB processes are commonly used in commercial cellular systems at mmWave carrier frequencies, and these design processes can be scaled down even to 10 GHz. A pictorial illustration of a patch array (in a 4×1 and 2×2 configuration) and a dipole array (in 2×1 configuration) are presented in Figure 3.9. Such antennas allow the compact design of dual-polarized arrays and realize the concomitant enhanced rates. However, a mmWave antenna is more significantly impacted by the materials in the UE (e.g., housing, battery, cameras, sensors, etc.) than a sub-7 GHz antenna.

Between dipoles and patches, dipole antennas are more affected by placement issues at the UE side than patch antennas since they are sensitive to metal, glass or plastic material distortions. Thus, beamforming with the dipole antennas can show a big deviation in real-world performance over that computed in freespace. This deviation

Table 3.2
Design parameters for different antenna types

Antenna type	Patches	Dipoles
Polarization	Dual-polarized	Single polarized
Achievable data rate	2X	1X
Impact of placement or tilting	Low	High
Freespace vs. real-world performance	Comparable	Distorted
Area or size	Small	Relatively bigger
Bandwidth coverage	Small	Relatively larger

requires a careful design of housing[7] in UE designs. Further, dipole antennas typically require more area (and a bigger size) than patch antennas. Thus, in thin UE designs, the antenna modules may need to be tilted or placed at an angle resulting in more complicated beamforming tradeoffs. Further, such a tilting may not even be possible below a certain carrier frequency as the radiator needs to have a sufficiently large ground plane for the radiator to function efficiently. On the other hand, dipoles allow a wider bandwidth coverage relative to patch antenna elements allowing the reuse of the same antenna design across different bands or geographies [65]. The wide bandwidth coverage property of dipole antennas makes them attractive for usage in sub-7 GHz frequencies to cover disparate bands across different geographies with a single antenna array. Table 3.2 captures a broad overview of the tradeoffs between patches and dipoles.

3.2.2 ARRAY GEOMETRY AND ARRAY SIZES

An antenna array can be arranged in a regular pattern with a fixed inter-antenna element spacing (e.g., linear, planar, circular, etc.). Regular arrangements (sometimes called as uniformly spaced arrays, uniform arrays, etc.) are easily amenable to application of antenna array theory for understanding their array gain patterns (main lobe, side lobes, nulls and beamwidths). That said, non-regular and sparse arrangements (sometimes called as thinned arrays [66, 67]) are of increasing interest as mmWave systems evolve into higher carrier frequencies allowing more degrees-of-freedom in terms of theoretical antenna capabilities and less degrees-of-freedom in effectively using them at the RF level, enabling antenna selection solutions [68–72].

Over the last few years, the evolution of premium-tier as well as high-tier smartphones has been toward thinner and sleeker designs with minimal bezel or no

[7]An antenna is enclosed by many components around it. Everything around the antenna, which impacts the antenna's radiation performance, can be said to constitute the housing. For a UE, antenna module placement impacts the distortion induced by the housing. At the UE side, housing is typically used for the back plane and the side/edge of the phone. This includes the whole piece of plastic or metal sheet that covers the phone, especially the battery, and which can be easily removed during maintenance.

Figure 3.10 Qualcomm® QTM052 antenna module.

bezel on the front display. For example, an Apple iPhone 15 has dimensions of $147.6 \times 71.6 \times 7.8$ mm, whereas a Samsung Galaxy S24 smartphone has dimensions of $147 \times 70.6 \times 7.6$ mm. Given that the ground plane of an antenna element is on the order of the wavelength for the antenna to radiate well, edge placement of antenna arrays operating at 20–50 GHz frequencies is confined to linear arrays.

Planar or circular arrays require more area for placement on the UE side and they can be mounted on the front or back face of the UE (face placement) at these frequencies. As will be discussed in antenna module placement tradeoffs in Chapter 3.2.3, linear arrays (and edge placement) appear to be more favorable over planar arrays (and face placement). However, as carrier frequencies increase, planar arrays can be mounted on the edge as well. In a commercial UE, a mmWave antenna array competes for space or real-estate with antennas at sub-7 GHz frequencies, WiFi, Bluetooth systems, near field communications/sensing, cameras, sensors, battery and circuit elements associated with all these components. With these constraints, the array sizes of typical linear arrays for edge placement range from 4×1 through 8×1 over the 20–50 GHz regime. Typical planar array sizes for face placement include 2×2, 4×2 and 4×4 at these frequencies. As an illustration, a practically designed mmWave antenna module solution with a dual-polarized 4×1 array in the form of Qualcomm®[8] QTM052 mmWave antenna module is compared with the dimensions of a 1 cent U.S. coin in Figure 3.10. In another illustration, a form factor UE with multiple dual-polarized 4×2 arrays on the front and back faces is illustrated in Figure 3.11.

The first-generation small cell mmWave base stations are modest in size with planar arrays of 8×4 and 8×8 being typically used. As mmWave technology evolves, larger array sizes such as 16×4, 32×4, 64×4, 32×8, 64×16, etc. are being actively considered and studied in terms of performance tradeoffs. For example, the remote radio head of a base station encompassing a dual-polarized 32×4 array is pictorially illustrated in Figure 3.12. Symmetric designs (e.g., 32×32, 64×64, etc.)

[8]Qualcomm RF modules are products of Qualcomm Technologies, Inc. and/or its subsidiaries.

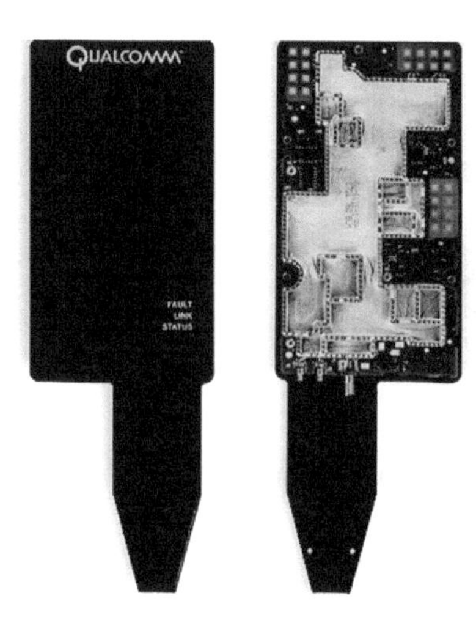

Figure 3.11 Prototype UE with multiple 4×2 arrays on the front and back faces.

are considered in deployments where users can be located at different elevation angles such as in shopping malls, stadiums, high-rise buildings, etc., as well as in CPEs where optimal installations cannot be guaranteed. However, as power and thermal overheads and beam management latencies become the fundamental issues limiting mmWave performance, practical array sizes are likely to be limited [73]. In particular, thermal management solutions due to increased RF power consumption and dissipation appear challenging in base station designs. As mmWave systems evolve to cater to a higher degree of densification, we are likely to see intermediate nodes serving as relays such as repeaters or reflectarrays, integrated access and backhaul (IAB) and reconfigurable intelligent surfaces (RIS). Such intermediate nodes can

Figure 3.12 Pictorial illustration of a Qualcomm® remote radio head with a 32×4 dual-polarized array.

use large arrays (e.g., 128×16, 256×16, 256×32, etc.) as they may tradeoff better performance with increased power or thermal overheads. Low-cost implementations of base stations, CPEs and intermediate nodes can also consider smaller array sizes along with apparatuses for focusing energy in steerable directions.

3.2.3 ANTENNA MODULE PLACEMENT

3.2.3.1 Spherical Coverage

In contrast to a base station, the notion of a sector makes less sense at the UE since such a design can lead to significant performance degradation if useful signals cannot be picked up from anywhere in the $360° \times 180°$ sphere in azimuth and elevation (respectively) around the UE. Since the base station and the UE can be in any specific environment (street level, indoor office, stadium, home, etc.) as illustrated in Figure 3.13, there is no reason to expect any specific prior over the sphere for the incoming signal corresponding to the dominant cluster direction(s); see Chapter 6.3 for a discussion on priors. Thus, good *spherical coverage* is expected in a good UE design. However, the impact of material properties at mmWave frequencies means that the scan angle of an antenna where acceptable gains can be seen is much smaller (typically a $\pm 45°$ to $\pm 60°$ region around the boresight). This necessitates achieving good spherical coverage by using multiple antenna arrays strategically placed in different locations of the UE.

Antenna elements are integrated with associated RF elements such as phase shifters, power amplifiers, low-noise amplifiers and variable gain amplifiers at mmWave frequencies. Thus, each RFIC chip can typically support only a small number of antenna elements, depending on the aperture allocated to the antennas and the range of carrier frequencies supported. Thus, the use of multiple antenna arrays is

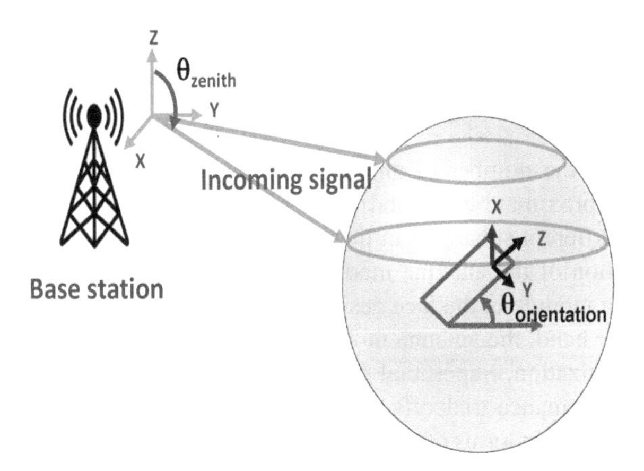

Figure 3.13 Illustration of arrival of path corresponding to the dominant cluster in the channel between the base station and UE.

practically realized with the use of multiple antenna modules (where each module is often made of different types of antenna elements such as dipoles and patches arranged in different geometries). Good antenna module placement is thus an important and *new* problem at mmWave frequencies (and beyond). In contrast, note that discrete antenna placement in sub-7 GHz systems has the objective of minimizing correlation (or ensuring maximal separation between antenna elements). Proper antenna module placements can lead to subarray diversity that leverages the richness in the channel multipath structure and can also provide robustness to hand or body blockage.

3.2.3.2 Face and Edge Designs

We now illustrate two popular commercial mmWave antenna module placements, which are pictorially described in Figure 3.9. These designs are:

- A predominantly *face* design with two antenna modules (on the front and back corners of the UE) with each module made of
  1. 2×2 dual-polarized patch subarrays
  2. 2×1 (single-polarized) dipole subarray on one edge of the module
  3. 1×2 (single-polarized) dipole subarray on a perpendicular edge of the module.

  Note that it is typical to count a dual-polarized patch antenna as two antenna elements since they are fed by two independent antenna feeds. Thus, for each antenna module in the face design, the 2×2 dual-polarized patch subarray counts for 8 antenna elements and the dipole subarrays count for 4 antenna elements leading to a total of 12 antenna elements per module. Since there are two modules, we have 24 antenna elements in this design.
- An *edge* design with three antenna modules (on three edges of the UE) with each module made of 4×1 dual-polarized patch subarrays alone. For each antenna module in the edge design, the dual-polarized patch subarray accounts for 8 antenna elements. Along with the use of three modules, we have 24 elements in all here.

Information on the number of antenna modules, number of antennas (in both polarizations), approximate elemental gains of the antennas, and number and description of the different subarrays in these designs are summarized in Table 3.3. A pictorial illustration of the antenna module design is provided in Figure 3.9. Note that each antenna module in the face design consists of three subarrays per polarization. On the other hand, the antenna module in the edge design consists of only one subarray per polarization. Superficial details on codebook construction and related beamforming performance tradeoffs for these designs are listed here with specifics provided in Chapter 4. In terms of performance metric, a CDF of the *spherical coverage* that captures the beamforming gain or EIRPs achievable with the UE's antennas in a sphere ($360^\circ \times 180^\circ$ in azimuth and elevation, respectively) around it becomes a paramount benchmark of UE design and performance [74, 75, 76]. In particular, a good spherical coverage CDF corresponds to good array gains *not only* in the top few

Table 3.3
Design parameters for the face and edge designs

Parameter of interest	Face design	Edge design
No. of antenna modules	2	3
No. of antenna elements	24	24
No. of subarrays per module	4 (2×2 dual-polarized patch subarrays, 2×1 and 1×2 dipole subarrays)	2 (4×1 dual-polarized patch subarrays)
No. of beams per subarray	4 for patch and 2 for dipole subarrays	4 for each patch subarray
Codebook size per module	12 (= 4 beams $\times$ 2 patch subarrays + 2 beams $\times$ 2 dipole subarrays)	8 (= 4 beams $\times$ 2 patch subarrays)
Total codebook size	$24 = 12 \times 2$	$24 = 8 \times 3$
Elemental gain	≈ 5.8 and ≈ 4.7 dBi for patches and dipoles, respectively	≈ 5.5 dBi

(e.g., top 30) percentile points, but also in the middle (e.g., 30-th to 75-th) percentile points.

3.2.3.3 Face and Edge Design Tradeoffs

- **Manufacturing/Mounting Constraints:** Broadly speaking, placement of antennas on the edge of the UE (edge design) appears to be the easiest from a practical implementation standpoint. Further, an edge placement is robust to the precise choice of location of the antenna modules. Given the real-estate constraints in commercial devices, this robustness adds a significant level of versatility to UE design. However, as premium-tier UEs come in reduced thickness, mounting an edge antenna module requires tilting of the module within this reduced thickness leading to loss in gains over one polarization. Additionally, an edge placement can significantly reduce penetration losses associated with planar arrays placed underneath the display. Specifically, planar arrays require more space and can be placed on the face (front or back) of the UE. However, displays that are almost bezel-less have become popular in the current generation of UEs and will be increasingly used in future designs. This constraint renders the use of a planar array questionable at least on the front face of the UE. This is because in addition to finding sufficient real-estate within the display unit of the UE for the antenna module to be mounted, careful mounting of the antenna modules can also lead to manufacturing complexities and cost/time overruns.
- **Penetration Losses:** A face design can incur significant additional radiation losses due to penetration of mmWave signals through typical display materials (e.g., glass, plastic, ceramic, etc.). In particular, works such as [32, 77] and [78],

as well as Chapter 2, point out that the loss is material-dependent (depending on permittivity and loss tangent), depends on the antenna type (dipole or patch), and clearance between display and cover.

The cover or display acts as a lens/dispersive medium and scatters the signal relative to the baseline case of no cover. A glass cover will scatter more energy than a plastic cover resulting in attenuation of signals in certain directions and comparable performance or even amplification of signals in certain other directions (all relative to the case with no cover). Despite the losses experienced by antennas on display, some sensing applications propose the use of such designs as they can better detect user gestures in sensing [79]. Thus, face placements are being commercially studied.

On the other hand, analogous to display-related losses for the face design, frame-related losses can accrue for the edge design. Note that typical frame materials include plastic and metal. The typical impact of these materials is to decrease the beam's strength and/or to tilt or steer the beams away from their intended directions. From prior works such as [77] and [78], it is known that additional losses are a function of the permittivity and loss tangent of the material, antenna type, clearance between frame and antenna substrate, beam steering direction, and other details pertaining to specific placement of various materials around the antenna in the phone. In the case of plastic frames, commonly used in a broad range of UEs at the medium- and high-tier, these losses are usually minimal. In contrast, metallic frames can lead to further losses.

- **Exposure Constraints:** Another issue with the use of planar arrays on the front face of the UE is exposure of sensitive body parts (e.g., eye, skin, etc.) to the beamformed signal with high energy. Some subarrays can steer energy toward the body of the user with minor signal energy peaks. Thus, relative to the face design, the edge design is expected to have rather minor exposure-related concerns. As an illustrative example, consider the form factor UE design in Figure 3.14 with two

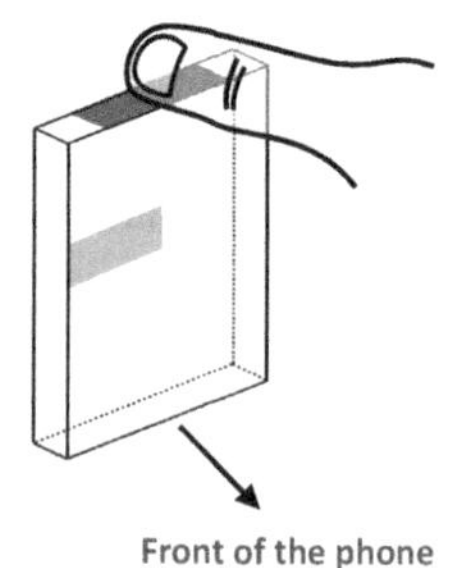

Figure 3.14 Pictorial illustration of a form factor UE with two antenna modules on the top edge and back face with parts of the hand exposed to transmit radiation from the top edge antenna module.

antenna modules where transmitted radiation from the top edge antenna module impacts the finger placed on/above this antenna module.

- **Spherical Coverage:** The use of linear arrays in the edge design necessitates as many antenna modules as possible for good spherical coverage in freespace, resulting in increased cost, power consumption as well as beam management overhead (see discussions in Chapter 4.4). In general, the edge design tries to appropriate the good features of the face design such as a small number of antenna modules by adding a layer of robustness to design. By compromising on performance over some parts of the sphere, the number of antenna modules supported by the UE can be reduced. This poor performance could be due to the edge pointing away from the serving base station(s) and toward the ground plane in portrait mode, or due to hand blockage in the landscape mode. On the other hand, the use of planar arrays with each antenna module (instead of linear arrays) allows two-dimensional beam scanning leading to a better parsing of the clusters in the channel. Further, this limits signal leakage (or interference) in unintended directions possible with one-dimensional beam scanning. Thus, a reasonable spherical coverage can be expected with the use of only two antenna modules (on the front and back) which can minimize cost, power consumption, as well as beam management overhead. In Chapter 4.4, the spherical coverage performance for the face and edge designs are illustrated with a codebook of size 24 showing that both designs are fairly competitive with each other.
- **Higher-Rank:** While patch elements allow dual-layer transmissions and receptions over the same set of beam weights, dipole elements allow only a single layer. On the other hand, a mixed mode dual-layer scheme involving some/all patch elements on one layer and some/all dipole elements on another layer (albeit with different beam weights on the two layers) is also possible. Such scenarios are important to maintain beam diversity, to support inter-band carrier aggregation and higher-rank schemes within and across antenna modules.

A table summarizing these broad tradeoffs is provided in Table 3.4.

3.2.3.4 More Advanced Antenna Module Designs

The face and the edge designs can be viewed as basic building blocks for mmWave systems. Using these blocks, we now explain some other popular designs that are proposed for use in practice. These designs are illustrated in Figure 3.15 and include:

- **Design 3:** A *maximalist edge* design with four antenna modules (on four sides of the UE) with each module made of 4×1 dual-polarized patch subarrays and a 4×1 dipole subarray. While full spherical coverage can be obtained with patch elements alone, the use of dipole elements provides complementary coverage since they point along different directions of the sphere. Thus, we can obtain better robustness at the expense of cost associated with more antenna elements as well as the control circuitry for these elements. A certain version of this design has been

Table 3.4
Relative tradeoffs between the face and edge designs

Parameter of interest	Face design	Edge design
Mounting constraints	More	Relaxed
Penetration losses	More through front display	Limited
Flexibility for MPE issues	Less	More
Spherical coverage	Comparable in freespace Tradeoffs possible in blockage mode	
Higher-rank	Possible	Possible

proposed in [80, 81] with dipole antennas alone instead of dual-polarized patches and dipoles (which is a minor design enhancement).

- **Design 4:** An *L-shaped edge* design with four antenna modules (on four sides of the UE) with each module being L-shaped and spanning two adjacent sides of coverage. Each side of coverage is made of 4×1 dual-polarized patch subarrays alone. A number of features of this design can be seen in other designs such as [82, 83, 84, 85, 86, 86, 87, 88]. A major advantage of this design is good coverage over the entire sphere. But a disadvantage of this design is the need to perform beam scanning over all modules leading to increased training latencies and power consumption. Another disadvantage of this design is the increased cost associated with the design of antenna flexible structures to allow signals from one side of the L to the other side.

The readers are further referred to [89, 90, 91, 76] for some recent studies on design tradeoffs of 5G antenna arrays with form factor considerations for antenna

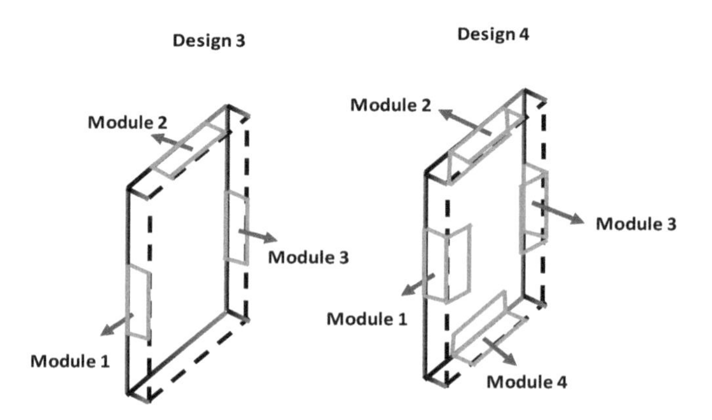

Figure 3.15 Pictorial illustration of a maximalist edge (Design 3) and L-shaped edge (Design 4) designs.

module placement. A number of reference UE architecture designs have been introduced at 3GPP for the purpose of developing testing and conformance requirement specifications for EIRP with mmWave transmissions. For example, the TR 38.803 specification has a number of potential UE reference architectures for the high bands (> 24 GHz) [92, Sec. 6.2.1.1, pp. 107–108]; also, see [82]. In compliance and testing studies considered at the 3GPP Working Group 4 level, a number of companies have proposed and considered diverse UE designs. These designs include the proposals in [84, 93, 83, 94, 95, 96, 85, 97, 81, 86, 80, 98]. Antenna array modeling and spherical coverage issues for 5G-NR systems (especially the UE) can be found in [89, 76, 99]. Antenna modeling for the 116–260 GHz range is studied in [64].

3.3 CONSIDERATIONS AT THE RF LEVEL

3.3.1 BEAMFORMING ARCHITECTURE TRADEOFFS

3.3.1.1 Digital Architecture

At sub-7 GHz frequencies in 4G systems, each of the possibly multiple antennas is equipped with an RF chain of an independent low-noise amplifier (LNA), down-converter(s)/mixer(s) and analog-to-digital converter (ADC) for the downlink or reception path, and a digital-to-analog convertor (DAC), up-converter(s)/mixer(s) and power amplifier (PA) for the uplink or transmission path. This is possible due to the low cost, complexity, area and power consumption of an RF chain at sub-7 GHz frequencies, as well as the smaller number of antennas or RF chains necessary at these frequencies for meeting a certain link budget. Note that the RF chain operates at a significantly lower bandwidth than at mmWave. Thus, beamforming at sub-7 GHz frequencies has traditionally been a *digital* architecture, as illustrated in Figure 3.16, where as many data-streams/layers as the number of antennas can (in theory) be either transmitted or received. On the receive path with this architecture, the in-phase and quadrature (I/Q) signals from each antenna element are down-converted and digitized. The baseband I/Q bits from the ADCs are processed using a digital signal processor (DSP) where different sets of beam weights can be used to null out interfering signals. Since the interference gets nulled only after the digital signal processing, the dynamic range of the mixer and ADC/DAC of each antenna element must be versatile to handle the interference. Since the DSP can perform complex combinations of beam weights, they can be used gainfully to handle a wide variety of interference scenarios across a broad frequency range.

Using the digital beamforming architecture at mmWave carrier frequencies can be disadvantageous since the power consumption of different components can be quite significant as the components' supported bandwidth increases. Such an architecture can also be onerous in terms of area on the chip, thermal overheads generated, cost, and in meeting regulatory or compliance requirements due to increased EIRPs possible with this architecture. Thus, for making such beamforming systems viable, it becomes a necessity to have substantially fewer RF chains than antenna elements.

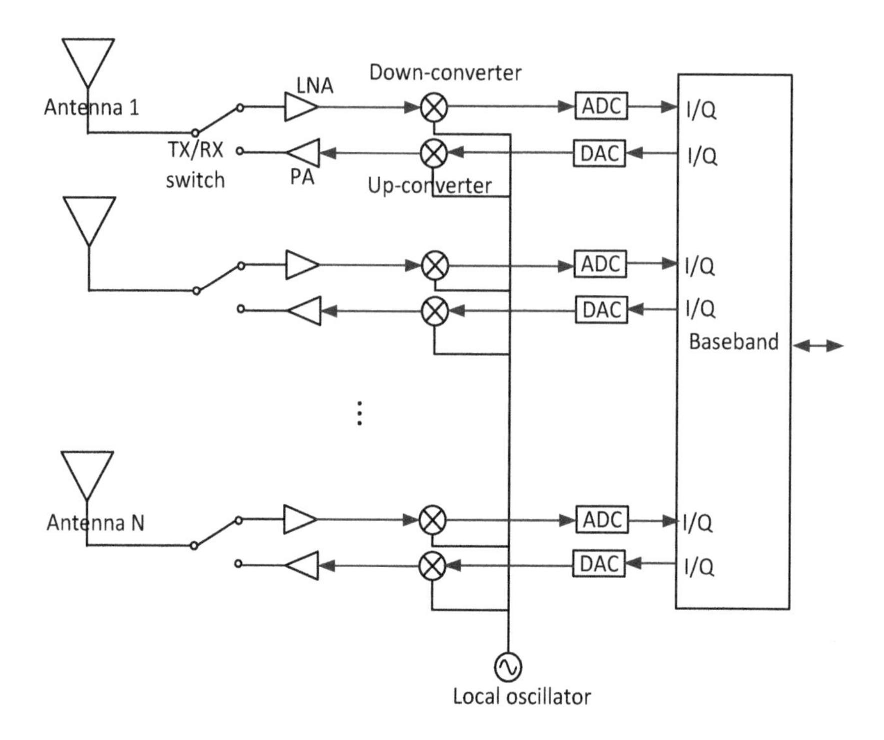

Figure 3.16 Conventional/traditional digital beamforming architecture.

3.3.1.2 RF Architecture

At one extreme of this thinking is an *RF/analog* beamforming architecture that uses only one RF chain[9] for the entire antenna array, independent of the number of antennas. In a particular rendition of analog beamforming, an RF phase shifting architecture is illustrated in Figure 3.17(a). Here, on the receive path, signals from the different antennas are phase shifted and combined at the RF domain. The combined symbol is then down-converted and digitized to enable baseband processing. A proper choice of beam weights can balance the objectives of amplifying a weak signal and suppressing a strong interference signal. Since the interfering signal is suppressed prior to the down-/up-converter or mixer, the requirements on the mixer and the ADC/DAC can be relaxed. The use of a variable gain amplifier (VGA) along with the phase shifter on the RF path can be challenging since a change in the gain state of the VGA is accompanied by a change in the phase response of the phase shifter.

[9]In practice, since dual-polarized antenna elements can be designed within the same aperture as single-polarized antenna elements (at these frequencies), the RF/analog beamforming architecture corresponds to two RF chains with one RF chain per polarization.

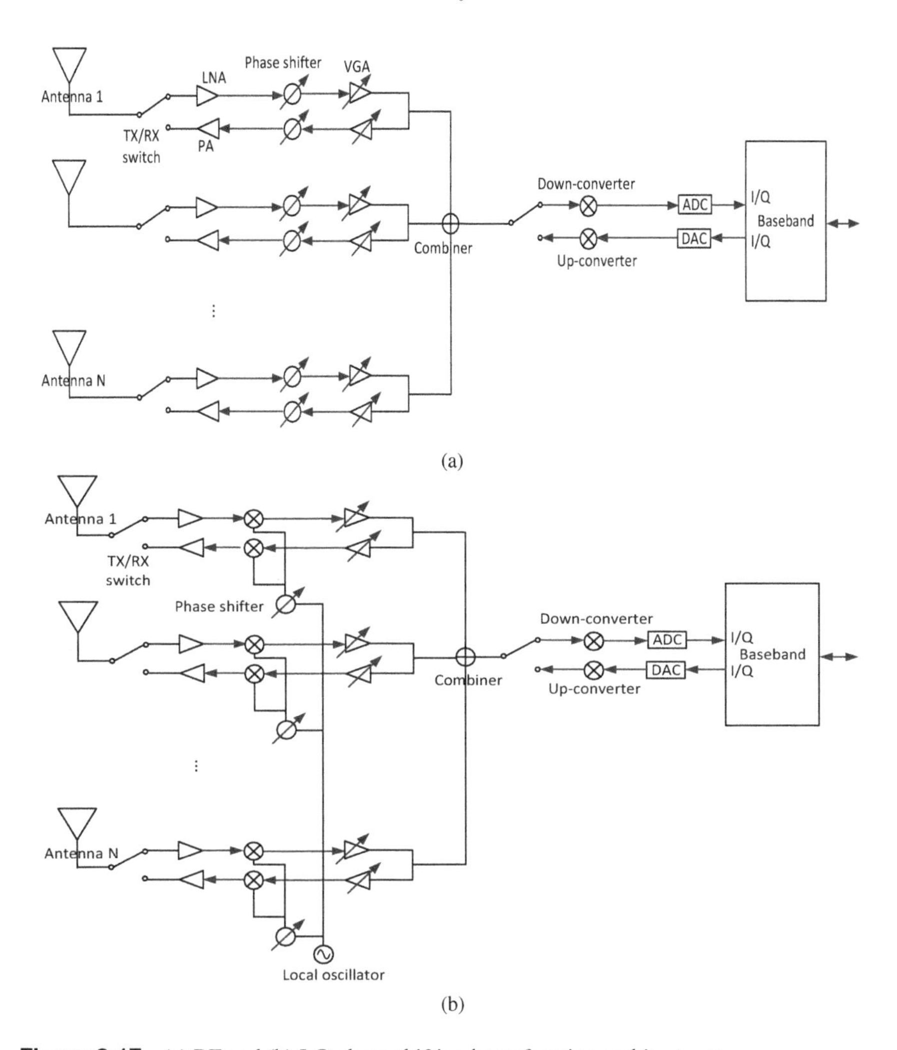

Figure 3.17 (a) RF and (b) LO phase shifting beamforming architectures.

In another rendition, a local oscillator (LO)-based phase shifting architecture is illustrated in Figure 3.17(b). Here, on the receive path, the RF signal from each antenna is mixed with a potentially different phase shifted version of the LO. The combined signal is then baseband processed after a mixer down-converts the intermediate frequency (IF) signal to baseband if needed, and an ADC digitizes the signal. Similar processing can be assumed on the transmit path. The main advantage with this architecture is the relaxed requirements on the phase shifter linearity since they are not on the RF path. In particular, a phase shifter on the IF path is relatively easier to design for narrowband systems. The disadvantage with this architecture is that a mixer is needed for each antenna element and interference is mitigated only after the down-/up-conversion operations. This imposes a strong dynamic range on the mixers and also subjects the array to intermodulation products (see discussion in Chapter 3.3.3) that can lead to interference. Distributing the LO over a large array can also be problematic in practice with feedline losses and jitter.

While the analog beamforming architecture is cost effective, it comes at the price of beam steering along only one beam. This limits the achievable rate since full spatial multiplexing gain in a channel is not realized. The beam weights used in analog beamforming typically correspond to steering energy along a single direction, although this is not always the case since beam weights can be designed appropriately to steer energy along multiple directions. However, this approach requires careful design of adaptive beam weights, which can be complicated in terms of measurements and processing (see Chapter 4.2.4 for a discussion).

3.3.1.3 Hybrid Architecture

A *hybrid* beamforming architecture is more realistic and traverses between the extremes of analog and digital beamforming architectures [100]. Here, a smaller number of RF chains are used relative to the number of antennas allowing beam steering along multiple directions or transmission of multiple layers at a time. Similar to the analog beamforming architecture, multiple transmit and receive paths are created with the same/different set of antenna elements via the use of appropriate transmit and receive RF circuitry.

Two popular practical implementations of a hybrid beamforming architecture are now described. In one approach, a distinct RF chain is used for a distinct subset of the antennas with the union of the distinct subsets being the entire antenna array. This approach is also known as the *subarray* or *sub-connected* hybrid architecture [101, 102]. This is illustrated in Figure 3.18 where two subsets of N antenna elements each form a size $2N$ antenna array. One *independent* RF chain is available over each subset of N antenna elements. In a second approach, multiple RF chains are used over all the antennas and this approach is also known as the *fully connected* or *shared* hybrid architecture. This approach is illustrated in Figure 3.19 where, for example, two RF chains are used over the entire antenna array of size N.

The sub-connected architecture makes the design modular or easier as it is analogous to multiple analog or RF beamforming constructions over distinct subsets of antennas. Different such modules can be placed on different parts of the UE to result in diversity to combat blockage. However, it leads to smaller array gains than possible with the fully connected architecture. Further, as the antenna array size increases, feedline length can increase leading to loss in signal strength at mmWave frequencies for the fully connected architecture. Feedline crossings of different RF components can also increase leading to enhanced interference and performance degradation, necessitating complex mitigation mechanisms at the circuit level. On the other hand, the sub-connected architecture avoids real-estate constraints as antennas can be used without coexistence concerns with other circuits or components. The array gain (including feedline losses) vs. ease of design tradeoffs determine the choice of whether to use a sub-connected or a fully connected architecture in practice.

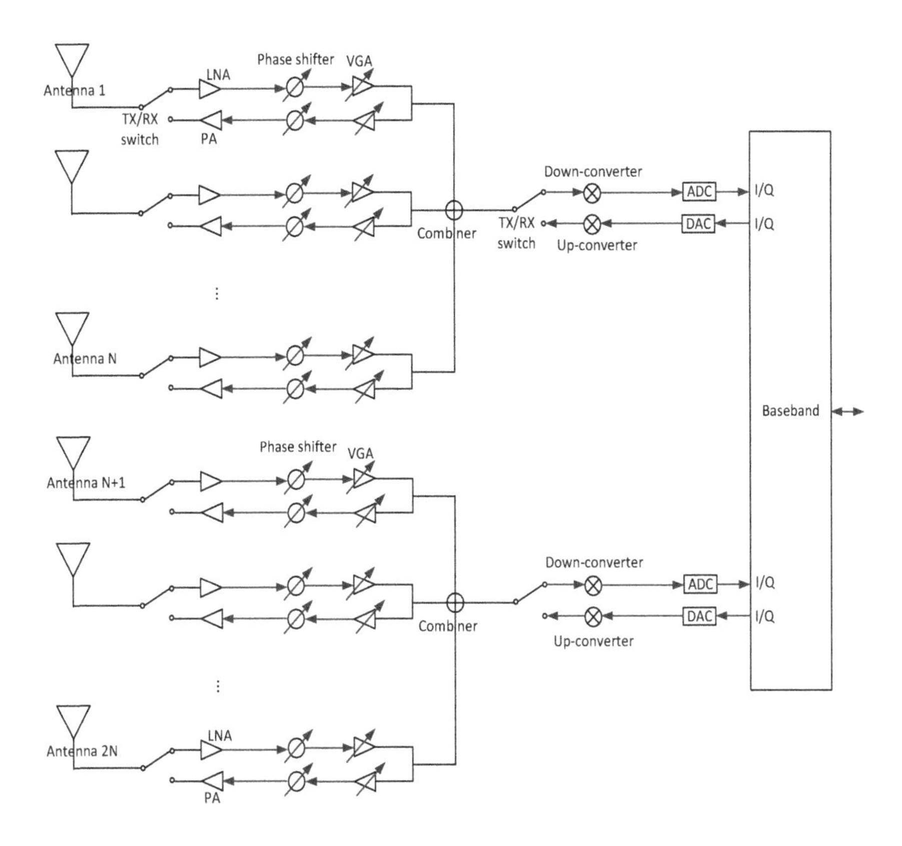

Figure 3.18 A sub-connected hybrid beamforming architecture with two RF chains.

Table 3.5 summarizes a list of tradeoffs between the sub-connected and fully connected architectures. These tradeoffs lead to a higher preference for the sub-connected hybrid architecture in practical implementations especially at the UE. On the other hand, since area constraints are minimal and the base station is a centrally shared entity, the fully connected architecture could be preferred at the base station.

3.3.1.4 System Model for Hybrid Architecture

The system model of the hybrid beamforming architecture is captured by the equation:

$$y(k) = G_{\mathsf{dig}}(k)^H \, G_{\mathsf{RF}}^H \cdot \left[H(k) \, F_{\mathsf{RF}} \, F_{\mathsf{dig}}(k) \, s(k) + n(k) \right], \tag{3.28}$$

where $H(k)$ denotes the channel matrix, $n(k)$ denotes the additive white Gaussian noise vector, $s(k)$ denotes the transmitted vector and $y(k)$ denotes the received vector, all over the k-th subcarrier. Assuming that the transmitter and the receiver have $N_{\mathsf{RF},t}$ and $N_{\mathsf{RF},r}$ RF chains, but with N_t and N_r antenna elements (satisfying $N_{\mathsf{RF},r} \leq N_r$ and

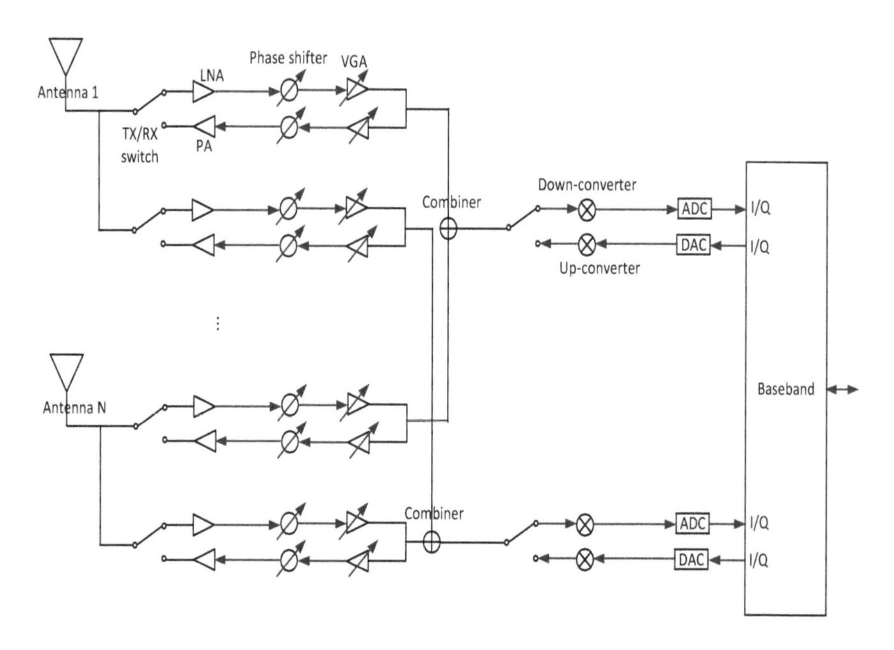

Figure 3.19 A fully connected hybrid beamforming architecture with two RF chains.

$N_{\mathsf{RF},t} \leq N_t$), the analog precoders F_{RF} and G_{RF} are of dimensions $N_t \times N_{\mathsf{RF},t}$ and $N_r \times N_{\mathsf{RF},r}$, respectively. On the other hand, the subcarrier-dependent digital precoders $F_{\mathsf{dig}}(k)$ and $G_{\mathsf{dig}}(k)$ are of dimensions $N_{\mathsf{RF},t} \times s$ and $N_{\mathsf{RF},r} \times s$, respectively, where

$$1 \leq s \leq \min\left(N_{\mathsf{RF},t}, N_{\mathsf{RF},r}\right). \tag{3.29}$$

In a sub-connected hybrid architecture, the analog precoders are block diagonal. Assuming that M and K antennas are excited at the transmit and receive sides with a

Table 3.5

Tradeoffs between sub-connected and fully connected hybrid beamforming architectures

Objective	Sub-connected	Fully connected
Beamforming gain	Smaller	Larger
Feedline losses	Smaller	Larger
Interference and coexistence issues	Easier	More difficult
Real-estate considerations	Can be mitigated	More difficult

single RF chain (that is, $N_t = N_{\text{RF},t}M$ and $N_r = N_{\text{RF},r}K$), we have

$$F_{\text{RF}} = \begin{bmatrix} \underbrace{\mathsf{f}_1}_{M\times 1} & \underbrace{0}_{M\times 1} & \cdots & \underbrace{0}_{M\times 1} \\ \vdots & \vdots & \ddots & \vdots \\ \underbrace{0}_{M\times 1} & \underbrace{0}_{M\times 1} & \cdots & \underbrace{\mathsf{f}_{N_{\text{RF},t}}}_{M\times 1} \end{bmatrix}, \; G_{\text{RF}} = \begin{bmatrix} \underbrace{\mathsf{g}_1}_{K\times 1} & \underbrace{0}_{K\times 1} & \cdots & \underbrace{0}_{K\times 1} \\ \vdots & \vdots & \ddots & \vdots \\ \underbrace{0}_{K\times 1} & \underbrace{0}_{K\times 1} & \cdots & \underbrace{\mathsf{g}_{N_{\text{RF},r}}}_{K\times 1} \end{bmatrix}. \tag{3.30}$$

As a generalization of the sub-connected and fully connected hybrid architectures, one can consider D groups (of antennas) with each group consisting of M antennas (thus, $N_t = DM$). Each of the D groups is activated with S RF chains for $N_{\text{RF},t} = DS$. Such an architecture is sometimes labeled as the partially connected hybrid architecture. Note that with $S = 1$, we have $N_{\text{RF},t} = D$ and $N_t = N_{\text{RF},t}M$ or the sub-connected hybrid architecture. Similarly, with $S = N_{\text{RF},t}$, we have $D = 1$ and hence $N_t = M$ or the fully connected hybrid architecture. While such a generalization is mathematically interesting from a signal processing optimization perspective, from a practical point-of-view, a simple sub- or fully connected architecture shall suffice to tradeoff cost, area and complexity with performance. In all the four architectures discussed here, multi-carrier/OFDM signaling can be easily supported.

The readers are referred to [100, 103] for a more detailed circuits-oriented discussion on different beamforming architectures and their tradeoffs.

3.3.2 SIMPLIFIED BEAMFORMING NETWORKS

While phased arrays and their renditions via digital, analog, or hybrid architectures as described in Chapter 3.3.1 lead to no specific constraints on the beam weights (phases and amplitudes) generated, constrained beam weight generation with simplified beamforming networks has been well studied in the literature.

If a device intends to transmit toward N directions simultaneously (where $N > 1$), a hybrid beamforming architecture with N RF chains can be considered. Such an approach can be sub-optimal from the perspectives of power/thermal, cost, and area. An alternative approach is to use a Butler matrix. The Butler matrix is a beamforming network with N input ports where the signal is applied and N output ports where the antenna elements are connected (typically, N is a power of 2). A Butler matrix architecture produces a spatial Fourier transform and for this, it requires the following circuit elements in implementation:

- $(N/2) \cdot \log_2(N)$ quadrature (or 90°) hybrid couplers
- $(N/2) \cdot (\log_2(N) - 1)$ fixed value phase shifters.

A quadrature (or 90°) hybrid couple splits the input port's power equally across two output ports and produces a relative phase offset of 90°. As an example, the "in-phase" output port has a 90° phase and the "out-of-phase" output port has a 180° phase. A Butler matrix architecture produces N fixed, orthogonal and simultaneously steerable beams allowing multi-layer transmissions. Fixed and simultaneously

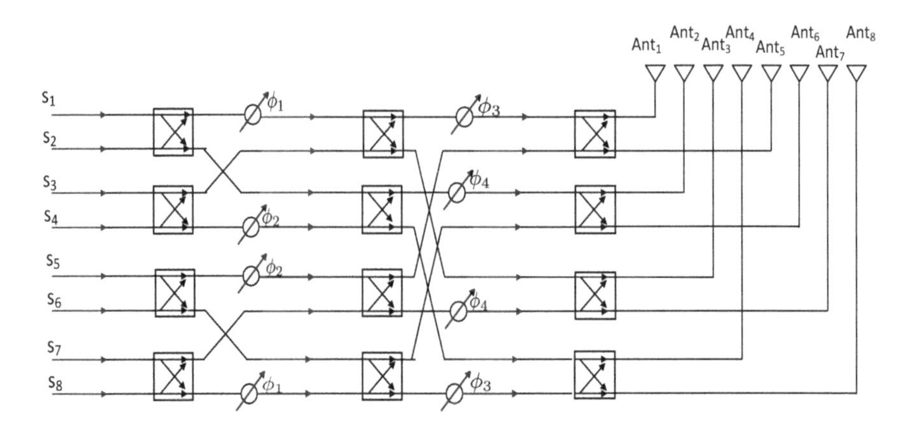

Figure 3.20 Butler matrix architecture for $N = 8$.

steerable means that each of the N input ports of a Butler matrix can produce a distinct beam, all of which can be realized at any instant by selecting all the input ports. Since multiple beams can be generated with a compact architecture, a Butler matrix is advantageous in area and power savings, especially as carrier frequency increases. However, the main disadvantage of this architecture is that typically it can only generate beams with the narrowest beamwidth possible for that array dimension along specific sets of steerable directions, thereby losing the flexibility or adaptability possible with a phased array architecture.

As an example, an 8-input 8-output Butler matrix architecture with four fixed value phase shifters (denoted as ϕ_1, ϕ_2, ϕ_3 and ϕ_4) is illustrated in Figure 3.20. Note that only four independent phase shifters are used for symmetry reasons. Here, the path from signal s_1 to the first antenna element requires a 90° "in-phase" phase shift followed by ϕ_1, 90°, ϕ_3 and 90° for a net phase shift of $-270° + \phi_1 + \phi_3$. Similarly, the path from s_1 to the second antenna element requires a 180° "out-of-phase" phase shift followed by 90°, ϕ_4 and 90° for a net phase shift of $-360° + \phi_4$. Following this recipe, the net phase shifts seen by the signal s_1 across the eight antenna elements are given as

$$\Delta\mathsf{s}_1 = \begin{bmatrix} -270° + \phi_1 + \phi_3, & -360° + \phi_4, & -360° + \phi_1, & -450°, \\ -360° + \phi_1 + \phi_3, & -450° + \phi_4, & -450° + \phi_1, & -540° \end{bmatrix} \quad (3.31)$$

$$= \begin{bmatrix} 90° + \phi_1 + \phi_3, & \phi_4, & \phi_1, & -90°, \\ \phi_1 + \phi_3, & -90° + \phi_4, & -90° + \phi_1, & -180° \end{bmatrix}. \quad (3.32)$$

A similar phase description across the eight antenna elements can be obtained with signals s_i, $i = 2,\ldots,8$. The beam pattern of these eight fixed beams (corresponding to signals s_i, $1 \leq i \leq 8$) with fixed value phase shifter choices $\phi_1 = -67.5°$, $\phi_2 = -22.5°$ and $\phi_3 = \phi_4 = -45°$ are illustrated in Figure 3.21. Clearly, the beam patterns are symmetrical, orthogonal and correspond to fixed directions with the

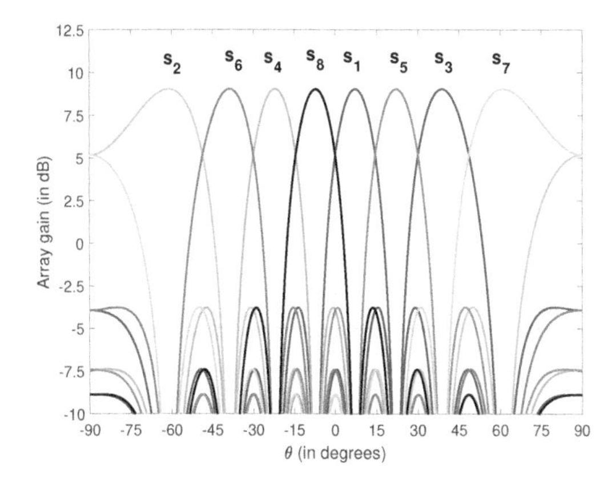

Figure 3.21 Array gain with the $N = 8$ fixed beams generated with a size 8 Butler matrix architecture.

overlap point between adjacent beams being ≈ 3.9 dB below the peak array gain. Such a design can be used at base stations for transmissions to multiple users (via MU-MIMO) or to a single user over multiple layers (via SU-MIMO). Alternately, if the Butler matrix architecture can incorporate a set of tunable reflective loads or varactors, a tunable phase shifter can be emulated leading to the design of beam weights with beam pattern peaks in between or intermediate to those that can be generated with a standard Butler matrix architecture. Sometimes, these beams may be called as intermediate beams.

Other versions of simplified beamforming networks include a Blass matrix that uses transmission lines (instead of phase shifters) and directional couplers (that divide power into two streams of which a special case is a quadrature hybrid coupler) to form multiple beams by means of true time delays. This architecture is thus suitable for broadband operation. More interesting lens array architectures include a Luneberg or a Rotman lens that operate based on the principle of geometrical optics. In this network, multiple beams are formed without the use of phase shifters and/or switches. Instead, phase shifting and power splitting or combining operations are handled by the careful design of a lens structure or cavity. In particular, a geometrically configured waveguide of a carefully chosen shape and appropriate length of the transmission lines leads to passive phase shifts of the inputs with desirable beam patterns.

Tunable Butler matrix architecture designs are pursued in [104, 105]. Beamforming design with a lens array architecture and associated design tradeoffs are discussed in [106–111].

3.3.3 POWER AMPLIFIERS

Beamforming circuits should be designed so as to provide high EIRP transmissions, adaptive and highly directional gains, low signal leakages in undesired directions, operation over large bandwidths, and interworking across multiple frequency bands. The design of a good quality PA driving each antenna element of the antenna array is thus a key stepping stone toward this goal. An alternate design where multiple antenna elements are fed by a common PA can also be considered. However, since the power from the common PA is divided across the antenna elements, a higher PA rating is needed for the common PA complicating its area, cost and power consumption profile.

The objective with a PA design is to generate high output power by efficiently converting direct current (DC) power to RF via linear amplification. The quality of the PA design is evaluated by the realization of maximum power gain under stable operating conditions with a minimum number of amplifier stages, independent of the requirements on linearity, reliability, efficiency, cost or area. Overall, the design of PAs needs to incorporate the following major constraints:

- PAs drive large voltages or currents into small load impedances. Thus, matching networks are critical. Any loss in the matching network has a severe impact on the efficiency of the PA.
- Due to increased current consumption, heat generation can be high. Thus, we need to carefully provide heat sinks to keep the junction temperatures as low as possible.
- Due to the interface with the external off-chip environment, packaging and board parasitics are very important.
- The spectral leakage and harmonics generation in a PA must be kept to a minimum in order to minimize interference to other users.

The quality of a PA design is captured using the following figures-of-merit:

- PA gain: The typical circuit level model of a PA is described in Figure 3.22(a) with an output power-to-input power relationship described in Figure 3.22(b). It consists of a linear region where the input power P_{in} is amplified by G dB with output power given as

$$10 \cdot \log_{10}(P_{\text{out}}) = 10 \cdot \log_{10}(P_{\text{in}}) + G. \tag{3.33}$$

From (3.33), with $P_{\text{in}} = 1$, we get $P_{\text{out}} = 10^{G/10}$ and this operating point is marked in Figure 3.22(b). It also consists of a non-linear or compression region where P_{out} saturates (or is clipped) even as P_{in} increases leading to a degraded error vector magnitude (EVM) or a poor adjacent channel leakage ratio (ACLR) performance. The gain of the PA is the rate at which output power from the PA increases for a unit increase in the input power. Mathematically, it is defined in dB scale as $10 \cdot \log_{10}\left(\frac{dP_{\text{out}}}{dP_{\text{in}}}\right)$ where $\frac{d\bullet}{d\bullet}$ is the derivative. This gain in the linear and non-linear regions are called the small signal and large signal gains, respectively. The maximum P_{out} obtainable from the PA is called the saturated output power (denoted as P_{sat} and sometimes as P_{max}). P_{sat} can be seen to be the product of the maximum supportable voltage (directly related to the DC supply voltage) and the maximum

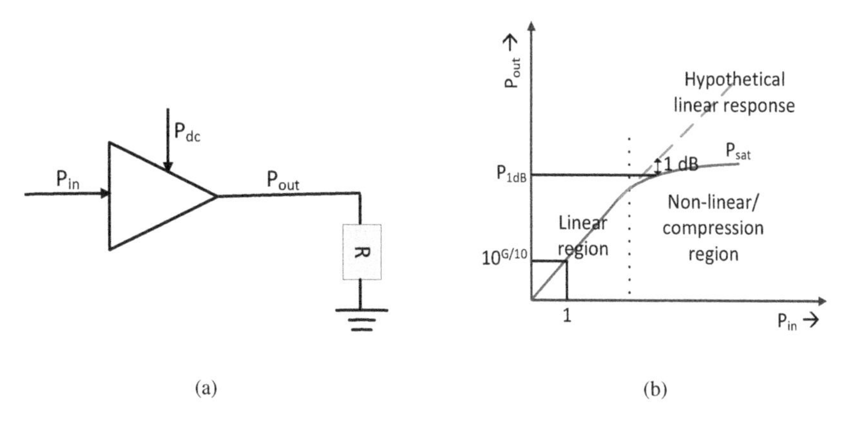

Figure 3.22 (a) Typical circuit-level representation of a PA. (b) Typical P_{in}-P_{out} (in linear scale) relationship for a PA.

supportable current. Note that the gain of the PA (the derivative) approaches zero (in absolute scale) as $P_{\text{out}} \to P_{\text{sat}}$. The gain of a PA is frequency-dependent and hence its bandwidth is defined as the frequency regime over which the gain is relatively flat, which is appropriately defined. Since mmWave systems cover a broad frequency regime, a broad PA bandwidth is a useful design metric.

- Efficiency: A number of efficiency metrics associated with a PA can be defined. Power added efficiency (PAE) quantifies the effectiveness of a PA in converting DC power to RF power. It is defined as

$$\eta_{\text{PAE}} = \frac{P_{\text{out}} - P_{\text{in}}}{P_{\text{dc}}}. \tag{3.34}$$

While η_{PAE} can range from 0% to 100%, a good PA has a high value of η_{PAE}. On the other hand, drain efficiency is defined as the efficiency of the PA relative to the DC power (the primary input DC power is fed to the drain of a field-effect transistor) and is given as

$$\eta_{\text{d}} = \frac{P_{\text{out}}}{P_{\text{dc}}}. \tag{3.35}$$

From (3.33), in the linear region, we have

$$\eta_{\text{PAE}} = \eta_{\text{d}} \cdot \left(1 - 10^{-G/10}\right). \tag{3.36}$$

Note that if the gain G is large, then $\eta_{\text{PAE}} \approx \eta_{\text{d}}$. Any power that is not converted to useful signal in a PA is dissipated as heat. Thus, PAs with low efficiencies have high levels of heat dissipation, which could be a limiting factor in practical implementation.

- $P_{1\text{dB}}$: The 1 dB compression point (denoted as $P_{1\text{dB}}$) is the output power level at which the true gain decreases by 1 dB relative to the gain corresponding to the hypothetical linear response. In many PA data sheets, $P_{1\text{dB}}$ can be referenced to the input power level as well. $P_{1\text{dB}}$ captures the notion that as the PA approaches saturation, wherein P_{out} nears $P_{1\text{dB}}$, a non-linear behavior is observed leading to signal distortion, higher-order harmonics and intermodulation. Thus, to avoid these deleterious effects, PAs should be operated below $P_{1\text{dB}}$ with a power backoff. Since the PAE is largest at P_{sat}, power backoff trades off linearity of the device with PAE.

Distortions induced by power amplifiers are broadly classified into three types:

- *Intermodulation (IM) distortion*: When a multi-tone signal (such as an OFDM signal) is input to a non-linear device, the device produces IM distortion or signals at integer linear combinations of the individual tones of the multi-tone signal. For example, consider a two tone input at carrier frequencies f_1 and f_2. In addition to the fundamental frequency components (f_1 and f_2), second-order intermodulation components ($2f_1, 2f_2, f_1 \pm f_2$), third-order intermodulation components ($3f_1, 3f_2, 2f_1 \pm f_2, 2f_2 \pm f_1$), fourth-order intermodulation components ($4f_1, 4f_2, 2f_1 \pm 2f_2, 3f_1 \pm f_2, 3f_2 \pm f1$), etc. can be produced. If f_1 and f_2 are comparable frequencies, the second-order intermodulation components are either closer to DC frequency, or at significantly higher carrier frequencies which can then be easily filtered. On the other hand, the first odd-order intermodulation components/products such as $2f_1 - f_2$, $2f_2 - f_1$, etc. become comparable with the frequency content of the two-tone input and hence cannot be easily filtered leading to distortion. These components, sometimes called IP_3[10], are an important non-linearity metric of the PA along with $P_{1\text{dB}}$. Figure 3.23 illustrates the fundamental and intermodulation components with the two tone input example as described above.
- *Adjacent channel leakage ratio (ACLR)*: In general, in the linear region of the PA, intermodulation products of the input signals do not interfere with the true signals. However, as the PA approaches saturation, intermodulation products can be fed back to the input of the PA and further contribute to the output signal's poor quality. Intermodulation products can also leak into adjacent channels, which is often measured by the ACLR or adjacent channel power ratio (ACPR). Specifically, amplitude variations arising from temperature and power supply variations as well as multipath fading can non-linearly distort both the amplitudes and phases which are captured by the amplitude-to-amplitude (AM-to-AM) and amplitude-to-phase (AM-to-PM) distortion components, respectively. These components capture the change in output power and phase for a 1 dB increment in the power applied to the amplifier's input. An ideal amplifier would have no interaction between its amplitude and phase response and the power level of the input signal. AM-to-AM

[10]The intercept points of order n (or IP_n for short) capture the power level at which the power of the desired tone and the n-th order intermodulation product are equal.

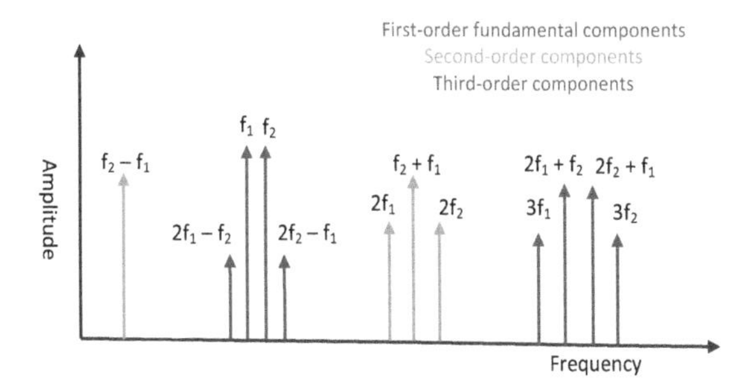

Figure 3.23 Fundamental/first-order and intermodulation components for the two tone example.

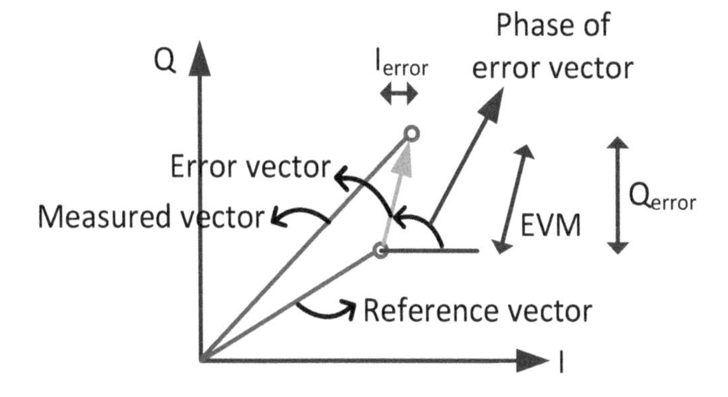

Figure 3.24 Pictorial illustration of EVM in a communications system.

and AM-to-PM components are important as they cause distortions in amplitude-modulated signals (e.g., QAM) and phase-modulated signals (e.g., FM, QPSK), respectively.

- *Error vector magnitude* (*EVM*): Another figure-of-merit[11] capturing the fidelity of the output signal relative to the input signal (or, in general, a measured vector relative to a reference vector) is the EVM which measures the in-band signal quality. EVM is an RF metric that captures the RMS of the magnitude of the error vector defined as the difference between the transmitted symbol vector (in the I/Q plane) and the closest ideal constellation location. A pictorial illustration of the EVM is presented in Figure 3.24 with the magnitude and phase of the error

[11]Measuring the performance of a system with multiple impairments is difficult since each impairment can impact the bit error rate differently. In this context, the EVM is a simpler metric that captures the impact of these different impairments with a single value.

vector being useful metrics of interest. Similarly, errors in the individual I and Q components can also be defined.
Low EVM values are typically difficult to be realized at high throughputs since the use of higher-order modulation means that the constellation points are relatively close and, are therefore more susceptible to noise and PA non-linearities.

In general, the designs of integrated circuits (ICs) for PAs at sub-7 GHz and mmWave frequencies differ in the following ways:

- At sub-7 GHz, the transistors have more usable gain and therefore, there is a wider range of tradeoffs available between size, output power and efficiency
- The lower the frequency, the better we are able to model parasitic resistances and capacitances to design inductors or transformers
- Measurements are more accurate at lower frequencies which implies that models are more accurate and thus, more academic/research work has been done in the IC field allowing us to better explore the design tradeoffs.

Some aspects of power amplifier design for 5G-NR systems with form factor constraints are discussed in [112] and [113]. The readers are referred to a comprehensive PA survey in [114] for comparison of benchmarks across frequencies. Other references to look for PA design include [115] and [116].

3.3.4 PHASE SHIFTERS

A phase shifter is an important component in RF circuitry that introduces the correct phase response to align signals across a multi-antenna phased array on either the transmit or receive side. Phase shifters can be controlled electrically, magnetically or mechanically and can be realized in various technologies (e.g., diode-based, microelectromechanical systems (MEMS)-based, ferrite-based, etc.) with different topologies.

A number of metrics are associated with the design of a phase shifter:

- Insertion loss is the loss in signal power (from input to output) when inserting a phase shifter in a system. This loss is determined by the number of stages needed for the phase shift operation and the frequency of coverage. Insertion loss can change with phase shift delivered and this effectively requires the use of VGAs for specific phase shifts for a flat gain response over the phase shift range.
- Dynamic range is the difference between the largest and the smallest phases that can be obtained by a phase shifter. A wide dynamic range can be obtained by cascading multiple phase shifter cells at the cost of larger chip area and more insertion loss.
- Similar to the case of PAs, IP_n and in particular, IP_3 measure the linearity of the phase shifter response.
- In a typical transceiver, the phase shifter is inserted either before or after other blocks such as antenna elements, LNAs and mixers. Return loss (or equivalently, VSWR) is a measure of the impedance of the phase shifter and whether it is

matched to the network for maximum power transfer or not. In a good design, the reflected power due to impedance mismatch should be low.

- In the most common implementation, the phase shifter is located before a PA in a transmit path and after an LNA in a receive path. A phase shifter needs to work in the linear region of the PA/LNA to guarantee an output response independent of the signal power and MCS used.

The figure-of-merit of a phase shifter circuitry (in degrees/dB) is defined as the ratio of the maximum phase shift to the maximum insertion loss incurred.

A continuous phase shift can be generated by analog phase shifters such as the voltage-controlled phase shifter operated by varactor diodes. On the other hand, digital phase shifters discretize the absolute phase into pre-determined phase states with the phase states changed by digitally controlling each phase shifter bit. In a digital phase shifter, the step-size between values is called the phase resolution and is determined by the number of bits. For good beam steering capability and side lobe control, adequate phase resolution is required which increases the cost and area of phase shifter cells. While 3 bit (or better granularity) phase shifters are sufficient for achieving peak directivity without significant performance degradation, higher resolution phase shifters are important for interference management via side lobe control (see the discussions in Chapter 4). In an analog phase shifter, the phase resolution depends on the quantization accuracy of the control voltage of the DAC.

In addition, a phase shifter can be an active or a passive RF component. An active analog phase shifter operates based on a programmable weighted combination of I/Q signals using a vector modulator including two VGAs in the I/Q paths. Here, two orthogonal vectors are summed with varying ratios allowing for different phase shifts to be generated. The gains in the VGAs are selected using digital bits (DACs) as demonstrated in Figure 3.25(a). Figure 3.25(b) shows the different gain settings and input bits for a 3 bit DAC to produce different phases. While an active vector modulator design can realize a wide dynamic range, relatively low insertion loss, better integration and better gain/phase calibration, it comes at the cost of increased power consumption, requirement of several control voltages, increased area on the chip, non-linearity in performance and limited operating frequency range or bandwidth.

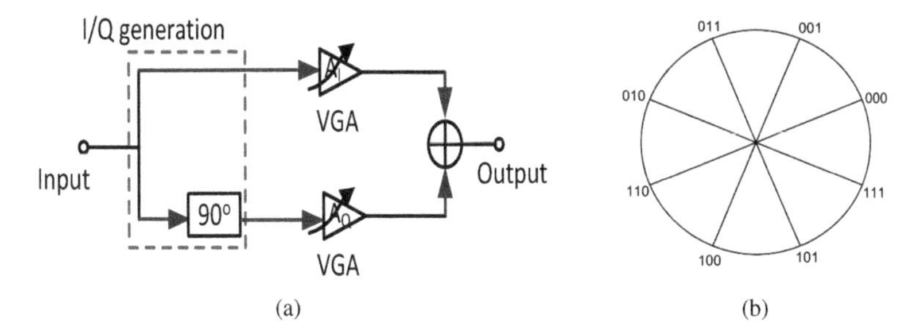

Figure 3.25 (a) Block diagram of a vector modulator phase shifter. (b) Phase selection diagram as a function of digital gain selection input bits.

A passive phase shifter can be of many types including reflection type, switched line, varactor-loaded transmission line and high- or low-pass. In a reflection type phase shifter, a hybrid coupler divides the input signal into I/Q components and a reflective load varies the phase by changing the impedance of the load. A reflection type phase shifter can provide a continuous phase shift. However, its major disadvantages include the use of couplers that occupy a large area, and high and variable insertion loss for different phase shifts. In a switched line phase shifter, a desired phase shift is realized by changing the length of the transmission line through switching among transmission lines with different electrical lengths. A switched line phase shifter offers a wide discrete phase shift range, whereas its disadvantages include high switching losses and poor switch isolation. A transmission line-based phase shifter has zero DC power consumption and offers a continuous phase shift. However, it has a limited phase shift range and difficulty in achieving proper matching over the entire phase shift range. In a high- or low-pass phase shifter type, transmission lines with different lengths are replaced with different filters having specific phase responses. Its major disadvantages include increased area for a filter structure, increased cost in fabrication and high insertion loss.

Phase shifter design has the same general implementation challenges as a PA design. At higher frequencies, we have less gain, higher losses due to the limited $R_{\text{on}}C_{\text{off}}$ figure-of-merit of switches, skin depth issues and more uncertainty in modeling due to reduced accuracies in measurements. More specific phase shifter design tradeoffs for mmWave frequencies are discussed in [117, 118, 119] and [120].

3.3.5 PHASE NOISE

Phase noise is an impairment that introduces a random phase drift on the communication symbols across different carriers and can be effectively seen as a multiplicative noise. As systems operate at higher carrier frequencies, the impact of phase noise typically becomes stronger. Further, it increases with the use of higher bandwidths and higher-order MCSs.

LO signals are typically generated by a phase locked loop (PLL) driven by a reference oscillator[12]. The different LO frequencies are generated by selecting different values for the frequency divider of the PLL. Noise in the reference oscillator is transferred to phase error/noise in the output signal. The accumulation of phase error makes the phase $\phi(t)$ behave like a random walk leading to phase instabilities. IEEE defines terms such as phase stability and Allan deviation (a measure of frequency stability in clocks/oscillators) in [121, Appendix A.2]. In general, the variance of $\phi(t)$ increases with time. Thus, $\phi(t)$ is non-stationary[13]. However, the oscillator signal

---

[12]A reference oscillator, often called as temperature compensated crystal oscillator (TCXO), is a crystal oscillator with a temperature-sensitive reactance circuit in its oscillation loop to compensate the frequency-temperature characteristics inherent to the crystal unit.

[13]An alternate view is to consider the phase noise only over a small time-period; e.g., a few slots. Over this period, the drift of the phase noise is limited and it can be treated as quasi-stationary over this time-period.

$w'(t) = e^{j(2\pi f_c t + \phi(t))}$ is stationary, as the following computation shows:

$$R_{w'w'}(\tau) = \mathsf{E}\left[w'(t+\tau)w'(t)^\star\right] \tag{3.37}$$
$$= e^{j2\pi f_c \tau} \cdot \mathsf{E}\left[e^{j(\phi(t+\tau)-\phi(t))}\right] \tag{3.38}$$
$$= e^{j2\pi f_c \tau} \cdot e^{-\frac{\mathsf{E}\left[|\phi(t+\tau)-\phi(t)|^2\right]}{2}}. \tag{3.39}$$

The last equality in (3.39) follows from the closed-form computation of the characteristic function [19] of a Gaussian random variable $X \sim \mathcal{N}(\mu, \sigma^2)$:

$$\mathsf{E}\left[e^{jtX}\right] = e^{jt\mu - \frac{t^2\sigma^2}{2}} \tag{3.40}$$

and noting that $\phi(t+\tau) - \phi(t)$ is a zero mean process. Also, the right-hand side of (3.39) is a function only of τ even if $\phi(t)$ contains random walk terms.

The common form of representing the LO's phase noise is as a "phase noise mask," as illustrated in Figure 3.26. This characterization is typically performed with a spectrum analyzer that measures the device's phase noise. In this characterization, a single sideband power density function $L(f)$ is plotted as a function of f. This metric captures the ratio of the single sideband noise power in a bandwidth of 1 Hz at a frequency offset of f from the carrier relative to the power in the carrier. This metric is typically denoted in units of dBc, where "c" denotes the reference to the carrier. The phase noise mask or the power density of the phase noise is the Fourier transform of $R_{w'w'}(\tau)$ [122, Section 2.1].

To understand the impact of phase noise, we consider the discrete-time case where information symbols $s(k)$ are transmitted over N subcarriers of an OFDM system leading to the transmitted discrete-time signal

$$x(n) = \sum_{k=0}^{N-1} s(k) e^{\frac{j2\pi kn}{N}}, \ n = 0, \ldots, N-1. \tag{3.41}$$

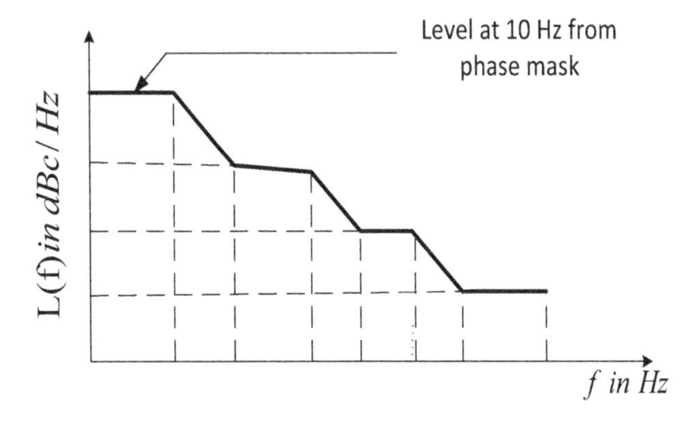

Figure 3.26 A typical phase noise mask as a function of frequency.

Here, n and k denote the indices in time and frequency domain, respectively. We will focus on the impact of phase noise on performance assuming that a cyclic prefix or guard interval is used to address multipath delay and inter-symbol interference, and that the cyclic prefix is removed at the receiver. Assuming a flat channel response, the signal $x(n)$ is affected by the phase noise $\phi(nT) \triangleq \phi(n)$ (where T is the symbol period) as

$$r(n) = x(n) \cdot e^{j\phi(n)}. \tag{3.42}$$

Note that $\phi(n)$ captures the sum phase noise contributions from the transmitter and the receiver. Assuming that channel estimation is done over a slot using a demodulation reference signal (DMRS), the average of phase noise over that slot is estimated and already incorporated in the channel estimate. Thus, we can assume that $\phi(n)$ captures the random variation around this average and hence its contribution is small. With this assumption, we can approximate $e^{j\phi(n)}$ with the first-order Taylor series expansion and we have

$$r(n) \approx x(n) \cdot (1 + j\phi(n)). \tag{3.43}$$

The demultiplexed signal $y(k)$ is then given as

$$\begin{aligned} y(k) &= \frac{1}{N}\sum_{n=0}^{N-1} r(n) e^{\frac{-j2\pi kn}{N}} && (3.44)\\ &\approx \frac{1}{N}\sum_{n=0}^{N-1} x(n) e^{\frac{-j2\pi kn}{N}} + \frac{j}{N}\sum_{n=0}^{N-1} x(n)\phi(n) e^{\frac{-j2\pi kn}{N}} && (3.45)\\ &= s(k) + \underbrace{\frac{j}{N}\sum_{k'=0}^{N-1} s(k') \cdot \sum_{n=0}^{N-1} \phi(n) e^{\frac{j2\pi(k'-k)n}{N}}}_{\triangleq e(k)} && (3.46)\end{aligned}$$

where $e(k)$ is the error term. We can break down the above summation over k' into $k' = k$ part and $k' \neq k$ part. Thus, we have

$$\begin{aligned} y(k) &\approx s(k) + s(k) \cdot \frac{j}{N}\sum_{n=0}^{N-1}\phi(n) + \frac{j}{N}\sum_{k'=0,k'\neq k}^{N-1} s(k') \cdot \sum_{n=0}^{N-1}\phi(n) e^{\frac{j2\pi(k'-k)n}{N}} && (3.47)\\ &= s(k) \cdot \left(1 + \frac{j}{N}\sum_{n=0}^{N-1}\phi(n)\right) + \frac{j}{N}\sum_{k'=0,k'\neq k}^{N-1} s(k') \cdot \sum_{n=0}^{N-1}\phi(n) e^{\frac{j2\pi(k'-k)n}{N}}. && (3.48)\end{aligned}$$

Thus, the impact of phase noise can be seen in terms of two major contributing components:

- *Common phase error term*: Sometimes called CPE, but not to be confused with a customer premises equipment as used in many chapters. The common error term (corresponding to $k' = k$ in the above expansion) added to *every* subcarrier is proportional to the average of the phase noise, $\frac{1}{N}\sum_{n=0}^{N-1}\phi(n)$, and results in a rotation of the symbol $s(k)$. Since the CPE is constant for all subcarriers, it can be corrected by an appropriate phase de-rotation algorithm.

- *Inter-carrier interference (ICI) component*: The contribution corresponding to $k' \neq k$ component in the above expansion leads to a time-varying impairment that comes from an average of phase noise with a spectral shift. This leads to a reciprocal mixing of adjacent subcarriers on to a desired subcarrier which induces ICI. The I/Q symbols show a Gaussian noise-type smearing effect around each symbol and a consequent loss of orthogonality. This component is caused by the phase noise from the reference oscillator. This component can also be corrected, but correcting it is computationally more expensive.

The CPE component is proportional to the average of the phase noise over the entire symbol and it typically affects the $[-\mathsf{SCS}/2,\ \mathsf{SCS}/2]$ part around the bandwidth where SCS denotes the subcarrier spacing. The noise power corresponding to the CPE part is similar to a sinc function profile, a majority of whose power is within $[-\mathsf{SCS}/2,\ \mathsf{SCS}/2]$. The ICI component affects frequencies beyond the subcarrier spacing through the phase noise bandwidth. Which of these two components dominate depends on the relationship between the LO's bandwidth and subcarrier spacing. If the LO bandwidth is less than half the subcarrier spacing, then the CPE term dominates; otherwise, the ICI component dominates. We could also have a situation where both components are comparable.

Let σ^2 denote the sum of the EVMs in the CPE and ICI components (that is, $\sigma^2 = \mathsf{EVM}_{\mathsf{CPE}} + \mathsf{EVM}_{\mathsf{ICI}}$). The SNR degradation induced by the phase noise in the $\sigma^2 \ll 1$ regime is approximated as [123]

$$\Delta\mathsf{SNR} = 10 \cdot \log_{10}\left(1 + \sigma^2 \cdot \mathsf{SNR}\right) \tag{3.49}$$

where SNR denotes the signal-to-noise ratio of the transmitted or received OFDM symbol with respect to additive white Gaussian noise. Thus, we observe that smaller phase noise degradation ($\Delta\mathsf{SNR}$) corresponds to a smaller value of σ^2. This, in turn, requires more expensive RF circuitry to meet a better phase noise mask.

A basic discussion on the impact of phase noise for multi-carrier systems can be found in [122, 124, 125]. SNR degradation with phase noise is studied in [123] and [126]. Design of time-frequency interleaving of Phase Tracking Reference Signal (PTRS) signals for 5G-NR systems and associated performance tradeoffs are discussed in [127] and [128].

3.4 IMPLICATIONS OF ANTENNA AND RF CONSTRAINTS ON SYSTEM DESIGN

3.4.1 SELECTION OF BEAMFORMING ARCHITECTURE

The question of what is the most appropriate beamforming architecture (analog, hybrid, or digital) arises a lot in system design. An analog architecture significantly reduces cost, area on the chip and power consumption, but comes with the limitation that only one beam (per-polarization layer) can be scanned at a time. Such a constraint can limit the achievable rates in a rich mmWave channel. It can also significantly slow down initial beam acquisition and mission-mode beam failure recovery (in case of blockage or fading).

At the other extreme, while a digital beamforming architecture can significantly speed up the beam search process, the associated cost, area, and power consumption can make it significantly less attractive. Given that the performance over the data channel is limited by the number of clusters in the channel, which is typically small at mmWave frequencies, the use of a digital architecture cannot improve rates substantially over a hybrid architecture. Nevertheless, there have been arguments made for the digital architecture in [129]. A digital architecture may be important in defense applications or in applications with extremely low coherence durations (e.g., high speed trains, air-to-ground links, etc.), but its importance in a cellular communications application requires further careful consideration. Its utility at different carrier frequencies of interest (e.g., mmWave or sub-THz bands) also needs constant reappraisal as technology drivers evolve. In general, a hybrid architecture often makes sense both at the base station and UE.

In this context, most initial commercial implementations (at 28 and 39 GHz) at the UE/CPE use two RF chains over two polarizations. Further evolution of this trend in commercial designs at the UE could be toward 4–8 layers, but with more layers at the CPE. Power consumption can be significant for the uplink part of the RF chain (due to the increased contribution of power amplifiers) than for the downlink part (low-noise amplifiers). Thus, asymmetric designs with more RF chains for reception than for transmission may also be useful as it is at sub-7 GHz frequencies (e.g., 2TX/4RX, 2TX/6RX, 2TX/8RX, 4TX/8RX, etc.).

At the base station, significantly more layers can be accommodated compared to that affordable at the UE/CPE. While base stations can enjoy network level spectral efficiency improvements with multi-user transmissions by amortizing the infrastructure cost across multiple UEs/CPEs, considerations such as size/area, cost, power, thermal management and weight are still not ignorable at the base station end. For example, small cell base stations are often deployed on lamp posts, traffic lights and low-weight accommodating fixtures, and considerations such as size, weight, and right-of-way take prominence. Thermal management in CPEs and base stations is also a serious problem that requires special cooling/heat sink solutions that can quickly become expensive and bulky. Further, in terms of network level performance improvement, multiplexing more than 4 UEs over the same time and frequency resource can become onerous from a user scheduling perspective. This is because of the need for a high user density deployment to determine appropriate users for simultaneous scheduling and the need for high-precision CSI feedback overhead to mitigate inter-user interference in multi-user schemes. Thus, the RF chain evolution at the base station may also mirror the trends seen in this evolution at the UE/CPE. This is especially the case as carrier frequencies increase beyond 28 and 39 GHz.

3.4.2 DOWN-/UP-CONVERSION TRADEOFFS

Signals are typically transported between the baseband and RF using a *superheterodyne* interface that takes the RF signal to an intermediate frequency (IF), and then from the IF to the baseband. On the other hand, a *direct conversion* interface takes the RF signal straight to baseband without down-conversion to IF. In a superheterodyne interface, the circuit-level bottlenecks are typically at the IF level with the addition of each port/layer meaning the addition of an IF cable/connector or multiplexing of another IF chain to the same cable. The choice of the IF is driven by two factors:

- It should be non-overlapping with as many other frequencies at the UE as possible to avoid interference. Similar constraints on IF also hold for other devices in the network
- It should be routable with minimum loss on the UE using cost-effective techniques.

While interference with other carrier frequencies (around the IF stage) may be avoided by careful design, coexistence issues arise with the opening of the bands between 7.125 and 24.25 GHz (called Frequency Range 3 in 3GPP terminology) to commercial cellular band operations. This makes interference avoidance a more complicated problem. Placing the ADC/DAC on the antenna chip enables a more tractable handling of the coexistence problem, thus making a digital architecture more palatable.

While a hybrid architecture can allow a tradeoff between these issues, it also includes phase shifters for each RF chain that take up area and often have a limited RF bandwidth. In such scenarios, phase shifting can be performed with the LO or at IF or at baseband (as in Figure 3.17(b)). Such choices become naturally attractive at upper mmWave bands (and beyond). On the other hand, increased insertion loss with the IF cable at higher carrier frequencies suggests a direct conversion architecture to reduce power and improve performance (as the gain stage is at baseband). The power consumed by the ADC scales exponentially with sampling rate and effective number of bits (ENOB) and thus, an analog beamforming architecture has a significant power savings advantage over a digital beamforming architecture when the highest bandwidth and MCS operations are considered.

There are also many scenarios where a phased array design intended for the UE/CPE is re-purposed and reused for base station designs. In such scenarios, the higher performance required at the base stations limits and drives the beamforming architecture selection. While an unconstrained vector modulator architecture may be useful for adaptive beamforming, constrained architectures such as Butler matrix, Blass matrix, Rotmans lens, etc., may be useful in applications with more predictable channel environments (e.g., in repeaters, CPEs, RIS nodes, etc.). Such tradeoffs become more complex as future generations evolve into the upper mmWave and sub-THz bands.

3.4.3 POWER CONSUMPTION

Traditional sub-7 GHz as well as mmWave systems consume significant power especially with higher-rank transmissions or receptions. Among the RF components, assuming a superheterodyne architecture, the PA and the variable gain amplifier at IF (IF VGA) typically consume the most power on the transmit path. The receive path power consumption is also typically dominated by the IF VGA. The voltage-controlled oscillator (VCO) driving the frequency synthesizer is another power hungry component on both the transmit and receive paths. This power consumption can significantly drain out the battery at the UE side impacting user experience and QoS.

Thus, physical and higher layer procedures optimized for power savings can have a major impact on the latency, overhead and performance (both at an individual user level as well as at the network level). For a UE operation, energy efficiency is captured in two ways:

- Energy efficiency when there is no data to be transmitted to the UE since the UE continues to monitor the network for arrival of data traffic and
- Energy efficiency in connected mode operations (measured as bits transmitted per unit energy consumption).

Such notions have been well-studied from a physical layer viewpoint in [130, 131] and in many other works in the traditional communications literature. These studies have started impacting 3GPP specifications from a network energy savings perspective.

Given the bursty nature of traffic, communications systems benefit from the use of a *connected mode discontinuous reception* (CDRX) signaling protocol which allows the UE to toggle back and forth between sleep and wake-up periods. Note that wake-up signaling based on CDRX mode operation is common to LTE as well as sub-7 GHz frequencies. What is unique to mmWave frequencies is the beam state in CDRX mode. That is, the choice of the beams to be used in reception or transmission once the UE wakes up from sleep state depends on the level of stationarity of the channel. Further, the scheduling offset between PDCCH and PDSCH/PUSCH can be configured by the base station (K_0 and K_2 in 3GPP terminology [132]) to allow for micro, light or deep sleep.

The network triggers the UE when there is data available to send which allows the UE to wake up to receive the data. Otherwise, the UE is in (or goes back to) a sleep state. Clearly, the longer the sleep period, more power savings can be realized. However, this power savings comes at a cost of missing the control signal from the base station over PDCCH triggering the transmission of a burst of data packets. To address this issue, the base station and the UE coordinate the use of ON and OFF periods in either a short DRX or a long DRX cycle where the UE turns off some or most of the RF components to save power.

Typically, there are three identified sleep states (deep, light and micro sleep) which differ in terms of which RF components are turned off and thus the net energy consumption at the UE. In a deep sleep state, the UE is idle or inactive and does not monitor PDCCH or PDSCH. The UE turns off most of the power hungry RF

Table 3.6
Model for relative power consumption in different operational states at the UE

State	Relative power
Deep sleep	1
Light sleep	20
Micro sleep	45
PDCCH monitoring only	175
SSB or Channel State Information-Reference Signal (CSI-RS) measurements	175
PDCCH and PDSCH reception	350
Uplink transmission	350

components to save considerable power and it enters the connected mode only when the data arrives. In contrast, a micro sleep state is limited to a fraction of a slot and the UE can have a customized PDCCH monitoring pattern with monitoring skipped across certain slots. After a PDCCH decode has been attempted, if the UE does not have a grant, then it enters a micro sleep state until the next PDCCH monitoring occasion (or pre-scheduled grant). As a result, very few RF components are turned on leading to a relatively higher energy consumption. An intermediate sleep state called light sleep corresponds to turning off RF components for reuse after a short period of time (e.g., a few slots later).

TR 38.840 [133] provides a relative power consumption model for mmWave systems which is captured in Table 3.6 for different operational states at the UE. In addition to an estimate of the power consumed with these three sleep states, power consumption estimates for radio resource management (RRM) measurements and transmissions are also provided. Note that the relative power consumption in uplink transmissions assumes a certain nominal EIRP and has to be adapted based on the actual EIRP. These relative power estimates are specific to mmWave systems. As mmWave systems evolve, such models need to be revisited and adapted appropriately. In these systems, there are differences arising from wider operating bandwidths and efficiency differences across RF circuitry, reduced slot durations and commensurate increase in PDCCH decoding operations per symbol, increased beam management/RRM measurements to maintain link connectivity, etc.

3.5 SYSTEM LEVEL SOLUTIONS TO ADDRESS RF CONSTRAINTS

3.5.1 POLARIZATION VS. SPATIAL MIMO

With two RF chains, a broad question of interest is whether these can be gainfully used over a single beam/direction across two orthogonal polarizations (polarization MIMO transmissions), or over two independent beams/directions (spatial

MIMO transmissions). Polarization MIMO transmissions are preferred since a dual-polarized set of antenna elements can be packed within the same area as a uni-polarized set of antenna elements (e.g., dipoles) at mmWave frequencies. Such schemes allow beam management with the feedback overhead or latency associated with a single beam and could lead up to a doubling of the rate. Note that this doubling of rate implicitly assumes that the gains across the polarizations are comparable. Further, if the XPR is not large, it is implicitly assumed that the mixing of energy across polarizations can be separated by the appropriate use of precoding and combining matrices (at the digital level).

In contrast, spatial MIMO transmissions over two RF chains lead to focusing of energy on the dominant and the second dominant cluster in the channel (necessarily a weaker one since the stronger cluster is used over the first layer) and therefore a less than doubling of the rate as well as an increase in beam management overhead and latency. Inter-beam interference between beam directions that are not necessarily orthogonal can also lead to a reduction in spectral efficiency. Spatial MIMO inherently enjoys the advantages of beam diversity (to blockage, for example). In the case of 4–8 layers, this can correspond to spatial MIMO over 2–4 distinct beams controlled by one/many RFIC chip(s) and polarization MIMO per beam. An example case would be two spatial directions with polarization MIMO in each direction leading to rank-4 transmissions (see more discussions in Chapter 4.5.1).

3.5.2 PA NON-LINEARITY MITIGATION

To mitigate the PA distortions in the input signal to reduce leakage and to enhance the linearity of the PA, a pre-distortion (or inverse filtering) step is often introduced into the input of the PA, thereby cancelling any non-linearity that the PA has. The pre-distortion step can be carried out in analog or digital domains. Digital pre-distortion (DPD) techniques can be broadly classified as open or closed loop-based. Open loop systems use a lookup table that contains correction values for amplitude and phase derived from AM-to-AM and AM-to-PM measurements. Thus, the overall combination response of the PA and the pre-distorter becomes a linear system. In contrast to this static approach, in a closed loop system, the PA's output signal is compared with an ideal signal to find the correction values. Since the approach is adaptive, it is useful to address wide bandwidth systems which are systems with memory. Memory arises in these systems due to thermal constants of the active devices or components in the biasing network that have frequency-dependent behavior. As a result, the current output of the PA not only depends on the current input, but also on the past input values. Digital post-distortion (DPoD) solutions at the receiver that mitigate the distortion of the PA are also actively pursued in 5G-NR systems. In particular, DPD and DPoD solutions can be used together. Since DPoD cannot address the ACLR impairment, in one particular rendition, the transmitter can use DPD to primarily mitigate ACLR and the receiver can use DPoD to mitigate in-band EVM.

3.5.3 PHASE NOISE MITIGATION

A poorer phase noise mask or profile arising from low-cost chip designs can lead to significant phase drifts over typical symbol period lengths. These impairments need wider subcarrier spacing for amelioration. Managing such issues at low cost and yet achieving high rates appears feasible with 64-QAM constellation. Extending such developments to 256- or 1024-QAM constellations needed in high data rate systems is a topic of active research. Further, the impact of phase noise increases with carrier frequency. Thus, in mmWave systems, the use of PTRS becomes necessary.

The CPE component of phase noise is either partially or totally compensated by the use of a frequency tracking loop (FTL) and in particular, by how fast the FTL is operated relative to the symbol duration. In 5G-NR, an optionally available PTRS allows the tracking of the phase error trajectory or drift over time. Error estimation from PTRS can be used for phase compensation and correction of the CPE especially with higher-order MCSs. Once the CPE is removed, the residual error is due to the ICI component. In the scenario where the effect of this component is not small (e.g., higher-order MCS in upper mmWave and sub-THz bands), we need improved receiver architectures or we need to increase the overhead associated with PTRS. This is realized by time-interleaving the occurrence of the PTRS with the data symbol in an appropriate time-pattern (e.g., every symbol, every other symbol, every fourth symbol, etc.). The ICI component can also be addressed in practice with the use of wider subcarrier spacings (e.g., 60 or 120 kHz for data/control symbols and 120 or 240 kHz for SSB/synchronization symbols).

3.5.4 POWER CONSUMPTION REDUCTION

Since power consumption at mmWave carrier frequencies can seriously hinder the user experience, a number of system level adaptations can be considered to save power in connected mode operations. These include:

- Adaptation of the transmit power/EIRP
- Adaptation of the number of used carriers in stand-alone or dual connectivity mode operations
- Rank (number of RF chains) and MCS adaptation
- Adaptation of the number of used antenna elements
- Adaptation of the time-period over which the UE stays active and a dynamic transition to sleep mode
- Switching active antenna modules based on channel, power and thermal conditions, etc.

Other power savings mechanisms such as cross-slot scheduling and secondary cell dormancy has been standardized in 3GPP specifications. In cross-slot scheduling, a guaranteed minimum time interval of K_0 slots between PDCCH and PDSCH allows the UE to skip unnecessary RF operations. In secondary cell dormancy, non-monitoring or dormancy of a secondary cell in a carrier aggregation mode and wake-up of secondary cell as needed using the PDCCH of primary cell is enforced.

3.6 APPENDIX

3.6.1 JACOBIAN OF COORDINATE TRANSFORMATIONS

Let $G_{\text{total}}(x,y,z)$ denote the total beamforming gain (over the two polarizations) with a certain beamforming scheme at a point (x,y,z) represented in the XYZ Cartesian coordinate system. Then, the CDF of spherical coverage evaluated at α over a sphere of radius R is given as

$$F(\alpha) = \frac{\iiint \mathbb{1}\left(G_{\text{total}}(x,y,z) \leq \alpha\right) dxdydz}{\iiint dxdydz} \tag{3.50}$$

where $\mathbb{1}(\bullet)$ denotes the indicator function of the underlying variable. Let a coordinate (x,y,z) in the Cartesian coordinate system be transformed to (r,θ,ϕ) in the spherical coordinate system via:

$$x = r\sin(\theta)\cos(\phi) \tag{3.51}$$

$$y = r\sin(\theta)\sin(\phi) \tag{3.52}$$

$$z = r\cos(\theta). \tag{3.53}$$

The differential element in the Cartesian coordinate system is transformed to the differential element in the spherical coordinate system as

$$dxdydz = \mathscr{J}\, drd\theta d\phi \tag{3.54}$$

where

$$\mathscr{J} = |\det(J)| \tag{3.55}$$

with J denoting the Jacobian matrix of the transformation:

$$J = \begin{bmatrix} \frac{\partial x}{\partial r} & \frac{\partial x}{\partial \theta} & \frac{\partial x}{\partial \phi} \\ \frac{\partial y}{\partial r} & \frac{\partial y}{\partial \theta} & \frac{\partial y}{\partial \phi} \\ \frac{\partial z}{\partial r} & \frac{\partial z}{\partial \theta} & \frac{\partial z}{\partial \phi} \end{bmatrix} \tag{3.56}$$

$$= \begin{bmatrix} \sin(\theta)\cos(\phi) & r\cos(\theta)\cos(\phi) & -r\sin(\theta)\sin(\phi) \\ \sin(\theta)\sin(\phi) & r\cos(\theta)\sin(\phi) & r\sin(\theta)\cos(\phi) \\ \cos(\theta) & -r\sin(\theta) & 0 \end{bmatrix}. \tag{3.57}$$

This computation results in

$$\mathscr{J} = r^2|\sin(\theta)| = r^2\sin(\theta). \tag{3.58}$$

With this, (3.50) transforms as follows:

$$F(\alpha) = \frac{\int_{r=0}^{R}\int_{\theta=0}^{\pi}\int_{\phi=0}^{2\pi} \mathbb{1}\left(G_{\text{total}}(\theta,\phi) \leq \alpha\right) r^2\sin(\theta) drd\theta d\phi}{\int_{r=0}^{R}\int_{\theta=0}^{\pi}\int_{\phi=0}^{2\pi} r^2\sin(\theta) drd\theta d\phi} \tag{3.59}$$

$$= \frac{\int_{\theta=0}^{\pi}\int_{\phi=0}^{2\pi} \mathbb{1}\left(G_{\text{total}}(\theta,\phi) \leq \alpha\right) \sin(\theta) d\theta d\phi}{4\pi}. \tag{3.60}$$

It is critical to note the scaling factor $\sin(\theta)$ in (3.60) reduces the weightage of points at the poles (where $\theta = 0°$ and $180°$) and increases the weightage of points at the equator (where $\theta = 90°$). In an analogous example, the non-use of the $\sin(\theta)$ weighting factor leads to issues observed with the Mercator projection where the area of regions far away from the equator are inflated.

4 Design at the Link Level and Performance

We start this chapter with a summary of well-known optimality results on MIMO precoding[1] that implicitly assume a digital beamforming architecture relevant for systems operating at lower carrier frequencies. We then explore how these results map to mmWave frequencies with an analog/hybrid beamforming architecture. We explore the scope and context of the popular directional transmission and reception strategies at mmWave frequencies and explain how the traditional singular value decomposition (SVD)-based approaches do not exhibit a property of strong robustness to small channel estimation errors. We then focus on the analog beamforming architecture and consider the most natural structure of directional beamforming codebooks consisting of progressive phase shift beam weights. After establishing the tradeoffs of array gain vs. beamwidth for such beam weights, we then consider the more general case of beam broadening to optimize these tradeoffs. We also describe other approaches with which beam weights can be learned in the analog domain including the notion of adaptive beam weights that mimic the SVD structure of optimal beamforming.

After this background, we consider practical implementation issues such as impact of phase shifter granularity/precision and calibration error on the performance of progressive phase shift beams and broad beams. We show that while the peak gain and beamwidths of progressive phase shift beams remain fairly robust with limited phase shifter granularity, those of broad beams can see significant variation. These observations can impact how SSB beams are deployed at base stations in a densified network. We then consider how antenna placements in a form factor UE can impact the spherical coverage performance of beamforming in a freespace and blockage scenario. From this study, we showcase good antenna module placements for practical system design.

We then extend the studies to the case where the systems allow a hybrid beamforming capability consisting of 4 layers with these layers used for SU- or MU-MIMO at the base station. We show that how information learned in beam training such as the best/second best spatial directions can be leveraged to produce substantial performance improvement with 4 layers over the analog beamforming scheme provided such an RF capability is available. We also expand the same approach

[1]In the classical signal processing for communications literature, the term *precoding* is used for rank-r signaling with the term *beamforming* reserved for the $r = 1$ case. In both mmWave technology and practice, the terms *hybrid beamforming* and *analog beamforming* are used to denote the same notions. We will use both sets of terminologies in this book as the situation warrants, with clarity obtained from the context of their usages.

DOI: 10.1201/9781032703756-4

for coordinated transmissions from multiple transmission reception points (TRPs) or base stations.

4.1 MIMO PRECODING

Let $H(k)$ denote the $N_r \times N_t$ channel matrix over the k-th subcarrier between a transmitter (e.g., base station, IAB node, etc.) with N_t transmit antennas and a receiver (e.g., a UE, a CPE, a side link node, etc.) with N_r receive antennas. For simplicity, we assume a narrowband setting where $H(k) = H$ over all the subcarriers of interest. We intend to communicate r (where $r \geq 1$) independent data-streams corresponding to an $r \times 1$ vector s over this channel. Let the components of s be independent and identically distributed (i.i.d.) with zero mean and variance σ_{s}^2. The channel matrix realization can be time- and frequency-varying. Hence, a possible solution to mitigate fading is *precoding* the r streams of data (r is also denoted as the *rank* of the precoder) with an $N_t \times r$ precoding matrix F without changing the average transmit power constraint (from the no precoding case). The simplest case of rank-1 precoding is often termed as *beamforming* in the classical signal processing literature. Let n denote the $N_r \times 1$ zero mean proper [134] complex white Gaussian noise vector added at the receiver. We assume that the covariance matrix of n is $\mathbf{\Sigma}_n = \sigma_{\mathsf{n}}^2 \cdot I_{N_r}$.

The system model for the $N_r \times 1$ received vector y is given as

$$y = HFs + n. \tag{4.1}$$

At the receiver, the data stream s is typically estimated (the estimate is denoted as $\widehat{s}$) with a *linear* processing scheme to reduce the complexity by using an $N_r \times r$ combining matrix G as follows:

$$\widehat{s} = G^H y = G^H HFs + G^H n. \tag{4.2}$$

With the above assumptions, Appendix 4.6.1 shows that the achievable spectral efficiency R (in bits per channel use) using the precoding matrix F and the combining matrix G is given as

$$R = \log_2 \det \left(I_r + \frac{\sigma_{\mathsf{s}}^2}{\sigma_{\mathsf{n}}^2} \cdot \left(G^H G\right)^{-1} G^H HFF^H H^H G \right). \tag{4.3}$$

We are interested in selecting the matrix pair (F, G) to maximize R as quantified in (4.3). Under the linear minimum mean squared error (LMMSE) structure, for a given F, the optimal choice of G is given as:

$$G_{\mathsf{opt}} = \frac{\sigma_{\mathsf{s}}^2}{\sigma_{\mathsf{n}}^2} \cdot HF \cdot \left(I_r + \frac{\sigma_{\mathsf{s}}^2}{\sigma_{\mathsf{n}}^2} \cdot F^H H^H HF \right)^{-1}. \tag{4.4}$$

With this choice, it can be seen that R reduces to

$$R = \log_2 \det \left(I_r + \frac{\sigma_{\mathsf{s}}^2}{\sigma_{\mathsf{n}}^2} \cdot F^H H^H HF \right), \tag{4.5}$$

which is the *classical* achievable rate expression assumed in most MIMO works [135, 136, 137, 50, 138, 139, 140]. The optimization problem of interest can be seen to be

$$F_{\mathsf{opt}} = \arg \max_{F\,:\,\mathsf{Tr}\left(F^H F\right)=r} \log_2 \det \left(I_r + \frac{P}{r\sigma_{\mathsf{n}}^2} \cdot F^H H^H H F \right) \tag{4.6}$$

where $P = \sigma_{\mathsf{s}}^2 \cdot r$ is the average transmit power.

Let the SVD of H be written as $H = U_H D_H V_H^H$. From Appendix 4.6.2, we can establish that F_{opt} satisfies a waterfilling solution structure. That is,

$$F_{\mathsf{opt}} = V_H D_{\mathsf{opt}} \tag{4.7}$$

with D_{opt} being $N_t \times r$ diagonal and satisfying

$$\left(D_{\mathsf{opt},ii}\right)^2 = \frac{r\sigma_{\mathsf{n}}^2}{P} \cdot \left(\mu - \frac{1}{\lambda_i(H^H H)} \right)^+, \; i = 1, \ldots, r \tag{4.8}$$

where $x^+ = \max(x, 0)$ and μ is such that $\sum_{i=1}^r \left(\mu - \frac{1}{\lambda_i(H^H H)} \right)^+ = \frac{P}{\sigma_{\mathsf{n}}^2}$.

This solution can be interpreted as pouring water over the inverted channel eigenmodes of strength $1/\lambda_i(H^H H)$ with the water level at μ and the power over this eigenmode given by $(D_{\mathsf{opt},ii})^2$. In the extreme case of $\frac{P}{\sigma_{\mathsf{n}}^2} \to 0$, only the strongest eigenmode is excited (rank-1 signaling) with all the power, whereas in the other extreme of $\frac{P}{\sigma_{\mathsf{n}}^2} \to \infty$, all the eigenmodes are excited (rank-r) with equal power ($D_{\mathsf{opt},ii} \to 1$). The number of eigenmodes excited by the waterfilling solution is non-decreasing as $\frac{P}{\sigma_{\mathsf{n}}^2}$ increases.

Note that in the low transmit power setting, F_{opt} and G_{opt} reduce to

$$F_{\mathsf{opt}} = V_{H,1} \tag{4.9}$$

$$G_{\mathsf{opt}} = H F_{\mathsf{opt}} = H V_{H,1} = \sqrt{\lambda_1(H^H H)} \cdot U_{H,1} \tag{4.10}$$

where $V_{H,1}$ and $U_{H,1}$ denote the first/dominant column of V_H and U_H, respectively. In other words, transmitting and receiving, respectively, along the dominant right and left singular vectors of H is optimal. In the high transmit power setting, we have the following reductions:

$$F_{\mathsf{opt}} = \widetilde{U} \tag{4.11}$$

$$G_{\mathsf{opt}} = H F_{\mathsf{opt}} \cdot \left(F_{\mathsf{opt}}^H H^H H F_{\mathsf{opt}}\right)^{-1} = H \widetilde{U} \left(\widetilde{D} \widetilde{D}_{\mathsf{opt}} \right)^{-1}, \tag{4.12}$$

where $\widetilde{U}$ denotes the first r columns of V_H. Also, $\widetilde{D}$ and $\widetilde{D}_{\mathsf{opt}}$ denote the $r \times r$ principal submatrices of $D = \mathsf{diag}(\lambda_i(H^H H))$ and D_{opt}, respectively. Further, in these extremes, we have the following conclusions on R:

$$\lim_{\frac{P}{\sigma_n^2}\to 0} \frac{R}{\log_2\left(1+\frac{P}{\sigma_n^2}\cdot\lambda_1(H^H H)\right)} = 1 \tag{4.13}$$

$$\lim_{\frac{P}{\sigma_n^2}\to\infty} \frac{R}{\sum_{i=1}^{r}\log_2\left(1+\frac{P}{r\sigma_n^2}\cdot\lambda_i(H^H H)\right)} = 1. \tag{4.14}$$

In other words, (4.13) demonstrates that with the use of optimal precoding, at low-SNRs, the dominant eigenmode in the channel can be leveraged to create a single parallel channel between the transmitter and the receiver. Similarly, (4.14) demonstrates that the r dominant eigenmodes can be leveraged to create parallel channels between the transmitter and the receiver.

4.1.1 UNIQUE ASPECTS OF MMWAVE TRANSMISSIONS

We now provide physical interpretations for F_{opt} and G_{opt} in terms of the channel structure. For this, we consider the narrowband channel model representation in (2.15) for H, which is given as,

$$H = \sum_{\ell=1}^{L} \alpha_\ell \cdot u_\ell v_\ell^H \tag{4.15}$$

with receive and transmit array steering vectors $\{u_\ell\}$ and $\{v_\ell\}$ and complex gains $\{\alpha_\ell\}$ across L clusters/paths. We use this in expanding $H^H H$ as

$$H^H H = \sum_{i,j} \alpha_i^\star \alpha_j \cdot \left(u_i^H u_j\right)\cdot v_i v_j^H = VAV^H \tag{4.16}$$

where

$$V = [\alpha_1^\star v_1, \ldots, \alpha_L^\star v_L] \tag{4.17}$$

$$A(i,j) = u_i^H u_j, \quad i,j = 1,\ldots,L. \tag{4.18}$$

Let X be an $L\times L$ eigenvector matrix of $AV^H V$ with the corresponding diagonal matrix of eigenvalues denoted by $\mathbf{\Lambda}$. That is, the eigenvalue equation is given as:

$$\left(AV^H V\right)\cdot X = X\cdot\mathbf{\Lambda}. \tag{4.19}$$

Pre-multiplying both sides of (4.19) by V and regrouping the matrices, we have

$$VX\cdot\mathbf{\Lambda} = \left(VAV^H V\right)\cdot X = \left(H^H H\right)\cdot VX. \tag{4.20}$$

Reading (4.20) from right to left, we see that VX forms the eigenvector matrix for $H^H H$ with the diagonal eigenvalue matrix being the same as $\mathbf{\Lambda}$. In other words, all the eigenvectors of $H^H H$ (and thus all the columns of F_{opt}) can be represented as *linear combinations* of $v_1,\ldots,v_L$. These are important observations on which much of the analog/hybrid beamforming codebook designs rely on. We can expand G_{opt} also as follows:

$$G_{\text{opt}} = \frac{\sigma_s^2}{\sigma_n^2} \cdot \sum_{\ell=1}^{L} u_\ell \cdot \left(\alpha_\ell \cdot v_\ell^H U \widehat{D} \right) \tag{4.21}$$

$$= \frac{\sigma_s^2}{\sigma_n^2} \cdot \begin{bmatrix} u_1 & \cdots & u_L \end{bmatrix} \cdot \begin{bmatrix} \alpha_1 \cdot v_1^H U \widehat{D} \\ \vdots \\ \alpha_L \cdot v_L^H U \widehat{D} \end{bmatrix} \tag{4.22}$$

where

$$\widehat{D} = D_{\text{opt}} \cdot \left(I_r + \frac{\sigma_s^2}{\sigma_n^2} \cdot D_{\text{opt}}^H D D_{\text{opt}} \right)^{-1}. \tag{4.23}$$

In other words, all the columns of G_{opt} can be represented as *linear combinations* of $u_1, \ldots, u_L$.

To illustrate the above observations, in Figure 4.1(a), we plot the beam patterns corresponding to the first/dominant column of F_{opt} in a 4×16 (that is, $N_r = 4$ and

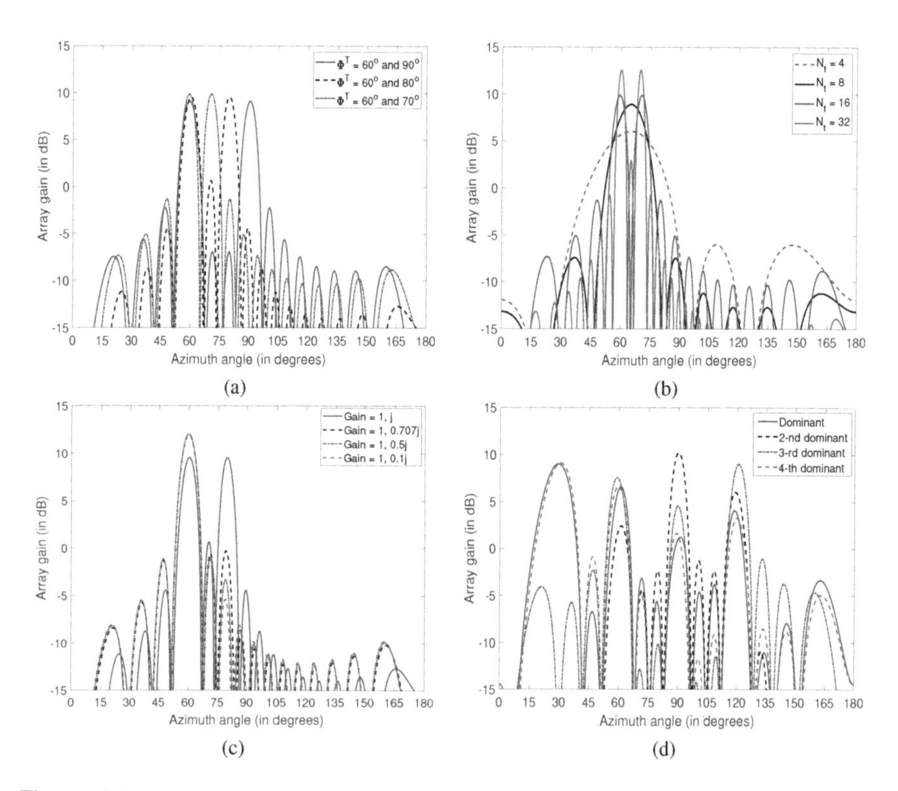

Figure 4.1 Beam patterns of dominant column of F_{opt} as a function of variations in (a) cluster angles, (b) antenna dimensions and (c) gains. (d) Beam patterns of all columns of F_{opt}.

$N_t = 16$) channel with $L = 2$ clusters (angular spread of 0° for each cluster) along the $\phi^{\mathsf{T}} = 60°$ and 90° directions[2], as well as $\phi^{\mathsf{T}} = 60°$ and 80°, and $\phi^{\mathsf{T}} = 60°$ and 70°. In all the three examples, the clusters have complex gains of 1 and j, respectively. From this plot, we clearly see that the beam pattern shows sharp peaks at/near the specific values of ϕ^{T}. As the cluster angles get closer to each other, the interaction between the part of the beam pattern due to each cluster leads to peaks around the ϕ^{T} values instead of exactly at the ϕ^{T} values. In Figure 4.1(b), the beam patterns are plotted as a function of azimuth angle for different values of N_t (where $N_t = 4, 8, 16, 32$) in the two cluster scenario with $\phi^{\mathsf{T}} = 60°$ and 70° corresponding to complex gains of 1 and j. From this plot, we observe that for smaller array dimensions, it becomes harder to parse the individual clusters (as they are close to each other) leading to a peak at an intermediate ϕ^{T} value. Resolvability of clusters improves as array dimensions increase. In Figure 4.1(c), we consider the $\phi^{\mathsf{T}} = 60°$ and 80° case as we change the gains of the two clusters from $\{1, j\}$ to $\{1, 1/\sqrt{2}j\}$, $\{1, 0.5j\}$ and $\{1, 0.1j\}$. Clearly, we see a dominant peak in the beam pattern at $\phi^{\mathsf{T}} = 60°$, whereas the peak at $\phi^{\mathsf{T}} = 80°$ appears to be sub-dominant as the relative gain of this cluster changes. In addition, secondary peaks are introduced at other angles depending on the relative gain values.

In Figure 4.1(d), we consider the case of $L = 4$ clusters at $\phi^{\mathsf{T}} = \phi^{\mathsf{R}} = 30°, 60°, 90°$ and 120° with complex gains of $1, j, -1$ and $-j$ (respectively) and plot the beam patterns of the four columns of F_{opt}. From the perspective of designing higher-rank precoding schemes, a directional approach would steer energy along the individual angles of ϕ^{T}. In contrast, the optimal higher-rank scheme relies on utilizing the intricate relationship across ϕ^{T} to construct orthogonal vectors by relying on infinite precision amplitude and phase control. The tradeoffs between these two approaches is studied in Figure 4.2. In the scenario of an angular spread of multiple rays per cluster, we observe that the beam patterns have a single peak or multiple peaks approximately around the central angular values depending on the relative gains and spreads of the rays of the cluster.

While the above interpretations on the structure of F_{opt} and G_{opt} are valid for any number L of clusters/paths, they are especially meaningful in sparse channels such as those seen at mmWave frequencies where L is small relative to the array dimensions. Since u_ℓ and v_ℓ in (4.15) denote the array steering vectors at the receive and transmit sides, based on the physical intuition of F_{opt} and G_{opt}, the search for a good F and G can be translated to a search for good linear combinations of the transmit and receive array steering vectors that make the channel matrix H. Unfortunately, the search for the weights in the *linear combinations* can be cumbersome; see Chapter 4.2.4 on how this can be done. Thus, a sub-optimal and simplistic approach is to search for the best set of transmit and receive array steering vectors, which leads to the class of *directional* precoder structures. Loss in performance with the directional schemes (relative to F_{opt} and G_{opt}) are small when the number of dominant paths in the channel L is small relative to the array dimensions (N_t and N_r). Further, in this

[2]From a notational perspective, $(\bullet)^{\mathsf{T}}$ and $(\bullet)^{\mathsf{R}}$ stand for the transmit and receive sides, respectively.

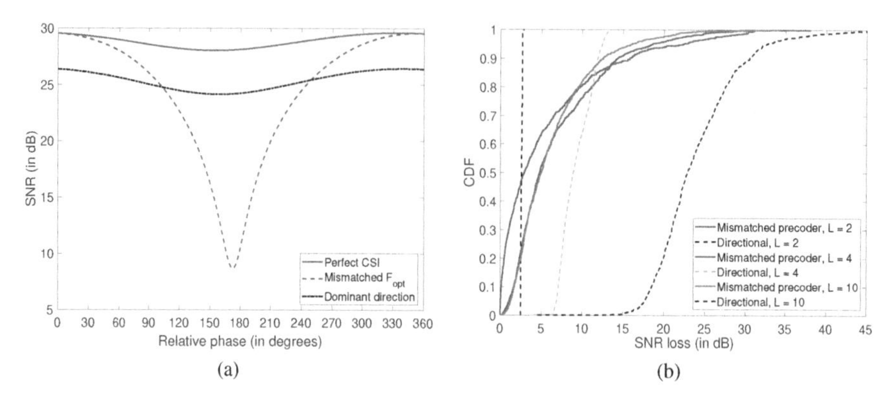

Figure 4.2 (a) Performance of a mismatched SVD and directional precoder for a single channel instantiation as a function of relative phase between clusters. (b) Performance gap between ideal SVD and mismatched SVD precoders, and ideal SVD and directional precoders as a function of the number of clusters in the channel.

scenario, the column vectors of V form a non-redundant basis for the eigenspace of $H^H H$.

In the optimal reception scenario, it is important to point out that the estimated vector $\widehat{s}$ in (4.2) depends on a *whitening* structure where the covariance matrix of the received vector y is first estimated:

$$\widehat{s} = G_{\mathrm{opt}}^H y \tag{4.24}$$

$$= \sigma_{\mathrm{s}}^2 \cdot F_{\mathrm{opt}}^H H^H \cdot (\mathbf{\Sigma}_y)^{-1} y \tag{4.25}$$

with

$$\mathbf{\Sigma}_y = I_{N_r} + \frac{\sigma_{\mathrm{s}}^2}{\sigma_{\mathrm{n}}^2} H F F^H H^H \tag{4.26}$$

where (4.25) follows from the matrix inversion lemma (see Chapter 4.6.3). In practice, the term $F_{\mathrm{opt}}^H H^H$ is estimated using pilot signals over specific time and frequency resources (resource elements or REs in 3GPP parlance). On the other hand, the covariance matrix $\mathbf{\Sigma}_y$ is estimated by sample averaging the outer product of the received vector over multiple resource elements in at least one resource block (multiple REs make a resource block or RB) with itself. Alternately, $\mathbf{\Sigma}_y$ can be estimated assuming different amounts of knowledge on $\mathbf{\Sigma}_n$ (only intra-cell signal, intra-cell signal and interference from other co-scheduled transmissions, inter-cell interference, etc.), which leads to different types of LMMSE implementations [141]. The simplest case of $\mathbf{\Sigma}_n = \sigma_{\mathrm{n}}^2 \cdot I_{N_r}$ is the most relevant scenario for mmWave deployments. This is because of the fact that given the heavy pathloss in mmWave systems, there is not much inter-cell interference for typical deployment ISDs. As elucidated in Chapter 5, as mmWave deployments get densified, this assumption may have to be revisited.

4.1.2 ROBUSTNESS OF F_{OPT} AND G_{OPT}

We now examine the SVD structure of F_{opt} and G_{opt} and study its robustness to errors/perturbations in the channel matrix H. Towards this goal, in Appendix 4.6.3, we note the criticality of the *eigen-gap* or separation between eigenvalues of the unperturbed channel matrix in how errors/perturbations reflect in terms of the robustness of the eigenvectors.

In the context of mmWave systems, we first consider a 4×16 channel with $L = 2$ clusters and 20 rays per cluster in a 5° angular spread around a cluster's central direction. The two clusters correspond to $\phi^{\text{T}} = \phi^{\text{R}} = 60°$ and 70° with gains of 1 and 0.95, respectively. For this single channel instantiation, we consider the case where the relative phase of the second cluster (and all the rays within that cluster) with respect to the first cluster is (are) varied across the 0° to 360° range. In an ideal setup, we consider the use of F_{opt} and G_{opt} for every channel realization as the relative phase changes. We then consider the use of a *mismatched* precoder where the F_{opt} and G_{opt} are designed for a relative phase offset of 0° between the two clusters while the true relative phase could be different. This scenario corresponds to the use of a hitherto optimal precoder structure while the channel (relative phase across clusters) drifts away due to fading leading to a potential loss in performance. Note that a 360° phase change across paths is possible with a relative movement of a wavelength (which is on the order of a few millimeters at mmWave frequencies). In a third scheme, we consider an ideal directional scheme where F and G are steered toward the dominant cluster (corresponding to $\phi^{\text{T}} = \phi^{\text{R}} = 60°$. From Figure 4.2(a), we observe that while the directional scheme is poorer than the optimal scheme (by about 3 dB) for any relative phase value, the mismatched scheme can lead to a significant drop in performance (even as high as 20 dB) with some relative phase changes. Thus, the mismatched precoder is an example of a non-robust precoder structure as relative phase changes can happen at much faster time-scales compared to gain changes.

While Figure 4.2(a) corresponds to an $L = 2$ cluster channel, the CDF of the loss in performance with a mismatched SVD precoder (SVD of the mismatched channel matrix) and a directional precoder, both relative to an ideal SVD precoder, are plotted in Figure 4.2(b) for the $L = 2, 4$ and 10 scenarios. Each cluster corresponds to 20 rays with a 5° angular spread around a central direction. The gains of the clusters are $\{1, 0.9\}$ for $L = 2$, $\{1, 0.9, 0.8, 0.7\}$ for $L = 4$ and $\{1, 0.9, \ldots, 0.1\}$ for $L = 10$. The CDF is over randomness in the relative phases of the rays within each cluster's angular spread. From this plot, we observe that, in general, while the mismatched precoder is not robust with a significant spread in loss relative to the ideal precoder, it is still better than a directional precoder for most of the channel realizations as the number of clusters in the channel increases. A directional precoder that selects the best cluster (out of many) in the channel can lead to a small loss in performance relative to the optimal precoder that it can maintain in a robust manner if the number of dominant clusters is small. On the other hand, while a mismatched precoder can quickly fall out of optimality with relative phase changes across clusters and rays within that cluster, it can remain better than a directional precoder that selects the best cluster. That is, even sub-optimal combining of the many clusters in the channel is often more useful than the selection of the best cluster with the performance gap being directly determined by the degree of sub-optimality in combining these clusters. This is the context in which approximation to F_{opt} and G_{opt} via a directional codebook-based approach makes sense for sparse mmWave channels, but not so much for richer channels at sub-7 GHz frequencies. See Appendix 4.6.4 for the class of feedback schemes useful at sub-7 GHz frequencies. Motivated by this development, the notion of adaptive beam weights (discussed in Chapter 4.2.4) that combine the energy across multiple clusters is also useful as the mmWave channel becomes richer (typically in indoor deployments with multiple reflections). Further, as Figure 4.1(b) illustrates, with smaller array dimensions, a directional precoder is relatively coarse (and interference-inducing) to excite the channel's cluster structure.

4.1.3 PARTIAL CSI SCHEMES AND CONNECTIONS TO HYBRID PRECODING

Perfect CSI at the transmitter node requires either obtaining perfect CSI at the receiver node followed by noise-free feedback of this CSI or CSI estimated perfectly at the transmitter node via uplink training (assuming a TDD system with channel reciprocity – that is, the same antenna configuration for transmitter and receiver). Both of these assumptions can lead to considerable overhead in practical systems. Thus, partial CSI-based schemes have become popular in practical implementations. A brief overview of the feedback-driven partial CSI based approaches commonly used in 3GPP Rels. 8 through 13 are presented[3] in Appendix 4.6.4.

Building on these developments, full- or three-dimensional (FD/3D) MIMO[4] that supports beamforming in *both* azimuth and elevation domains targeting users in high-rise/office buildings, stadium deployments, etc. was introduced in Rel. 13 [142] with support for up to $N = 16$ RF chains. This has been further extended to support $N = 32$ RF chains in Rels. 14 through 18, whereas Rel. 19 (and its evolution) could further expand support to the $N > 32$ RF chain case. In these settings, feedback codebooks started incorporating the possibility that base stations can use active antenna systems (AASs) over planar arrays. Further, the number of antennas, RF chains, and logical/CSI-RS ports in 5G-NR could be significantly more than that of legacy/LTE systems where each antenna is connected to an RF chain and which, in turn, is connected to a logical port. In particular, many antenna elements could be connected[5] to a port/RF chain for cost, power and thermal reasons.

---

[3]Note that while partial CSI acquisition is a sophisticated topic in its own right, the focus here is on setting the contextual connection between the past in terms of MIMO development and its utility in the mmWave context.

[4]Rel. 13 and 14 are also often called as LTE Advanced Pro.

[5]To be precise, the concept of *antenna virtualization*, or the notion that multiple antennas are treated as a single port/RF chain by the application of suitable precoding and can thus remain transparent to the UE, was introduced in Rel. 10. Thus, the notion of a one-to-one mapping between antenna elements and ports/RF chains has been loosened from Rel. 10 (and beyond). Nevertheless, the use of both FD-MIMO (in Rel. 13) and mmWave transmissions (in Rel. 15 and beyond) has led to a growing interest in this idea.

This evolution led to a considerable increase in the degrees-of-freedom in implementation (with all the different possibilities in terms of feedback codebooks difficult to be standardized). Hence, an abstraction has been introduced as follows [143, 144].

- A *transceiver unit (TXRU) virtualization* framework is introduced where an RF circuit[6] takes a set of digital inputs (corresponding to the TXRUs) and maps it into a set of analog outputs (corresponding to the antenna elements) and is denoted by an $N_t \times N_{\mathsf{TXRU}}$ virtualization matrix V_{T}. This transformation happens in RF/analog in time-domain and is wideband. In this specific context, wideband corresponds to transmissions over a component carrier (e.g., 20 MHz in FR1 or 100 MHz in FR2).
- A *port virtualization* framework is captured by an $N_{\mathsf{TXRU}} \times N_{\mathsf{P}}$ virtualization matrix V_{P} that corresponds to *digital* beamforming from the TXRUs to the N_{P} logical/CSI-RS ports. This transformation happens in the baseband domain and is frequency selective.

This is preceded by the traditional precoder matrix V_{precoder} of size $N_{\mathsf{P}} \times r$ (where $r \leq N_{\mathsf{P}}$) that takes the ports to the r layers[7] (in the baseband domain and is also frequency selective). Overall, the system model from the baseband to the antenna domain is given as

$$x\Big|_{N_t \times 1} = V_{\mathsf{T}}\Big|_{N_t \times N_{\mathsf{TXRU}}} V_{\mathsf{P}}\Big|_{N_{\mathsf{TXRU}} \times N_{\mathsf{P}}} V_{\mathsf{precoder}}\Big|_{N_{\mathsf{P}} \times r} s\Big|_{r \times 1} \tag{4.27}$$

where x is transmitted over the antenna array. In traditional/LTE evaluation assumptions that model passive antenna base stations, TXRUs are one-to-one mapped to logical antenna ports[8], and are also one-to-one mapped to passive antennas placed on a horizontal axis. That is, $N_t = N_{\mathsf{TXRU}} = N_{\mathsf{P}}$ and $x = V_{\mathsf{precode}}\, s$ as described in (4.1).

In the FD MIMO case, it is typical that $N_t \geq N_{\mathsf{TXRU}} \geq N_{\mathsf{P}}$. Two types of connection architectures are recognized as illustrated in Figure 4.3 and Figure 4.4. In a one-dimensional subarray connected architecture, the N_t antenna elements are divided into N_{TXRU} groups with N_t/N_{TXRU} antennas per group and where $N_{\mathsf{TXRU}} = N_{\mathsf{P}}$ with $V_{\mathsf{P}} = I$. The weights/connections across the antenna elements remain fixed as we move across the N_{TXRU} groups denoted by the vector w of size $N_t/N_{\mathsf{TXRU}} \times 1$. Let x_{TXRU} be the $N_{\mathsf{TXRU}} \times 1$ signal vector connecting the TXRUs. Then, assuming a single data-stream, the input into the antenna domain with this architecture is given

[6]Data is up-converted from the baseband to RF (for the transmitter path) and down-converted from RF to the baseband (for the receiver path) via a TXRU, which is functionally similar to an RF chain, as described in Chapter 3. A radio distribution network (RDN) connects the RF signals to the antenna array. The only difference between the hybrid beamforming framework of mmWave systems and TXRU framework of Rel. 13 is that the RF is different in these systems. However, this has massive practical implications on the cost, size, feasibility and other key performance indicators (KPIs) of these networks.

[7]The number of layers r is also the number DMRS ports.

[8]Typically, the ports are CRS-based or CSI-RS-based.

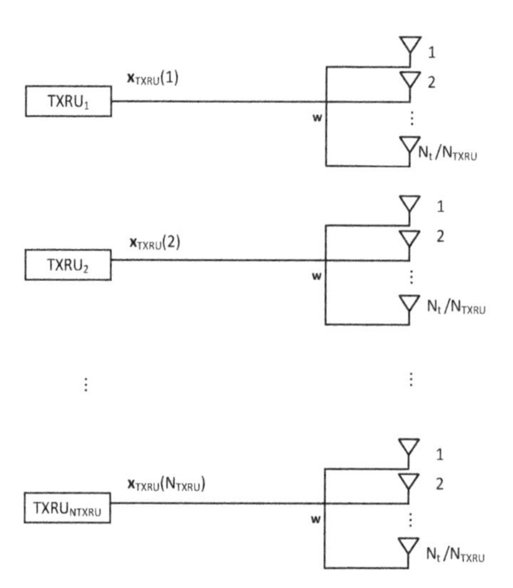

Figure 4.3 Subarray connected architecture with a power divider network.

as $x_{\text{TXRU}} \otimes w$ where $\otimes$ denotes the Kronecker product of the two vectors. Thus, we have

$$V_{\text{T}} = \begin{bmatrix} w & 0 & \dots & 0 \\ 0 & w & \dots & 0 \\ \vdots & \vdots & \ddots & \vdots \\ 0 & 0 & \dots & w \end{bmatrix}. \tag{4.28}$$

In a fully connected architecture, the signal from each TXRU is connected to all the N_t antennas leading to a linear combination of signals across the TXRUs. Here, the input into the antenna domain is given as Wx_{TXRU} where W is $N_t \times N_{\text{TXRU}}$ and x_{TXRU} is $N_{\text{TXRU}} \times 1$. Thus, we have $V_{\text{T}} = W$. These architectures capture the sub-connected and fully connected hybrid beamforming architectures explored in Chapter 3 at a more abstract level as used in 3GPP specifications.

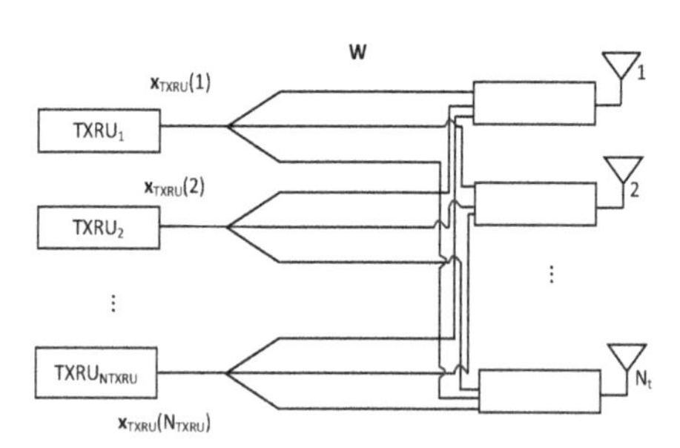

Figure 4.4 Fully connected architecture with power dividers and power combiners.

From a precoder perspective, 3GPP standard specifications assume that N_{tz} antenna elements in elevation are connected to N_1 TXRUs and N_{tx} antenna elements in azimuth are connected to N_2 TXRUs with a doubling of TXRUs to accommodate polarization MIMO transmissions for a total of $2N_1N_2$ TXRUs (which determines the number of RF chains or CSI-RS ports). Note that the antenna array is a dual-polarized array of size $N_{tz} \times N_{tx}$, which is quite flexible from an implementation standpoint. While codebook design for different choices of $\{N_{tz}, N_{tx}\}$ is difficult, the abstract virtualization/mapping considered here allows us to construct precoder matrices assuming $2N_1N_2$ RF chains.

4.2 ANALOG BEAMFORMING

The framework of hybrid beamforming is now well established from a channel state feedback perspective. The extreme case of hybrid beamforming with a single RF chain per-polarization (or two RF chains in all) is typically termed as analog/RF beamforming since all the antenna elements are co-phased at the radio frequency of interest. In this context, Chapter 4.1 has shown that the optimal rank-1 beamformer corresponds to tracking the dominant left and right singular vectors of the channel matrix, each of which in itself is a linear combination of the dominant steering vectors in the channel at both ends. One possibility here is to focus on a *directional structure* that tracks only the dominant steering vector at both ends and ignoring the sub-dominant steering vectors (or using them as fallback options for beam failure recovery). We start with a simple approach where we consider a fixed directional codebook of beam weights, where each set of beam weights steers energy toward a fixed spatial direction.

4.2.1 PROGRESSIVE PHASE SHIFT BEAM CODEBOOKS

To simplify the understanding of the performance of these directional beams, we now consider a one-sided beamforming setup as illustrated in Figure 4.5 (over the XY plane of a global coordinate system) with signal coming from a point source in the far field and collected by N antenna elements placed on the Y axis. The N antenna elements have a constant inter-antenna element spacing of d and are arranged in a linear array and the planar wavefront in the far field is at an angle θ_0 relative to the boresight of the antenna array (that is, boresight corresponds to $\theta_0 = 0^\circ$).

Let $s_1(t), \ldots, s_N(t)$ denote the signals observed by the N antenna elements. Due to the relative path difference of $(i-1)d\sin(\theta_0)$ at the i-th antenna (with respect to the first antenna) of the signal coming from the point source, we have

$$s_i(t) = s_1\left(t - \frac{(i-1)d\sin(\theta_0)}{c}\right), \; i = 1, \ldots, N. \tag{4.29}$$

The i-th antenna processes the received signal with an amplitude response and phase of $A_i e^{j\phi_i}$ to result in

$$r(t) = \sum_{i=1}^{N} A_i e^{j\phi_i} \cdot s_i(t) + n(t) \tag{4.30}$$

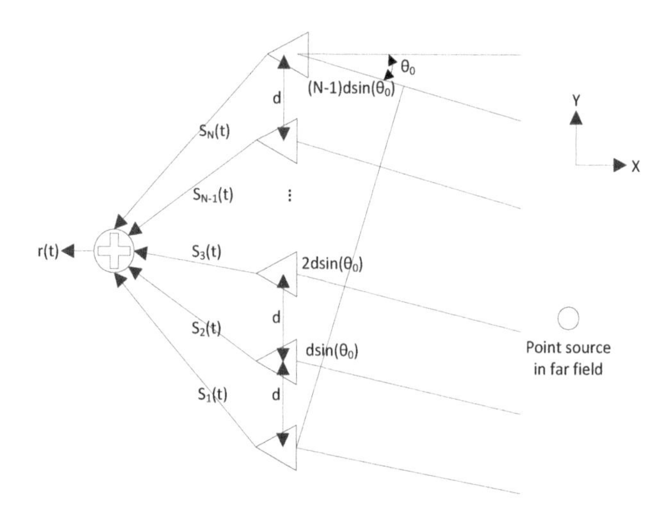

Figure 4.5 A simplified one-sided beamforming setup with N antenna elements.

where the additive noise $n(t) \sim \mathcal{CN}(0, N\sigma^2)$ with the factor of N in the noise variance coming from the collection of the signal (and just, the noise) from the N antennas. Let $s_1(t)$ be a tone at a carrier frequency f_c and we thus have

$$r(t) = \sum_{i=1}^{N} A_i e^{j\phi_i} \cdot e^{j2\pi f_c \left(t - \frac{(i-1)d\sin(\theta_0)}{c}\right)} + n(t) \tag{4.31}$$

$$= s_1(t) \cdot \sum_{i=1}^{N} A_i e^{j\phi_i} \cdot e^{\frac{-j2\pi(i-1)d\sin(\theta_0)}{\lambda}} + n(t). \tag{4.32}$$

The SNR after this beamforming operation is given as

$$\mathsf{SNR} = \frac{\left| \sum_{i=1}^{N} A_i e^{j\phi_i} \cdot e^{\frac{-j2\pi(i-1)d\sin(\theta_0)}{\lambda}} \right|^2}{N\sigma^2}. \tag{4.33}$$

It is clear that the SNR is maximized by aligning up the phases on any straight line; for example, by the choice

$$\phi_i = \frac{2\pi(i-1)d\sin(\theta_0)}{\lambda}. \tag{4.34}$$

With a constraint that $|A_i| \leq A$, this value of SNR is maximized by the choice $A_i = A$. This choice of beam weights corresponds to steering the energy of the antenna array toward the direction θ_0. This class of beam weights implements a *progressive phase shift* (PPS) or a *constant phase offset* (CPO) across the antenna array (or often simply described as DFT beam weights). With these beam weights, we have

$$\mathsf{SNR} = A^2 \cdot N/\sigma^2. \tag{4.35}$$

In other words, the use of N antennas leads to an SNR improvement of $10 \cdot \log_{10}(N)$ dB. Note that this is the best possible gain realizable with the use of N antenna elements since

$$\left| \sum_{i=1}^{N} A_i e^{j\phi_i} \cdot e^{\frac{-j2\pi(i-1)d\sin(\theta_0)}{\lambda}} \right|^2 \leq \left(\sum_{i=1}^{N} A_i \right)^2 \leq N^2 A^2. \tag{4.36}$$

In the general case where $A_i = A$ and ϕ_i corresponds to beam steering along a direction θ which could be different from θ_0 (that is, $\phi_i = 2\pi(i-1)d\sin(\theta)/\lambda$), we have

$$\mathsf{SNR}(\theta) = \frac{A^2}{N\sigma^2} \cdot \left| \sum_{i=1}^{N} e^{\frac{j2\pi(i-1)d\cdot(\sin(\theta)-\sin(\theta_0))}{\lambda}} \right|^2 \tag{4.37}$$

$$= \frac{A^2}{N\sigma^2} \cdot \left| \frac{1 - e^{\frac{j2\pi Nd\cdot(\sin(\theta)-\sin(\theta_0))}{\lambda}}}{1 - e^{\frac{j2\pi d\cdot(\sin(\theta)-\sin(\theta_0))}{\lambda}}} \right|^2 \tag{4.38}$$

$$= \frac{A^2}{N\sigma^2} \cdot \left| \frac{\sin\left(\frac{\pi Nd\cdot(\sin(\theta)-\sin(\theta_0))}{\lambda}\right)}{\sin\left(\frac{\pi d\cdot(\sin(\theta)-\sin(\theta_0))}{\lambda}\right)} \right|^2 \tag{4.39}$$

where (4.38) follows from the sum of a geometric sequence and (4.39) follows since $\sin(\theta) = \frac{\exp(j\theta)-\exp(-j\theta)}{2j}$.

Consider the scenario where $A = 1$, $d = \lambda/2$ and $\theta_0 = 0°$. Here, (4.39) reduces to

$$\mathsf{SNR}(\theta) = \frac{1}{N} \cdot \left| \frac{\sin\left(\frac{\pi N\cdot\sin(\theta)}{2}\right)}{\sin\left(\frac{\pi\cdot\sin(\theta)}{2}\right)} \right|^2. \tag{4.40}$$

First, note that $\mathsf{SNR}(\theta)$ is symmetric in θ around $\theta = 0°$. As $\theta \to 0°$, since $\frac{\sin(\theta)}{\theta} \to 1$, for any N, we have

$$\lim_{\theta\to 0°} \mathsf{SNR}(\theta) = \lim_{\theta\to 0°} \frac{1}{N} \cdot \left| \frac{\sin\left(\frac{\pi N\cdot\sin(\theta)}{2}\right)}{\frac{\pi N\cdot\sin(\theta)}{2}} \right|^2 \cdot \left| \frac{\frac{\pi\cdot\sin(\theta)}{2}}{\sin\left(\frac{\pi\cdot\sin(\theta)}{2}\right)} \right|^2 \cdot N^2 \tag{4.41}$$

$$= N. \tag{4.42}$$

The nulls in $\mathsf{SNR}(\theta)$ correspond to the θ values that make the numerator of (4.40) zero, but the denominator non-zero. In other words, we are looking for

$$\frac{\pi N\sin(\theta)}{2} = n\pi,\ n \in \mathbb{N},\ n \neq kN, \tag{4.43}$$

which is equivalent to

$$\theta = \sin^{-1}\left(\frac{2n}{N}\right). \tag{4.44}$$

The first null-to-null beamwidth is

$$2\sin^{-1}\left(\frac{2}{N}\right) \approx \frac{4}{N} \tag{4.45}$$

for large N. This beamwidth (in degrees) can be approximated as $229.2^\circ/N$.

While the first null in $\mathsf{SNR}(\theta)$ is at $\frac{\pi N\sin(\theta)}{2} = \pi$, the second null is at $\frac{\pi N\sin(\theta)}{2} = 2\pi$. Thus, the first/main side lobe corresponds to the case where

$$\frac{\pi N\sin(\theta)}{2} = \frac{3\pi}{2} \text{ or } \theta = \sin^{-1}\left(\frac{3}{N}\right). \tag{4.46}$$

With this choice, the numerator of (4.40) is 1 and we have

$$\mathsf{SNR}(\theta)\Big|_{\text{main side lobe}} = \frac{1}{N\cdot\sin^2\left(\frac{3\pi}{2N}\right)} \approx \frac{4N}{9\pi^2} \tag{4.47}$$

for large N. Thus, relative to the main lobe level of N, the first side lobe is at

$$10\cdot\log_{10}\left(\frac{9\pi^2}{4}\right) \approx 13.47 \text{ dB} \tag{4.48}$$

below for large N. We call this metric the *side lobe gap* in subsequent discussions. We now compute the angular spread over which $\mathsf{SNR}(\theta)$ is above $N/2$ (defined as the 3-dB or half-power beamwidth since the peak gain is N). Without explicitly computing this quantity, note that since the first null-to-null beamwidth is $2\sin^{-1}\left(\frac{2}{N}\right)$ and decreasing as N increases, the half-power beamwidth should also decrease as N increases due to the fact that it is smaller than the first null-to-null beamwidth. More precisely, the θ value corresponding to half-power satisfies

$$\frac{1}{N}\cdot\frac{\sin^2(Nt)}{\sin^2(t)} = \frac{N}{2} \Longleftrightarrow \frac{\sin(Nt)}{N\sin(t)} = \frac{1}{\sqrt{2}} \tag{4.49}$$

where $t = \frac{\pi\sin(\theta)}{2}$. Note that the t that solves for (4.49) is a function of N. For small values of N, this value t needs to be solved numerically. But for large N, since $\theta \to 0$, so does t. We have

$$\frac{\sin(Nt)}{N\sin(t)} \approx \frac{\sin(Nt)}{Nt} = \frac{1}{\sqrt{2}} \tag{4.50}$$

which leads to a choice of $Nt \approx 1.3916$. In other words, the half-power beamwidth is $2\sin^{-1}\left(\frac{2\cdot 1.3916}{\pi N}\right)$. In degrees, this beamwidth can be approximated as $101.5^\circ/N$. The ratio between the first null-to-null beamwidth and the half-power beamwidth is given as

$$\frac{2\sin^{-1}\left(\frac{2}{N}\right)}{2\sin^{-1}\left(\frac{2\cdot 1.3916}{\pi N}\right)} \approx \frac{\pi}{1.3916} = 2.2575 \tag{4.51}$$

for large N.

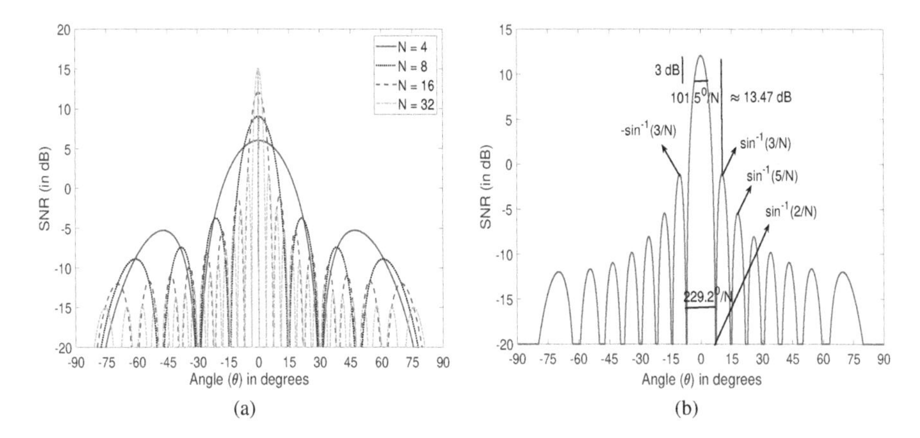

Figure 4.6 (a) $\mathsf{SNR}(\theta)$ vs. θ for different N and $d = \lambda/2$. (b) Beam properties with a beam implementing progressive phase shifts for $N = 16$.

In the case we have studied so far ($A = 1$, $d = \lambda/2$ and $\theta_0 = 0°$), Figure 4.6(a) plots $\mathsf{SNR}(\theta)$ as a function of θ for different values of N. From this plot, we observe the trends that the peak gain increases and the beamwidth of the beam decreases as N increases. The beam pattern becomes more concentrated/peaky in the area of coverage as the array dimension increases. This trend reversal is natural given that the total energy in the beam remains conserved, independent of N. Thus, a beam with a wider beamwidth (which spreads energy over a larger spatial area) should necessarily get less peakier and *vice versa*. Also, the location of the k-th side lobe can be parameterized as $\sin^{-1}\left(\frac{2k+1}{N}\right)$. The value $k = N/2 - 1$ is the largest k for which $\frac{2k+1}{N} < 1$ and hence $N/2 - 1$ side lobes are observable in the $[0°, 90°]$ region. Along with the main lobe, we have $2 \cdot (N/2 - 1) + 1 = N - 1$ peaks in the beam pattern in the $[-90°, 90°]$ region. This trend of $N - 1$ peaks in beam pattern is also observed in Figure 4.6(a). The beam properties associated with a typical beam are illustrated in Figure 4.6(b) for the case of $N = 16$.

A beam codebook of size K corresponding to progressive phase shifts along steered directions $\theta_k,\ k = 1, \ldots, K$ is of the form:

$$\mathscr{C}_{\mathsf{PPS}} = \left\{ w_k(i) = e^{\frac{j2\pi(i-1)d\sin(\theta_k)}{\lambda}},\ i = 1, \ldots, N,\ k = 1, \ldots, K \right\}. \tag{4.52}$$

By choosing $\{\theta_k\}$ to maximize the worst-case gain within the $[-90°, 90°]$ coverage region for different choices of K, Figure 4.7(a) shows SNR as a function of θ with $K = 16, 8$ and 4 beams in the $N = 16$ case. Clearly, we observe that as the number of beams K decreases, the codebook of progressive phase shifts is insufficient to provide good coverage over the $[-90°, 90°]$ region with deep coverage holes and poor performance. This trend is confirmed with the CDF of $\mathsf{SNR}(\theta)$ over the coverage region for the three codebook sizes, illustrated in Figure 4.7(b), with the median of $\mathsf{SNR}(\theta)$ being 11.0, 6.8 and -2.3 dB, respectively.

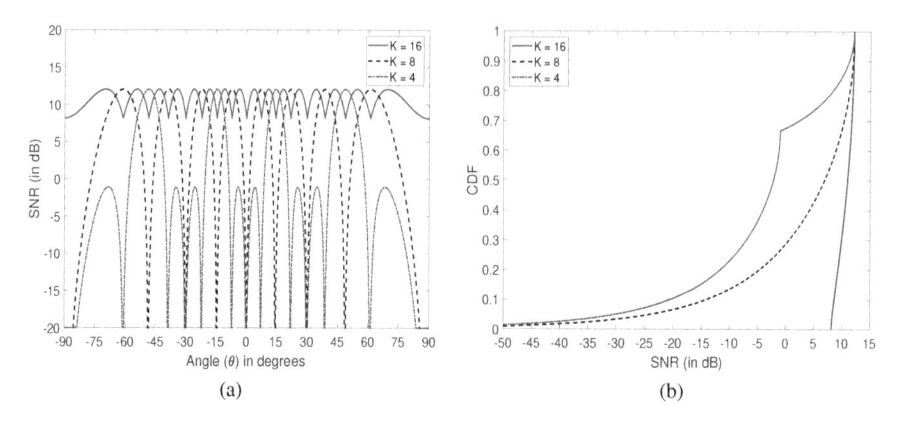

Figure 4.7 (a) SNR(θ) vs. θ for different progressive phase shift codebook sizes with $N = 16$. (b) CDF of SNR(θ) over θ for different codebook sizes with $N = 16$.

4.2.2 BEAM BROADENING

The reason for the observation in Figure 4.7(b) is that the null-to-null and half-power beamwidths of beams of progressive phase shifts are inversely proportional to the array dimension N. Thus, the number of beams required to cover a certain coverage area grows linearly with N and when a fixed or small codebook size is used, it invariably leads to coverage holes.

It is in this context that highlights the relevance of *beam broadening* or maximization of the worst-case array gain over the coverage region with a fixed number (K) of beams in the codebook. To keep ideas simple, consider a linear antenna array of size N on the Z-axis with inter-antenna element spacing of $d = \lambda/2$ leading to a beamspace transformation of the physical steering angle θ:

$$\Omega = \frac{2\pi d}{\lambda}\cos(\theta) = \pi\cos(\theta). \tag{4.53}$$

Since the boresight plane corresponds to $\theta = 90^\circ$ or $\Omega = 0$ and mmWave antennas are directional, we can define the goal of beam broadening to maximize the worst-case array gain over $\Omega \in [-\Omega_0/2, \Omega_0/2]$. A practically useful example would be 120° coverage over the $[30^\circ, 150^\circ]$ region with

$$\Omega_0 = \pi \cdot (\cos(30^\circ) - \cos(150^\circ)) = \pi \cdot \sqrt{3}. \tag{4.54}$$

Since K beams are expected to cover this beamspace, the focus area of each broadened beam is $\frac{\Omega_0}{K} = \frac{\pi\cdot\sqrt{3}}{K}$ or a region of $\left[\frac{-\pi\cdot\sqrt{3}}{2K}, \frac{\pi\cdot\sqrt{3}}{2K}\right]$. The approximate flat coverage region of a broad beam (in degrees) spans a one-sided beamspace of $\frac{\pi\cdot\sqrt{3}}{2K}$ and is given as

$$2\cdot\left(\frac{\pi}{2} - \cos^{-1}\left(\frac{\sqrt{3}}{2K}\right)\right). \tag{4.55}$$

Thus, the 3-dB beamwidth is a scaled version of the above term (let the scaling constant be denoted as γ for some $\gamma > 1$). Thus, the broadening factor relative to a progressive phase shift beam is given as

$$\text{Broadening factor} = \frac{2\gamma \cdot \left(\frac{\pi}{2} - \cos^{-1}\left(\frac{\sqrt{3}}{2K}\right)\right)}{2\sin^{-1}\left(\frac{2 \cdot 1.3916}{\pi N}\right)}. \quad (4.56)$$

We now use the approximations

$$\cos^{-1}(x) \approx \frac{\pi}{2} - x \text{ and } \sin^{-1}(x) \approx x \text{ as } x \to 0. \quad (4.57)$$

For N and K large, we thus have

$$\text{Broadening factor} \approx \frac{\gamma \cdot \pi\sqrt{3} \cdot N}{4 \cdot 1.3916 \cdot K} = \frac{\gamma \cdot N}{K}. \quad (4.58)$$

With $K = N/2$ beams to cover the 120° coverage region and $\gamma = 1.5$, the broadening factor realized with a broad beam is ≈ 3.

To design a broadened beam, we consider one constructive solution[9] wherein the antenna array of size N is partitioned into M *virtual subarrays* [147]. Each virtual subarray is used to beamform to a certain appropriately chosen virtual direction. The expectation from this approach is that the beam patterns from the individual virtual subarrays combine or constructively add to enhance the coverage area of the resultant beam with minimal loss in peak gain due to reduction in the effective aperture of the subarrays. As a specific example, in the $M = 2$ virtual subarray setting, half of the array is used to steer energy toward $\theta = \pi - \cos^{-1}\left(\frac{2\mathsf{f}}{N}\right)$ for an appropriately chosen parameter f and the other half of the array is used to steer energy toward

$$\theta = \cos^{-1}\left(\frac{2\mathsf{f}}{N}\right). \quad (4.59)$$

The broadened beam weights are thus given as

$$w(n) = \frac{1}{\sqrt{N}} \begin{cases} \exp\left(-\frac{j2\pi\mathsf{f}}{N}\left(n - \frac{N}{2} + \frac{1}{2}\right)\right) & \text{if } 0 \leq n \leq \frac{N}{2} - 1 \\ \exp\left(\frac{j2\pi\mathsf{f}}{N}\left(n - \frac{N}{2} + \frac{1}{2}\right)\right) & \text{if } \frac{N}{2} \leq n \leq N - 1. \end{cases} \quad (4.60)$$

Optimization over f is performed to maximize the worst-case gain over the $\left[\frac{-\pi \cdot \sqrt{3}}{2K}, \frac{\pi \cdot \sqrt{3}}{2K}\right]$ region or another beamspace coverage region.

This approach can be extended to $M = 3$ and $M = 4$ virtual subarrays. In the $M = 3$ case, we use a two parameter structure (f and $0 \leq L \leq \frac{N}{2}$) to propose beam weights

[9] Some other beam broadening solutions can be found in works such as [145] and [146]. A fundamental performance bound with which every constructive solution can be compared is also provided in [147].

of the form:

$$w(n) = \frac{1}{\sqrt{N}} \begin{cases} \exp\left(-\frac{j2\pi \mathsf{f}}{N}(n - \frac{N}{2} + \frac{1}{2} + L)\right) & \text{if } 0 \le n \le \frac{N}{2} - L - 1 \\ 1 & \text{if } \frac{N}{2} - L \le n \le \frac{N}{2} + L - 1 \\ \exp\left(\frac{j2\pi \mathsf{f}}{N}(n - \frac{N}{2} + \frac{1}{2} - L)\right) & \text{if } \frac{N}{2} + L \le n \le N - 1, \end{cases} \tag{4.61}$$

for three subarrays of length $N/2 - L$, $2L$ and $N/2 - L$, respectively. In the $M = 4$ case, we use a three parameter structure (f, $\delta\mathsf{f}$ and $0 \le L \le \frac{N}{2}$) to propose beam weights of the form:

$$w(n) = \frac{1}{\sqrt{N}} \cdot \begin{cases} \exp\left(-\frac{j2\pi(\mathsf{f}+\delta\mathsf{f})}{N}(n - \frac{N}{2} + \frac{1}{2}) - \frac{j2\pi \cdot \delta\mathsf{f}}{N}\left(L - \frac{1}{2}\right)\right) & \text{if } 0 \le n \le \frac{N}{2} - L - 1 \\ \exp\left(-\frac{j2\pi \mathsf{f}}{N}(n - \frac{N}{2} + \frac{1}{2})\right) & \text{if } \frac{N}{2} - L \le n \le \frac{N}{2} - 1 \\ \exp\left(\frac{j2\pi \mathsf{f}}{N}(n - \frac{N}{2} + \frac{1}{2})\right) & \text{if } \frac{N}{2} \le n \le \frac{N}{2} + L - 1 \\ \exp\left(\frac{j2\pi(\mathsf{f}+\delta\mathsf{f})}{N}(n - \frac{N}{2} + \frac{1}{2}) - \frac{j2\pi \cdot \delta\mathsf{f}}{N}\left(L - \frac{1}{2}\right)\right) & \text{if } \frac{N}{2} + L \le n \le N - 1, \end{cases} \tag{4.62}$$

for four subarrays of length $N/2 - L$, L, L and $N/2 - L$, respectively. Optimization or search over these parameters leads to a broadened beam construction.

The basic idea behind beam broadening is illustrated in Figure 4.8(a) where three different template beams corresponding to different beam broadening factors are designed using the virtual subarray technique described above for the case of $N = 64$. A set of beam weights of progressive phase shift steered toward the boresight direction leads to a 3-dB beamwidth of $\approx 101.5^\circ/64 = 1.59^\circ$ with a peak gain of

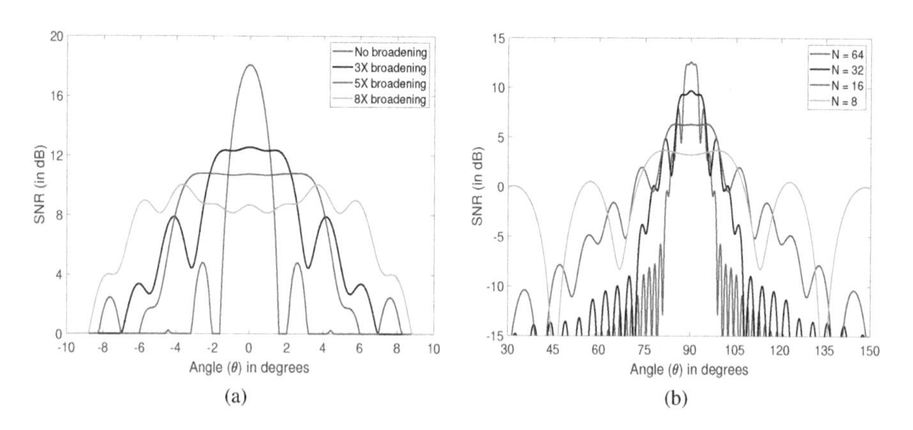

Figure 4.8 Beam broadening solutions for (a) $N = 64$ with different broadening factors and (b) for different array dimensions.

$10 \cdot \log_{10}(64) = 18.06$ dB. We then consider three beam broadening factors (of $\approx 3, 5$, and 8) leading to 3-dB beamwidths of $\approx 4.8°$, $\approx 8.4°$ and $\approx 14°$ by the careful design of beam broadening parameters. The beam patterns of these three template beams are also plotted in Figure 4.8(a), which shows an approximately flat gain response (with potential ripples) over the essential coverage area of the beam. The peak gains of these three beams are 12.55, 10.63 and 8.07 dB, respectively. The beamwidths of the template beams are increased. This comes at the necessary cost of reduction in peak gain since the total energy in all the beams remains the same. Thus, fewer beams can be used to cover a given coverage area by spatially shifting or modulating the template beams. Thus, beam broadening allows a tradeoff of the beam scanning latency with reduced gains realized with beam scanning.

Illustrating the design principle of beam broadening further, Figure 4.8(b) plots the beam patterns corresponding to the broadened beams for $N = 64$, 32, 16 and 8. The peak gains of these four beams are 12.55, 9.68, 6.33 and 3.69 dB (or an approximate 3 dB drop for a halving of array dimensions). Note that the peak gain of the broadened beam is $\approx 28\%$ of the peak gain of a progressive phase shift beam. On the other hand, the 3-dB beamwidths of these beams are 4.8°, 10°, 19° and 34° which can be compared with the 3-dB beamwidths of progressive phase shift beams of 1.59°, 3.17°, 6.34° and 12.69°, respectively. That is, the beamwidths of the broadened beams have approximately tripled relative to the beamwidths of progressive phase shift beams. In other words, only $\approx 75\%$ of the energy in a progressive phase shift beam is seen in the main lobe of a broadened beam. Since the total energy is conserved, a necessary outcome of beam broadening is that unlike the ≈ 13.47 dB gap between the main lobe and first side lobe of a progressive phase shift beam, we observe a gap of 4.68, 4.80, 4.36 and 3.15 dB for $N = 64$, 32, 16 and 8, respectively. In other words, beam broadening leads to a reduction of energy in the main lobe (reduced and approximately flat peak gain over a larger coverage region) with side lobe levels that are relatively more comparable to the main lobe.

We now consider a broad beam codebook of size K that steers the template beams (designed above) along directions θ_k, $k = 1, \ldots, K$:

$$\mathscr{C}_{\mathsf{broad}} = \left\{ w_k(i) = \mathsf{x}_K(i) \cdot e^{\frac{j2\pi(i-1)d\sin(\theta_k)}{\lambda}}, \; i = 1, \ldots, N, \; k = 1, \ldots, K \right\} \tag{4.63}$$

where $\{\mathsf{x}_K(i),\ i = 1, \ldots, N\}$ denotes the template beam weights. The realized $\mathsf{SNR}(\theta)$ over the coverage region with these three template beams are plotted in Figure 4.9(a). Unlike the trend in Figure 4.7(a) that shows a significant deviation from a flat response, $\mathscr{C}_{\mathsf{broad}}$ shows a near-flat beam pattern response over the coverage region. The CDF of the achieved $\mathsf{SNR}(\theta)$ over the coverage region with these three broad beam codebooks as well as different codebooks with progressive phase shift beam weights of different sizes are plotted in Figure 4.9(b). Note that while $\mathscr{C}_{\mathsf{PPS}}$ realizes a higher peak gain, the tails of the achieved SNR show significant distortion. On the other hand, $\mathscr{C}_{\mathsf{broad}}$ realizes a relatively flat CDF curve at the cost of reduced peak gain. Thus, a broadened beam codebook is useful from an initial beam acquisition perspective as there are no coverage holes within the intended coverage region.

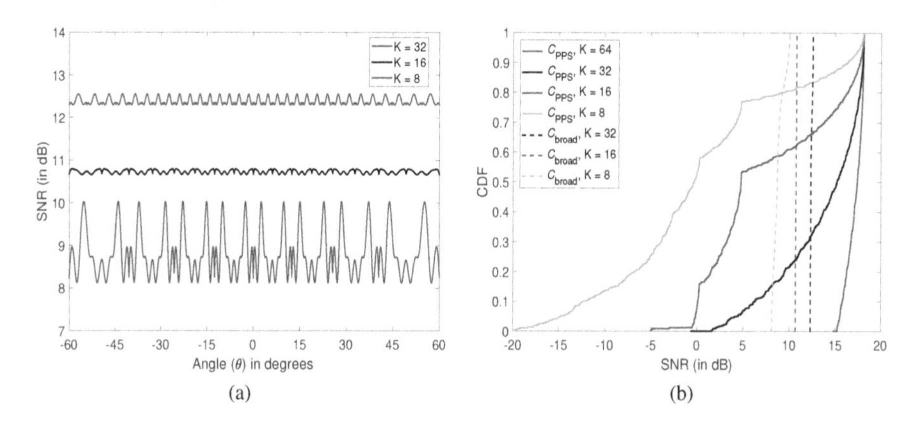

Figure 4.9 (a) SNR(θ) vs. θ for different broad beam codebooks with $N = 64$. (b) CDF of SNR(θ) over the coverage region with different codebooks with $N = 64$.

4.2.3 OTHER BEAM WEIGHT LEARNING APPROACHES

The optimal right singular vector structure of analog beamforming, described in Chapter 4.1, can be learned in practice via a simple scheme (applicable for TDD/reciprocal systems) known as *power iteration.* In the basic noise-less version of this scheme [148, Sec. 7.3], [149], a randomly initialized beamforming vector f_0 is beamformed over the forward channel H to obtain Hf_0 at the receiver end, which is then beamformed further over the reverse channel H^T to obtain $H^H Hf_0$. Iterating this procedure k times, the transmitter node can estimate the un-normalized beamforming vector

$$f_k = \left(H^H H\right)^k f_0. \tag{4.64}$$

If the eigenvectors of $H^H H$ (in decreasing order of dominance) are denoted as $v_1, \ldots, v_N$ with corresponding eigenvalues being $\lambda_1, \ldots, \lambda_N$, we have

$$f_k = \left(H^H H\right)^k f_0 = \sum_{i=1}^{N} (\lambda_i)^k \cdot v_i^H f_0 \cdot v_i. \tag{4.65}$$

Thus, as k increases, the first term corresponding to the dominant eigenvector dominates the contribution in f_k leading to convergence (at a speed of the order of $\left(\frac{\lambda_1}{\lambda_2}\right)^k$) to the desired beamforming vector. The more general noisy version of this scheme is studied in [147].

Given that the optimal right singular vector for each base station-to-UE link could be different, learning it has to be done in a unicast (that is, on a per-UE basis) and in a bidirectional manner (separate learning for the downlink and uplink) leading to an increase in system-level latencies and overheads. Further, the entries of f_k require high-precision amplitude and phase quantization (for convergence), which may not be available at one/both end(s) of the link. The efficacy of right singular vector learning depends on the link margin with poor learning at low SNRs, a typical scenario in practical deployments over large inter-site distances. Thus, the power iteration approach (or its variants) are typically not considered in practical systems.

While we considered codebook-based beam sweeping approaches for directional learning, this problem has had a long and illustrious history in the signal/array processing literature [150]. In the simplest case of estimating a single unknown source (signal direction) at the UE end as described in Figure 4.5 and with system equation:

$$y = \alpha_1 u(\phi_1) + n \tag{4.66}$$

where α_1 is known, $u(\cdot)$ denotes the array steering vector along a certain direction, and $n \sim \mathscr{C}\mathscr{N}(0, I)$. It can be seen that the maximum likelihood (ML) solution of maximizing the density function $f(y|\alpha_1, \phi_1)$ is equivalent to finding $\widehat{\phi}_1$ solving for

$$\widehat{\phi}_1 = \arg\max_{\phi} |u(\phi)^H y|^2. \tag{4.67}$$

In other words, correlation of the received vector y for the best signal strength results in the ML solution for the problem of signal coming from one unknown direction.

In general, if there are multiple (say, K) sources with system equation

$$y = \sum_{k=1}^{K} \alpha_k u(\phi_k) + n, \tag{4.68}$$

the density function of y is non-convex in the parameters resulting in a numerical multi-dimensional search over the parameter space. In this context, the main premise behind the MUltiple SIgnal Classification (MUSIC) algorithm [151] is that the signal subspace is K-dimensional and is orthogonal to the noise subspace. Furthermore, the K largest eigenvalues of the estimated received covariance matrix, R_y, correspond to the signal subspace and the other eigenvalues to the noise subspace (provided that the covariance matrix estimate is reliable). The MUSIC algorithm then estimates the signal directions by finding the (K) peaks of the pseudospectrum[10], defined as,

$$P_{\mathsf{MUSIC}}(\phi) \triangleq \frac{1}{\sum_{n=K+1}^{N} |u(\phi)^H \widehat{q}_n|^2} \tag{4.69}$$

where $\{\widehat{q}_{K+1}, \ldots, \widehat{q}_N\}$ denote the eigenvectors of the noise subspace of R_y. The principal advantage of the MUSIC algorithm is that the signal maximization task has been recast as a noise minimization task, a one-dimensional line search problem albeit at the cost of computing the eigenvectors of R_y. Nevertheless, since $\{\widehat{q}_1, \ldots, \widehat{q}_N\}$

[10]In general, the choice of K in (4.69) has to be estimated via an information theoretic criterion as in [152] or via minimum description length (MDL) criteria such as those due to Rissanen or Schwartz [153]. So the approach here consists of identification of the number of dominant clusters followed by the actual beam weight learning.

can be chosen to form a unitary basis, it can be seen that MUSIC attempts to maximize $\sum_{n=1}^{K} |u(\phi)^H \widehat{q}_n|^2$ (or in other words, it assigns equal weights to all the components of the signal subspace and is hence not ML-optimal).

Since the covariance matrices are unique for each link, as with the power iteration scheme, MUSIC also requires a unicast system design. MUSIC can suffer from poor performance in link margin constrained scenarios such as in initial acquisition, as consistent covariance matrix estimation becomes a difficult exercise with very few measurements. This is especially true as the array dimensions increase at both the transmitter and the receiver. It also suffers from a high computational complexity dominated by the eigen-decomposition of an $N \times N$ matrix in uplink training. In general, the computational complexity of MUSIC can be traded off by constraining the antenna array structure in various ways. Nevertheless, we expect the computational complexity of other such constrained AoA/AoD learning techniques such as Estimation of Signal Parameters via Rotational Invariance Techniques (ESPRIT) algorithm [154], Space-Alternating Generalized Expectation maximization (SAGE) algorithm [155, 41], higher-order singular value decomposition, etc., to be of similar nature to the MUSIC algorithm. All these reasons suggest that while the MUSIC algorithm (or its variants) may be useful for beam refinement after the UE has been discovered, their utility in initial acquisition is limited.

An alternative approach for beam weight learning uses compressive sensing techniques such as nuclear norm optimization [156, 58, 157] and machine learning-based compressive sensing [158]. In general, these approaches rely on a good dictionary of initial beam weights, corresponding to high-resolution amplitude and phase quantization, over which measurements are made followed by computations whose complexity is similar to that of convex optimization. Performance comparisons and a qualitative comparison of the tradeoffs across the different beam weight learning schemes can be seen in [147] and in Table 4.1, respectively. In addition to the complexity and quantization requirements of the considered solutions, we also focus on the nature of training (uni-directional vs. bi-directional) and overhead from a network-level training perspective (unicast or user specific vs. broadcast).

4.2.4 ADAPTIVE OR DYNAMIC BEAM WEIGHTS

For a given antenna array, the number of directional beams with varying beamwidths at the UE side needed for good performance is small[11]. This set of beams is useful for low-latency operational requirements such as those in initial acquisition, beam failure recovery and beam refinement. Due to the small number of beam weights considered, they are also *static* in the sense that they can be stored in the RFIC chip memory. This allows lower beam-switch latencies without the need for loading beam weights from a slower memory to the RFIC chip memory (which is limited by bus latencies).

[11]For example, with an N element array, $2N$ progressive phase shift beams may be sufficient for good performance—defined as max-to-min array gain deviation within the intended coverage area of the array being limited to ≈ 1.5 dB.

Table 4.1
Qualitative comparison between different beamforming approaches

Issue of interest	Codebook-based beam sweep	MUSIC/ESPRIT	Compressive sensing	Singular vector learning
Computational complexity	RSRP computation	Computing eigenvectors	Convex optimization or similar	Iterative method
Quantization needed	Phase-only Amplitude for adaptive beam	Phase for training	Both amplitude and phase for good dictionary	Both amplitude and phase
Performance robustness	Reasonable	Poor	Reasonable	Poor
Overhead with multiple users	Broadcast	Unicast	Broadcast	Unicast
Training direction	Uni-directional	Bi-directional	Uni-directional	Bi-directional

In contrast, as described in Chapter 4.1, the structure of the optimal beam weights is a linear combination of the array steering vectors in the different directions corresponding to the dominant clusters in the channel. Since a linear combination of progressive phase shifts or equi-gain beam weights does not preserve the equi-gain structure, implementing such a solution requires the use of high-precision phase and amplitude control. This observation motivates the treatment of beam weight design as an optimization problem over the space of available phase shifter and amplitude control combinations, instead of imposing a structural search (like with array steering vector-type beam weights such as those in a progressive phase shift or a broadened beam codebook). This generalization is also particularly useful for small array dimensions, typical at the UE, leading us to the notion of *adaptive* or *dynamic* beam weights.

Note that the space of potential candidate beams from this set of quantization possibilities is large and this set cannot be typically stored in RFIC chip memory. For example, if a B-bit phase shifter and a B_1-bit amplitude control are used per antenna element in an N element antenna array, the number of potential beam weights is $\left(2^B\right)^{N-1} \cdot \left(2^{B_1}\right)^N$. Even with modest values such as $B = B_1 = 3$, this leads to $\approx 2.1 \cdot 10^6$ and $\approx 1.3 \cdot 10^8$ beams for $N = 4$ and $N = 5$ arrays, respectively. Thus, the optimal set of beam weights has to be *adaptively* learned and loaded into RFIC chip memory in mission-mode operation leading to complexities in terms of beam weight settling times. These adaptive beam weights serve different purposes including:

- Combining energy across a wider angular spread of the dominant cluster(s) in the channel
- Polarization mismatches that require phase or amplitude compensation
- Mitigating hand or other blockages via beam weight-based solutions, etc.

We now provide a simple illustrative approach by which adaptive beam weights can be learned. For a given beamforming vector f at the base station and $H(k)$ being the channel matrix over the k-th subcarrier, $g = H(k)f$ is the optimal matched filter beamforming structure at the UE. Thus, if optimal beam weights can be designed on a per-subcarrier basis and phase responses can be estimated accurately at the UE, an estimate of $H(k)f$ can be used at the UE side for receive combining. However, we can only use a common set of beam weights (at the RF level) over a wideband (say, K subcarriers). Thus, considering the received power over a set of RS resources (e.g., SSB), we have

$$\text{RSRP over } K \text{ subcarriers} = \sum_{k=1}^{K} |g^H H(k) f|^2 = g^H \cdot \left(\underbrace{\sum_{k=1}^{K} H(k) f f^H H(k)^H}_{\triangleq R} \right) \cdot g. \tag{4.70}$$

In other words, the received power estimated with a beamforming vector g provides a window into the structure of R (the effective post-transmit beamformed covariance matrix seen at the UE). By using many such sampling beams, R can be estimated. The adaptive beam weights we consider for reception at the UE correspond to the dominant eigenvector of the estimated R, as quantized based on the phase shifter and amplitude control resolution available at the UE.

As an example illustration of how adaptive beam weights work, consider a 4×16 channel with linear arrays at both ends. This channel consists of two dominant clusters with complex gains of $\alpha_1 = -1.27 + 0.6j$ and $\alpha_2 = 0.69 - 1.22j$ leading to $|\alpha_1| = |\alpha_2| = 1.4$. The two clusters correspond to transmit angles of 109.2° and 112.2°. These clusters can be excited with the same directional beam given that the inter-cluster angular separation is smaller than the 3-dB beamwidth of a progressive phase shift beam at the transmitter with 16 antenna elements. At the receiver end, the angles are separated by 5°, 25°, 50°, 75° and 100° in five scenarios (each scenario is marked by distinct markers in Figure 4.10). The beam patterns corresponding to the adaptive beam weights generated for these five scenarios are plotted in Figure 4.10. These beam patterns show that the adaptive beam weights corresponds to exciting both the clusters with some fraction of the available power, thereby indicating the multi-beam property of the adaptive design.

In a second example, consider a channel matrix with multiple 4×1 dual-polarized arrays in different locations of the UE, as described in Chapter 3. In this example, there are four dominant clusters over a wide angular spread with the mean angles of arrival/departure and mean relative gains as described in Table 4.2. The locations of the four clusters are marked in Figure 4.11 with the cross, square, diamond and circle markers, respectively. Given that the third and fourth clusters are relatively the strongest, the best directional beam (illustrated in Figures 4.11(a, b)) selects the third cluster. The adaptive beam (illustrated in Figures 4.11(c, d)) selects the second and

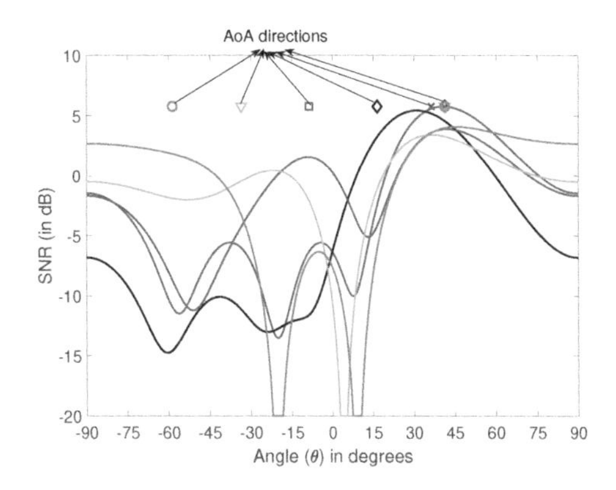

Figure 4.10 Illustration of beam patterns of adaptive beam weights with different AoA separation (AoAs of the clusters are marked with arrows).

fourth clusters in one polarization, and the second, third and fourth clusters in the second polarization via optimal choice of beam weight design. As a result, the sum spectral efficiency over two polarization layers improves by 2.68 bps/Hz (or ≈ 8.1 dB sum SNR improvement over the two polarization layers).

We now take one further step and illustrate the impressive performance improvement with adaptive beam weights in a hand blockage situation. For this, as Chapter 2.2.4 describes, depending on the angle of arrival of the dominant cluster at the UE, the phases and/or amplitudes seen across the antenna elements in the antenna array can be randomized/mixed up. Thus, a set of adaptive beam weight schemes that can de-randomize the phases and/or amplitudes can be expected to improve beamforming performance in blockage scenarios. To demonstrate this, we consider a hand blockage experiment where a hand phantom is placed directly on top of the antenna module of interest that contains a dual-polarized 4×1 array (with 0 mm air gap). Either one or two fingers of the hand phantom are near the antenna module leading to different blockage realizations.

Table 4.2
Channel structure for the adaptive beam weight example

Cluster	Gain	ZoA	AoA	ZoD	AoD
1	-7.96 dB	171°	80°	7°	133°
2	-6.02 dB	18°	89°	7°	265°
3	-4.08 dB	61°	251°	23°	254°
4	-3.64 dB	36°	337°	16°	194°

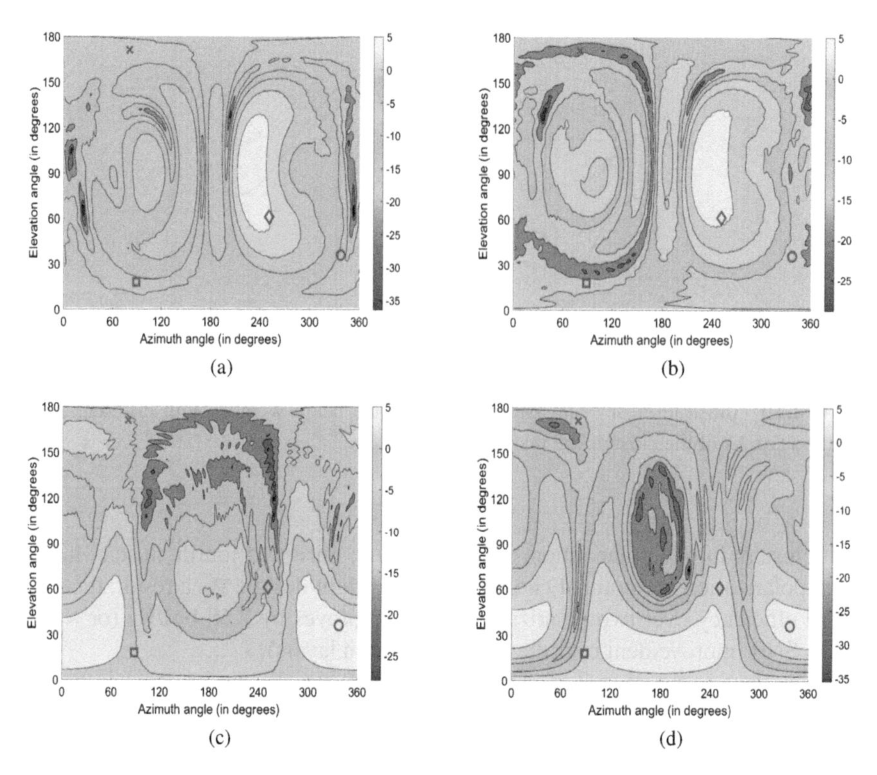

Figure 4.11 Beam patterns in the two polarizations associated with (a)–(b) directional beams and (c)–(d) adaptive beam weights.

For phase-based adaptive beam weights, we start by noting that a B-bit phase shifter can ideally produce 2^B phase possibilities:

$$\phi_k = \frac{2\pi \cdot k}{2^B}, \ k = 0, \ldots, 2^B - 1. \tag{4.71}$$

From this fact, we consider a codebook enhancement $\mathscr{C}_{\mathsf{enh,\,phase}}$ of size $(2^B)^{N-1}$ where

$$\mathscr{C}_{\mathsf{enh,\,phase}} = \left\{ u_{k_2,\ldots,k_N}, \ k_\ell = 0, \ldots, 2^B - 1, \ \ell = 2, \ldots, N \right\}, \tag{4.72}$$

with each set of beam weights being of the functional form:

$$u_{k_2,\ldots,k_N} = \frac{1}{\sqrt{N}} \cdot \begin{bmatrix} 1 \\ e^{j\phi_{k_2}} \\ \vdots \\ e^{j\phi_{k_N}} \end{bmatrix}. \tag{4.73}$$

Only the relative phases of the antenna elements with respect to the first antenna matters and thus without loss in generality, we can set the first phase term (ϕ_{k_1}) to be 0 for all the codebook entries. The basic motivation behind the structure of $\mathscr{C}_{\mathsf{enh,\,phase}}$ is to sample each antenna element with a B-bit phase shifter with the best set of beam weights from $\mathscr{C}_{\mathsf{enh,\,phase}}$ being the closest de-randomizer of the phase distortions induced by the hand. The effective role of the de-randomizer is to incorporate the impact of the hand distortions in the beam weights used, thereby matching the beam weights to the effective channel response as well as the hand effects and thus improving the realized array gains.

Since blockage induces both amplitude and phase distortions, the optimal beam weights for this scenario need to incorporate a search over *both* amplitudes and phases. Unlike phases with a limited range of 0° to 360°, approximating the amplitude information can lead to a quick increase in codebook size and therefore the overhead associated with learning these beam weights. Thus, to overcome this complexity, we consider a beam training procedure with N beams, each of which excites only one of the N antenna elements at any instant. Let S_i, $i = 1, \ldots, N$ denote the estimated RSRP with the i-th beam that excites only the i-th antenna. This beam training is performed *after* the introduction of hand blockage so that S_i can be estimated with the presence of the hand.

Based on these signal strengths, we consider a codebook enhancement where

$$\mathscr{C}_{\mathsf{enh,\,phase,\,amp}} = \left\{ v_{k_2,\,\ldots,\,k_N},\ k_\ell = 0, \ldots, 2^B - 1,\ \ell = 2, \ldots, N \right\}, \tag{4.74}$$

with each set of beam weights being of the functional form:

$$v_{k_2,\,\ldots,\,k_N} = \frac{1}{\sqrt{\sum_{i=1}^{N} \mathsf{S}_i}} \cdot \begin{bmatrix} \sqrt{\mathsf{S}_1} \\ \sqrt{\mathsf{S}_2} \cdot e^{j\phi_{k_2}} \\ \vdots \\ \sqrt{\mathsf{S}_N} \cdot e^{j\phi_{k_N}} \end{bmatrix}. \tag{4.75}$$

As before, we can set $\phi_{k_1} = 0$. In the above structure, instead of searching for the amplitude of the i-th antenna element, we approximate it by the normalized square root of the signal strength based on selecting the i-th antenna element. Note that instead of using the estimated S_i, if we used $\mathsf{S}_i = \frac{1}{N}$ for all i, then $v_{k_2,\,\ldots,\,k_N}$ reduces to $u_{k_2,\,\ldots,\,k_N}$.

To quantify the performance of de-randomizing the phases and/or amplitudes, we consider four adaptive beam weight schemes corresponding to:

- $B = 2$ bit phase-based
- $B = 3$ bit phase-based
- $B = 2$ bit phase- and amplitude-based
- $B = 3$ bit phase- and amplitude-based.

For the first scheme, with $B = 2$, note that the phases of each antenna element are of the form $\{\pm 1, \pm j\}$ and with $N = 4$, we consider a $\mathscr{C}_{\mathsf{enh,\,phase}}$ of size 64

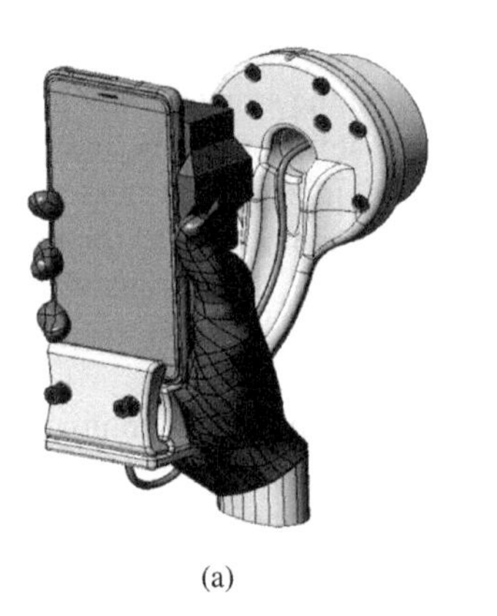

(a)

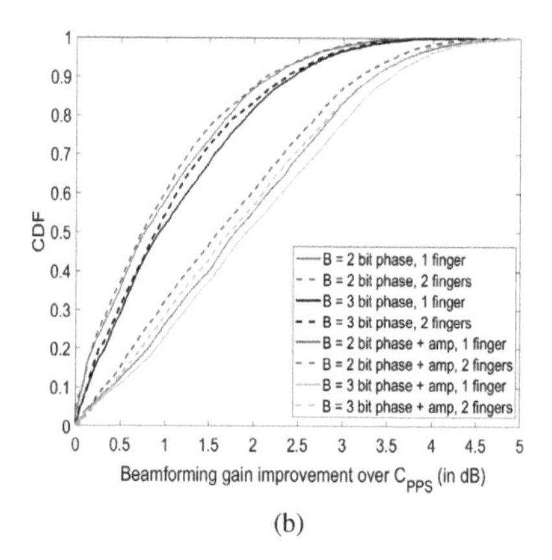

(b)

Figure 4.12 (a) Illustration of a hand phantom holding the UE. (b) Performance improvement over $\mathscr{C}_{\mathsf{PPS}}$ for different adaptive beam weight schemes.

$(= (2^B)^{N-1} = (2^2)^3)$. For the second scheme with $B = 3$, the size of $\mathscr{C}_{\mathsf{enh,\,phase}}$ is 512 $(= (2^3)^3)$. For these four schemes, Figure 4.12(b) plots the beamforming gain improvement with the codebook enhancements over $\mathscr{C}_{\mathsf{PPS}}$ of size 4 for the 0 mm air gap case with one and two fingers. From these plots, we observe the median, 80-th and 90-th percentile performance improvement of 0.7, 1.7 and 2.1 dB for the first scheme suggesting that the fingers of the hand do actually randomize the phases of different antenna elements which $\mathscr{C}_{\mathsf{enh,\,phase}}$ can de-randomize. Increasing B in the phase shifter selection approach only leads to a marginal performance improvement (comparable improvement of 0.9, 1.9 and 2.4 dB) suggesting that most of the gains with phase shifter selection are captured with the $B = 2$ bit phase shifter choice. On the other hand, addition of the signal strength to mirror a maximum ratio combining (MRC)-type solution can lead to significant gains (1.6 dB at the median and 3.2 dB at the 90-th percentile). Similar numbers for $B = 3$ phase and amplitude control over phase-only control are 1.7 dB gain at median and 3.3 dB at the 90-th percentile, again reinforcing that $B = 2$ is sufficient. Thus, it is important to consider a hand blockage mitigation strategy via adaptive beam weights (that mirror and account for the signal strength and phase variations seen across the antenna array commensurate with the hand position).

4.3 EFFECT OF PRACTICAL IMPAIRMENTS IN SYSTEM DESIGN

4.3.1 PHASE-ONLY VS. PHASE AND AMPLITUDE CONTROL

Like in digital beamforming where there are no specific constraints on the amplitudes and phases of the entries in the beamformer matrices, there are no specific constraints in analog beamforming also. The only constraints are the finite-precision quantization steps possible with the VGA and the phase shifter, respectively. Nevertheless, it is typically assumed in much of the literature that *only* the phases of the analog beamformers can be controlled. This assumption generally arises because the PAs/VGAs are typically set to operate at their peak ratings to avoid loss in EIRP that could happen with the use of amplitude control. However, there are many use cases where such amplitude control is necessary. Example use cases include:

- Side lobe control in interference management
- Multiple lobes/peaks in beam pattern design
- Handling polarization/blockage impairments with beam weights that mirror MRC operation
- Beam design for multi-user transmissions.

The use of amplitude control can lead to a loss in EIRP in transmit operations. However, it can help with improved array gain on receive operations as the optimal right singular vector (or MRC) performance can be better emulated with amplitude control. In receive operation, amplitude control adapts both the signal and noise and the SNR needs to be computed using amplitude control of the noise over the antenna array. The loss in beam correspondence between uplink and downlink due to the use of amplitude control is also a good justification for its non-use. However, as 3GPP specifications evolve, lack of beam correspondence has started being accommodated in some of the proposals. Nevertheless, the most general analog beamformer constraint is to impose *no* artificial equalization of amplitudes like it is often assumed in the literature. See [159] for further discussion of this issue.

4.3.2 BEAM CHARACTERIZATION VS. CALIBRATION

Beam characterization is a procedure by which we estimate the common cause variations (e.g., batch design of chip level components such as PCBs) in beam response. This procedure is performed once per device design and is applicable to all the devices within this design class. Characterization data is also used to design baseline beamforming codebooks for all the devices in the design class as a beam codebook is designed only once for the entire design class.

In general, beam codebook construction leads to the design of a targeted set of phase and amplitude responses. However, the impairments associated with the transmit and receive circuitries (e.g., due to mismatches in amplifiers, mixers, filters, couplers, etc.) imply that the mismatch between the targeted and observed phases and amplitudes needs to be estimated and compensated. In this context, *calibration* is a procedure by which the amplitude and phase response at every antenna of a multi-antenna array are tuned and corrected to ensure that they replicate the targeted response with a certain set of excitations [160, 161, 162, 163, 164, 165].

Without calibration, the beam weights used on the transmit or receive path do not produce the intended behavior. Calibration can help correct amplitude and phase mismatches between the circuitries in the transmit and receive paths. Since the beam weights on the transmit and receive paths are assumed to be reciprocal in TDD systems, the beam weights on the receive path cannot be reused for uplink transmissions without proper calibration. Calibration is a per-device procedure that captures per-part, special cause, or random variations. Since characterization is a one-time process and calibration is a per-unit process, significant performance improvements need to be demonstrated with it for enabling calibration. In the absence of calibration, a characterization look-up table can be used as a proxy to capture average performance of the antenna elements in the array.

To perform calibration, while a per-antenna procedure can be performed, it can be time-consuming, complex, and resource intensive. Thus, typically, calibration is performed over groups of antenna elements with certain pre-decided beam weight combinations (e.g., progressive phase shifts over the groups). When a UE antenna array is made of multiple panels and a group calibration is performed over tiles, less than perfect *a priori* knowledge of parameters such as inter-tile distances, phase offsets across tiles, etc. can lead to systematic errors. Further, calibration is performed separately per-frequency/subcarrier, per-temperature, and per-gain stage value. Due to the associated complexity and cost, only a finite number of points are sampled across the frequency, temperature and gain stage grid. Temporal stability of interconnects can also lead to gradual degradation that calls for built-in self tests (BISTs) and self calibration mechanisms, some of which can be absent in low-cost and low-complexity hardware.

Thus, even under the ideal assumption that calibration is performed on a per-antenna basis, there can be residual errors in amplitude and phase of each antenna (due to measurement precision and resources spent on calibration). Thus, it is reasonable to assume that the phase at the i-th antenna satisfies:

$$\widehat{\phi_i} = \phi_i + \varepsilon_i \tag{4.76}$$

where $\widehat{\phi_i}$ and ϕ_i are the measured and true phases with ε_i capturing the phase error in calibration. A reasonable model (assuming non-systematic errors) is to assume that the error is uniformly distributed over $\pm\delta$ degrees: $\varepsilon_i \sim \mathsf{Unif}\left([-\delta, \delta]\right)$ for an appropriate choice of δ. A truncated Gaussian model is also sometimes used with ± 3 times the standard deviation serving as the essential support of the error. Similarly, we can assume that the amplitude at the i-th antenna satisfies

$$\widehat{\alpha_i} = \alpha_i + \vartheta_i \tag{4.77}$$

where $\widehat{\alpha_i}$ and α_i are the measured and true amplitudes, respectively, with ϑ_i capturing the amplitude error in calibration. Typically, we assume that this error is uniformly distributed as $\vartheta_i \sim \mathsf{Unif}\left([-A, A]\right)$ for an appropriate choice of A.

4.3.3 IMPACT OF PHASE SHIFTER RESOLUTION

We now consider the impact of phase shifter resolution on the performance of different types of directional beams. Tradeoffs between cost, chip area and power/thermal on the one hand and performance on the other hand means that the lowest resolution phase shifter that is good for acceptable performance is necessary. Towards this goal, we consider the performance with $B = 2, 3, 4$ and 5 bit phase shifter quantizations of beam weights and compare the beam pattern response (and the corresponding beam properties) relative to the performance with an infinite precision phase shifter.

4.3.3.1 Progressive Phase Shift Beams

For a boresight beam of progressive phase shifts, the phases and amplitudes are equal and hence the beam pattern response with any phase shifter resolution is the same as that with infinite precision. However, as the beams get steered beyond the boresight direction, the impact of finite phase shifter resolution can be seen in terms of

- Main lobe gain
- Gap between the main lobe and the first side lobe (which we simply call as the "side lobe gap")
- Direction(s) of the side lobe(s) (both of which are important in network level interference considerations)
- 3-dB beamwidth of the main lobe (which translates to the number of beams needed to cover a given beamspace and thus initial acquisition latency).

For example, two typical beam pattern responses can be seen in Figure 4.13(a) for different values of B. These responses correspond to progressive phase shift beams steered along $-48°$ and $-28°$ with a linear array of size $N = 64$. These responses show that for a low phase shifter resolution, significant interference can be seen in some directions (red peaks), where no interference is expected if high-resolution phase shifters were used instead. The directions where interference is seen are deterministic (albeit complicated to quantify) and are a function of the beam weights used, the phase shifter resolution and the array geometry/size.

Taking this study further, Figures 4.13(b, d) plot the CDFs of these quantities for $K = 64$ beams designed to cover $\pm 60°$ around the boresight direction with the same $N = 64$ sized array. From these plots, we observe that while a $B = 2$ bit phase shifter resolution leads to a big offset in terms of the main lobe gain (almost a dB deviation for most steered directions) and the expected side lobe levels (expected asymptote at ≈ 13.47 dB). However, even a $B = 3$ bit resolution is sufficient to improve both these quantities significantly (main lobe to within 0.3 dB and the side lobe gap for over 90% of the beams to be > 12 dB). On the other hand, the 3-dB beamwidths of most beams appear to be similar, independent of the phase shifter resolution.

In Figures 4.13(e, f), we plot the CDFs of the main lobe gain and side lobe gap as a function of array dimensions ($N = 8, 16, 32$ and 64 with $K = N$ in each case) with the solid and dashed lines representing $B = 2$ and $B = 3$, respectively. These plots clearly show that as the phase shifter resolution increases for any N, both main lobe gain and side lobe gap converge to their asymptotes. However, speedier convergence is seen for any B as N increases. This is because as the array size increases, the positive variations (relative to a baseline) with some antenna elements are compensated by negative variations in other antenna elements, and the progressive phase shift structure is more robust to errors than more structured or arbitrary phase structures that lead to a broadened beam pattern. In terms of the 3-dB beamwidths of the main lobe, plotted in Figure 4.14(a), we observe no major dependence of this quantity for the

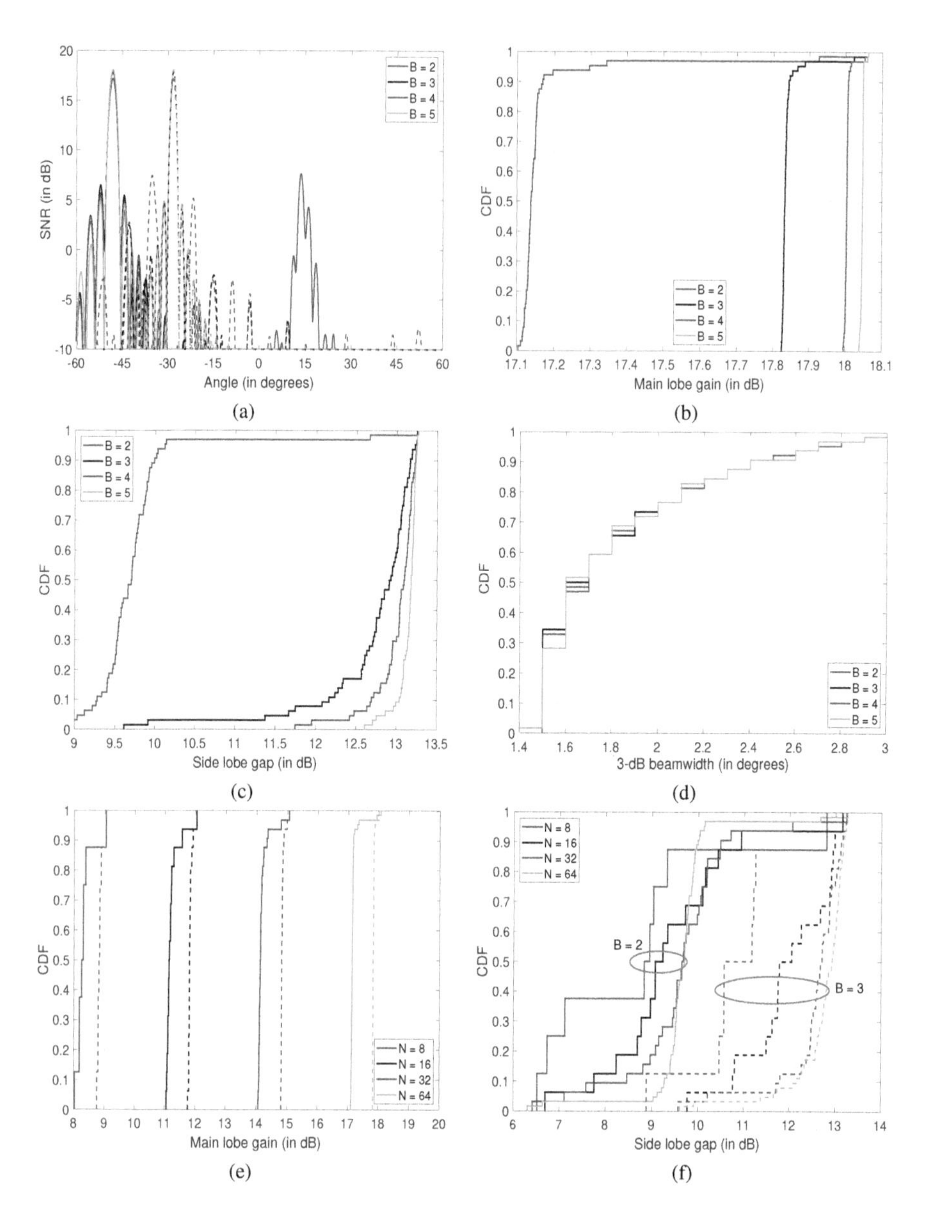

Figure 4.13 (a) Beam pattern response for different phase shifter resolution with two different progressive phase shift beams. CDFs of (b) main lobe gain, (c) side lobe gap and (d) 3-dB beamwidth for different progressive phase shift beams with $N = 64$ and different phase shifter resolutions. CDFs of (e) main lobe gain and (f) side lobe gap for different values of N with solid lines representing $B = 2$ and dashed lines representing $B = 3$, respectively.

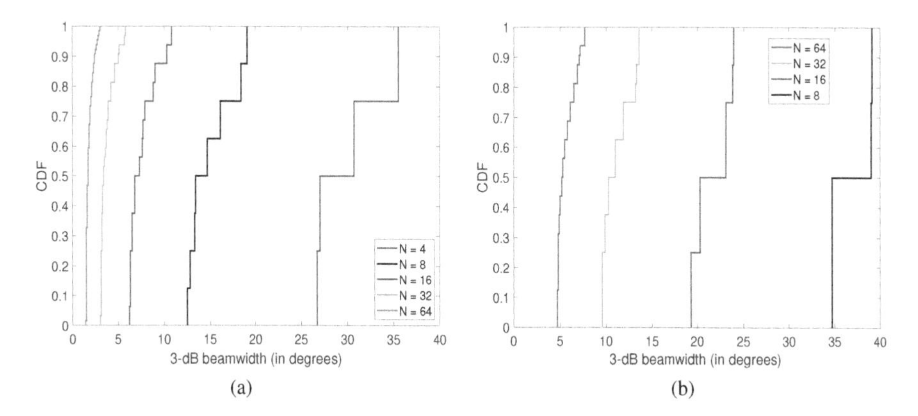

Figure 4.14 CDFs of 3-dB beamwidth for different array dimensions with (a) progressive phase shift beam weights and (b) broad beams.

different beams as a function of B (similar to the observation in Figure 4.13(d)), but we do observe a significant variation of the beamwidths for the different beams, especially for smaller array dimensions. It turns out that as the beam is steered past the boresight direction, the beamwidth increases in the azimuth domain, whereas it decreases in the elevation domain ensuring the conservation of the energy in the beam. Thus, while a one-dimensional azimuth beam scanning would require lesser number of beams as the beam is directed away from the boresight, a two-dimensional beam scanning would require the same number of beams as the beamwidth reduces in the elevation domain.

4.3.3.2 Broad Beams

We now extend this study to broadened beams whose design principle is outlined in Chapter 4.2.2. We consider two template beams designed for $N = 64$ (illustrated in Figure 4.8(b)), both of which are steered toward $-1.55°$, and plot their beam pattern responses for different phase shifter resolutions in Figures 4.15(a, b). We see that $B = 2$ leads to significant distortion in the beam pattern response with lesser beam broadening effect. However, with increased beam broadening, both $B = 2$ and $B = 3$ bit resolutions appear to induce some distortion in performance. In both cases, higher resolutions are necessary for acceptable performance relative to the infinite precision beam pattern response. We then use the first template beam to produce a size 32 codebook where the beams are shifted uniformly in azimuth to cover $\pm 60°$. Figures 4.15(c, d) plot the best $\mathsf{SNR}(\theta)$ observed with this codebook over the coverage region for $B = 2$ through $B = 5$ along with the infinite precision phase shifter performance. Clearly, we see that with low phase shifter resolutions, $\mathsf{SNR}(\theta)$ can vary considerably (over 3 dB at many angles), whereas the variation significantly reduces as the resolution increases. The CDFs of the main lobe gain and the side lobe gap are also plotted in Figures 4.15(e, f) for different phase shifter resolutions.

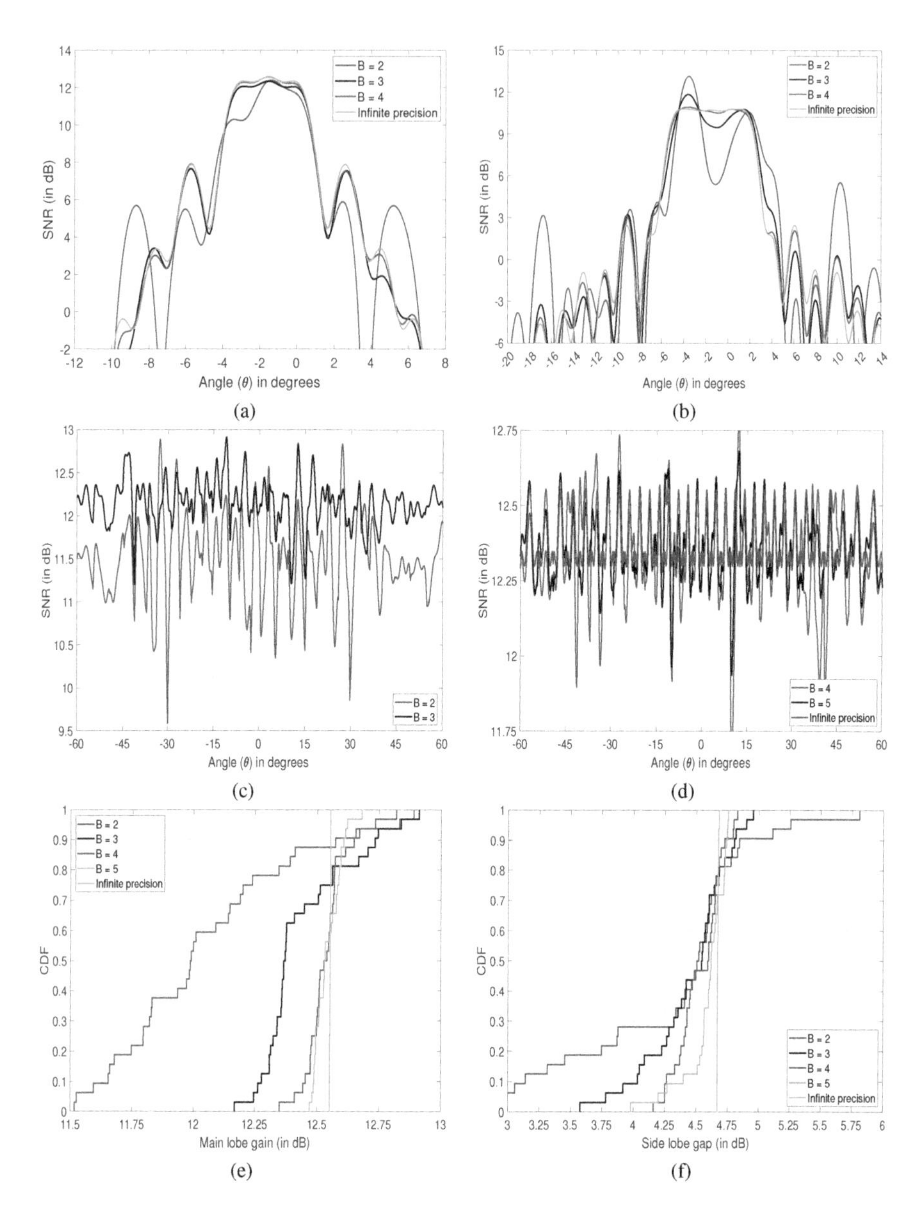

Figure 4.15 (a)–(b) Beam pattern response for two template broad beams of different broadening factors with $N = 64$ and different phase shifter resolutions. (c)–(d) Beam pattern response with a size 32 codebook of broad beams for different phase shifter resolutions: $B = 2$ to 5. CDFs of (e) main lobe gain and (f) side lobe gap for different phase shifter resolutions.

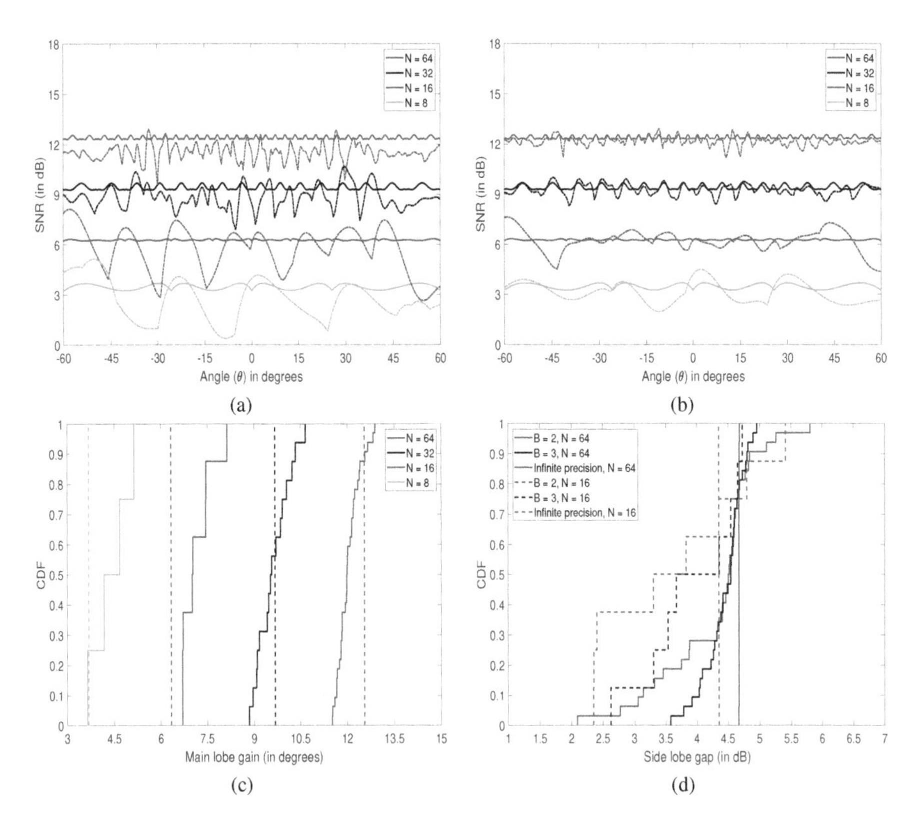

Figure 4.16 Beam pattern response with a size $N/2$ broad beam codebook for different choices of N with (a) $B = 2$ and (b) $B = 3$. Quantized beam weights are shown in dashed lines and infinite precision beam weights are shown in solid lines. (c) Main lobe gain with $B = 2$ in solid lines and infinite precision in dashed lines. (d) Side lobe gap with solid and dashed lines representing $N = 64$ and 16, respectively.

We now consider a size $K = N/2$ broad beam codebook covering $\pm 60°$ for $N = 64, 32, 16$ and 8 with $B = 2$ through $B = 5$. The template beams for the broad beam codebook for the four cases are illustrated in Figure 4.8(b). Figures 4.16(a, b) plot the beam pattern response for this codebook with different phase shift resolutions. From these plots, we observe that a max-to-min variation of 3.3, 3.8, 5.5 and 4.8 dB are seen for $B = 2$ with $N = 64, 32, 16$ and 8, respectively. The corresponding values for $B = 3$ are 1.7, 1.7, 3.3 and 2.5 dB, respectively. In general, this variation reduces as B increases with the infinite precision variation in the four cases being 0.3, 0.4, 0.1 and 0.5 dB, respectively, pointing out that beam broadening can be performed with minimal ripples in performance over the coverage region of each template beam. We also plot the main lobe gain and side lobe gap for the different N values and observe that both the main lobe gain and side lobe gap are generally larger

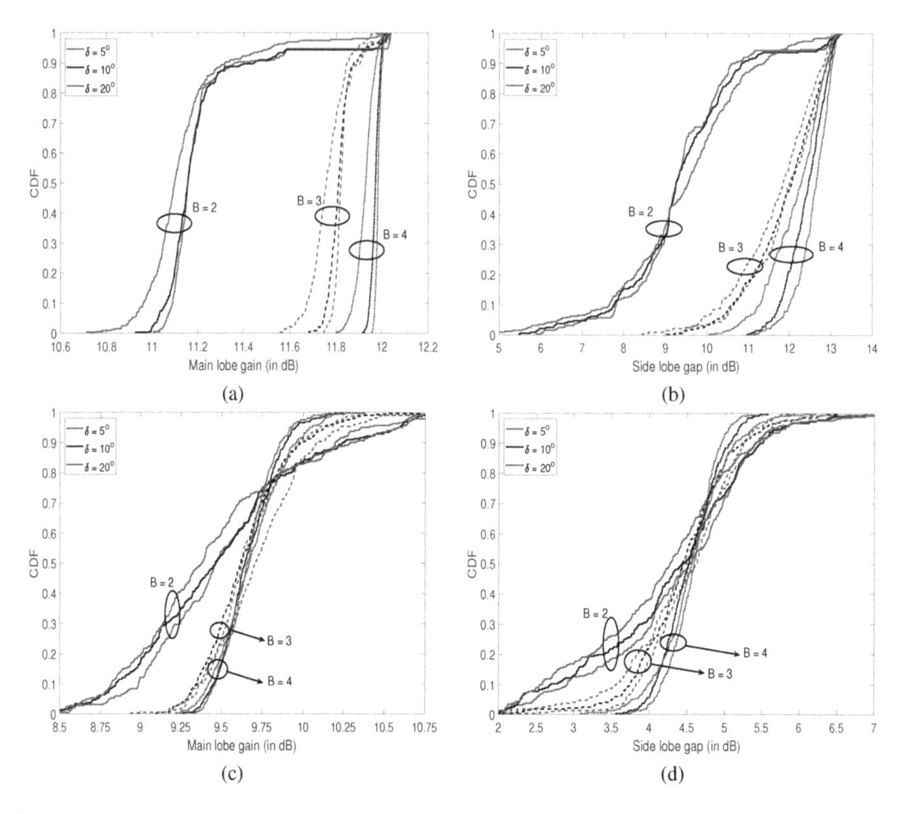

Figure 4.17 (a) Main lobe gain and (b) side lobe gap for a codebook of progressive phase shift beams (with $N = K = 16$) for different phase errors. Same metrics for $N = 32$ and $K = 16$ broad beams in (c) and (d).

as N increases. The CDF of the 3-dB beamwidth of the beams in the codebook are plotted in Figure 4.14(b) for different values of N. Comparing these CDFs with that of the progressive phase shift beam codebook in Figure 4.14(a), we observe that an approximate three-fold increase in the beamwidths are observed, as expected, since the template beams are designed according to this expectation.

4.3.3.3 Impact of Calibration Error

To understand the impact of calibration error, we consider the model where $\varepsilon_i \sim \mathsf{Unif}([-\delta, \delta])$ with $\delta = 5°, 10°$ and $20°$. Note that with a $B = 5$ bit phase shift resolution, in the ideal/no error scenario, the quantization step is $11.25°$. Thus, from this perspective, a $\pm 10°$ or a $\pm 20°$ error is a significant distortion from ideal behavior. In Figures 4.17(a, b), the main lobe gain and side lobe gap are plotted for a codebook of progressive phase shift beams with $N = K = 16$ for different values of δ and B. In Figures 4.17(c, d), the same two metrics are plotted for a broad beam codebook with

$N = 32$ and $K = 16$ for different values of δ and B. From these plots, we observe that the deterioration in performance with increased calibration error is typically small (less than a fraction of a dB even at the low percentile points), independent of the phase shift resolutions. In other words, the calibration error associated with a given phase shift resolution is not the main cause of loss in system-level performance, but poor quality phase shift resolution could be a determinant.

Thus, from the different studies presented here on main lobe and side lobe behavior relative to the asymptote values, the significant deviation seen with low resolution phase shifters (such as $B = 2$) appear to render them insufficient for optimal performance. In particular, increased side lobes in (essentially) unpredictable directions causes the use of such low resolution components to be questionable from a practical perspective. For most practical array sizes, a $B = 3$ bit phase shifter appears to optimize performance with cost and complexity, while a $B > 3$ bit phase shifter is useful for better performance if a cost increase is justified. Such higher resolution phase shifters may be useful at the base station since interference considerations in small cell settings become prominent.

4.4 SPHERICAL COVERAGE AND SPECTRAL EFFICIENCY TRADEOFFS WITH ANALOG BEAMFORMING

Chapter 3.2.3 motivated the notion of spherical coverage. This is important for mmWave systems since a single antenna module (or a set of antenna elements) cannot provide good coverage over the entire sphere, which is necessary for robust performance. Two popular UE designs where the antenna modules are placed on the front and back face of the UE (*face design*) and on the edges of the UE (*edge design*) are described in Figure 3.9 along with the design parameters in Table 3.3. We now study the spherical coverage comparisons across these two designs. For this, we need to introduce a set of beamforming schemes against which the two designs can be compared.

Given a subarray of N antenna elements, let

$$E_{\Theta}(\theta,\phi) = [\mathsf{E}_{\Theta,1}(\theta,\phi),\ldots,\mathsf{E}_{\Theta,N}(\theta,\phi)]^T \text{ and} \tag{4.78}$$

$$E_{\Phi}(\theta,\phi) = [\mathsf{E}_{\Phi,1}(\theta,\phi),\ldots,\mathsf{E}_{\Phi,N}(\theta,\phi)]^T \tag{4.79}$$

denote the antenna response functions in the Θ and Φ polarizations[12] along a certain direction (θ,ϕ) of the sphere. In the ideal case of an equi-spaced antenna array with no impairments from the UE housing, the array response function in (θ,ϕ) corresponds to the steering vector with the array in that direction (see the ideal response functions' description in (2.5) and (2.6)). To study the spherical coverage tradeoffs, we consider four different schemes described as below:

[12]Typically, antenna response functions in a high-frequency simulation software such as [166] are specified in the Θ and Φ polarizations to avoid unnecessary confusion with notations such as H- or V-polarizations that are explicitly associated with the point on the sphere where the antenna responses are computed.

- Maximum ratio combining (MRC) in every direction (θ, ϕ) of the sphere [167] without any phase or amplitude quantization constraints of the beamforming vector corresponds to performing the optimization:

$$G_{\mathsf{MRC}}(\theta,\phi) = \max_{\{\alpha_i\}} \left| \sum_{i=1}^{N} \alpha_i^{\star} \, \mathsf{E}_{\Theta,\,i}(\theta,\phi) \right|^2 + \left| \sum_{i=1}^{N} \alpha_i^{\star} \, \mathsf{E}_{\Phi,\,i}(\theta,\phi) \right|^2 \tag{4.80}$$

$$= \max_{\boldsymbol{\alpha}} \boldsymbol{\alpha}^H \cdot \left(E_{\Theta}(\theta,\phi) E_{\Theta}(\theta,\phi)^H + E_{\Phi}(\theta,\phi) E_{\Phi}(\theta,\phi)^H \right) \cdot \boldsymbol{\alpha} \tag{4.81}$$

$$= \lambda_{\mathsf{max}} \left(E_{\Theta}(\theta,\phi) E_{\Theta}(\theta,\phi)^H + E_{\Phi}(\theta,\phi) E_{\Phi}(\theta,\phi)^H \right) \tag{4.82}$$

$$= \frac{E_{\Theta}(\theta,\phi)^H E_{\Theta}(\theta,\phi) + E_{\Phi}(\theta,\phi)^H E_{\Phi}(\theta,\phi)}{2} + \frac{\sqrt{\left(E_{\Theta}(\theta,\phi)^H E_{\Theta}(\theta,\phi) - E_{\Phi}(\theta,\phi)^H E_{\Phi}(\theta,\phi) \right)^2 + 4 \left| E_{\Theta}(\theta,\phi)^H E_{\Phi}(\theta,\phi) \right|^2}}{2} \tag{4.83}$$

where $\boldsymbol{\alpha} = [\alpha_1, \ldots, \alpha_N]^T$, (4.82) follows from the characterization of $\lambda_{\mathsf{max}}(\bullet)$ in terms of quadratic forms and the equality in (4.83) is due to the closed-form computability of $\lambda_{\mathsf{max}}(\bullet)$ of a rank-2 matrix. The solution to the above optimization with $\boldsymbol{\alpha}$ such that $\|\boldsymbol{\alpha}\|^2 \leq 1$ is the dominant eigenvector of the underlying rank-2 $N \times N$ matrix. This solution requires infinite precision in both phase and amplitude as well as directional resolution (co-phasing beams are used in every direction (θ, ϕ)). Thus, this scheme serves as an optimistic (unachievable in practice) upper bound on the spherical coverage performance of the UE design.
- Equal gain combining (EGC) is similar to the MRC scheme except that the beam weights are constrained to have equal (maximal) gain for all the antennas. In the single polarization case where $\mathsf{X} \in \{\Theta, \Phi\}$, it can be easily seen that

$$\sum_{i=1}^{N} |\mathsf{E}_{\mathsf{X},\,i}(\theta,\phi)|^2 = G_{\mathsf{MRC}}(\theta,\phi) \geq G_{\mathsf{EGC}}(\theta,\phi) = \frac{1}{N} \left(\sum_{i=1}^{N} |\mathsf{E}_{\mathsf{X},\,i}(\theta,\phi)| \right)^2. \tag{4.84}$$

In the dual-polarized case, since the EGC optimization is more complicated, we reuse the phases from the optimizing MRC solution but constrained to have equal amplitudes as a potential candidate set of beam weights for the EGC scheme.This approach serves as an achievable or realizable EGC solution.
- Towards the goal of comparing the peak performance of the face and edge UE designs with a practical codebook-based beamforming scheme, it is important that the comparison be *fair* across these designs. With different codebook sizes, the beam acquisition latencies associated with each UE design can be different thereby making head-to-head comparisons difficult. To address these concerns, for the face design, 4 beams are used for each polarization of the 2×2 dual-polarized patch subarrays, and 2 beams are used for each dipole subarray leading

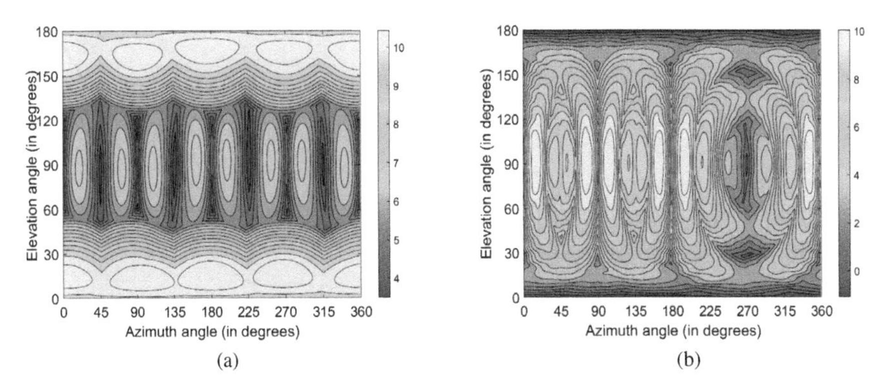

Figure 4.18 Array gain performance over the sphere for the (a) face and (b) edge designs with a size 24 codebook across all antenna modules.

to 12 beams per antenna module as well as 12 beams per polarization. For the edge design, 4 beams are used for each polarization of the 4×1 dual-polarized patch subarray for 8 beams per module and 12 beams per polarization. Since the codebook sizes are 24 for both the designs, the performance of these two UE designs can be compared fairly. Note that while more complicated and different-sized codebooks can be considered for the two designs and their performance can be compared with some performance penalty function (e.g., a 3 dB penalty for a doubling of the codebook in one design relative to the other, etc.). However, the method proposed here is reasonable for practical implementations.

The individual beam weights in the codebook can be optimally designed to cover certain angular regions over the sphere. While the beam design process in itself can be implementation-specific, general design principles are exposed in the previous chapters and [147]. In this work, the beams for the 4×1 subarrays are designed such that each beam results in a beamwidth of $\approx 25°$ to $30°$ in accordance with the observations for similar array sizes in Figure 4.14(a). Similarly, the beams for the 2×2 and 2×1 subarrays have a beamwidth of $\approx 55°$. All the beam weights (for either design) are constrained to meet an equal amplitude and a $B = 5$ bit phase shifter resolution. Figures 4.18(a, b) present the array gain performance with the codebooks over the sphere (represented over a plane) for the face and edge designs, respectively. For each point (θ, ϕ) over the sphere, the best representative from each design's codebook is used here. Clearly, from Figure 4.18, we observe that the codebooks are designed to meet good array gain performance over a significant fraction of the sphere and are hence sufficient to perform a fair comparison.

- We then finally compare all these schemes by selecting the best single antenna element (from amongst all the possible antennas across all the antenna modules

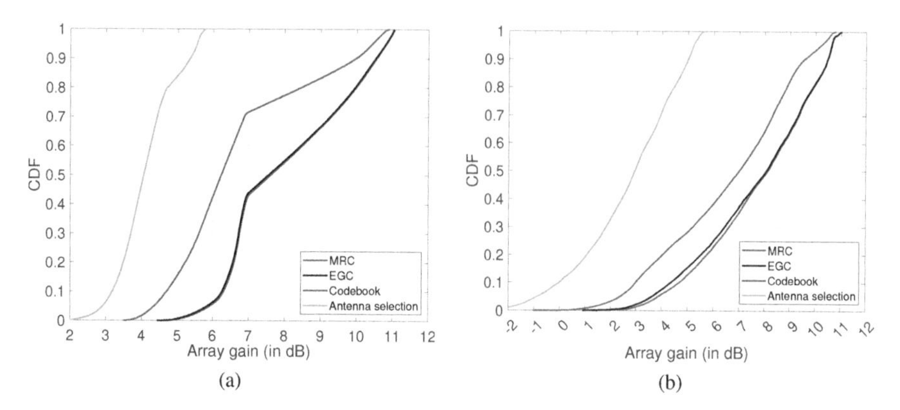

Figure 4.19 Beamforming performance of different beamforming schemes in freespace for the (a) face design and (b) edge design.

at the UE side) for a direction (θ,ϕ) to result in:

$$G_{\text{ant sel}}(\theta,\phi) = \max_{i=1,\ldots,N} \left|\mathsf{E}_{\Theta,i}(\theta,\phi)\right|^2 + \left|\mathsf{E}_{\Phi,i}(\theta,\phi)\right|^2. \tag{4.85}$$

Since no array gain is realized with this scheme and the gains are purely from antenna selection, this scheme is pessimistic in terms of the available antenna capabilities and corresponds to a legacy beamforming solution such as those available in prior generations (e.g., 3G or 4G) of most wireless devices.

4.4.1 SPHERICAL COVERAGE IN FREESPACE

In our first study, in Figures 4.19(a, b), we describe the array gain tradeoffs for the UE designs with these four schemes in freespace/no blockage operational mode. With both the designs, we observe that the EGC scheme performs as well as the MRC scheme over the entire sphere. This conclusion implies that phase-only control is sufficient to obtain the optimal spherical coverage and the cost associated with amplitude control can be forsaken with minimal performance penalties. This also motivates the design of directional beam codebooks with only phase shifter control instead of full-blown phase and amplitude control. This conclusion stems from the fact that all the antennas that make a certain subarray have similar amplitudes over the whole sphere or at least[13] over the main coverage regions of the antenna elements. No specific antenna sees an anomalous behavior (relative to others) necessitating amplitude control. On the other hand, if adaptive beam weights or multi-beams [159] are considered, amplitude control becomes important.

[13]Note that this is not the case with hand blockage as shown in Chapter 4.2.4.

For both the UE designs, the directional beam codebook schemes are within 1–2 dB of the MRC/EGC performance. This suggests the goodness of the codebook design principles as well as of the idea that a codebook can well approximate the optimal performance. However, the worst-case points of the codebook's performance are 7 and 10 dB away from the peak gain for the face and edge designs, respectively. While this observation could suggest that there are significant gaps relative to MRC/EGC performance, this is a naïve conclusion since with blockage, the tail performance of all the schemes are comparable and are significantly deteriorated.

With the edge design, single antenna selection is approximately 5–6 dB worse than MRC/EGC. This gap can be explained as the co-phasing gain from four antenna elements in the subarray. On the other hand, with the face design, this gap reduces from 6 dB at the peak to 3 dB at the tail corresponding to the switch from a 2×2 patch subarray to a 2×1 dipole subarray. Regarding codebook performance relative to MRC/EGC, the edge design shows a near-constant gap over the CDF curve (≈ 1 dB). On the other hand, the face design appears to have a gap that increases from the peak to the tail. This can be attributed to:

- Loss in array gain as we move from the beams' boresight steering direction to the edge of coverage of each beam
- Switch from a four element subarray to a two element subarray.

From a view of the codebooks in Figure 4.18, we observe that the edge design has coverage holes mostly over the poles, whereas the face design has coverage holes at random points over the sphere accounting for the degradation in codebook performance from the peak to the tail.

4.4.2 SPHERICAL COVERAGE WITH BLOCKAGE

We now study the performance of the face and the edge design with hand blockage, which as described in Chapter 2.2.4, can lead to significant performance deterioration at the UE side at mmWave frequencies. In this pursuit, in the 3GPP channel modeling document TR 38.901, a blockage model is proposed [18, pp. 53–57] to capture these detrimental effects under two variants: a stochastic variant (Model A) and a map-based/ray tracing-based variant (Model B). The stochastic variant proposes a methodology tailored to the hand in portrait or landscape orientations around a UE modeled to form factor considerations. As illustrated in Table 4.3, this model

Table 4.3
Hand blockage model for spherical coverage studies

Scenario	ϕ_1	x_1	θ_1	y_1	Blockage loss (in dB)
Portrait	260°	120°	100°	80°	Model 1 [18] : Flat loss of 30 dB
Landscape	40°	160°	110°	75°	Model 2 [168] : $\mathcal{N}(\mu = 15.3 \text{ dB}, \sigma = 3.8 \text{ dB})$

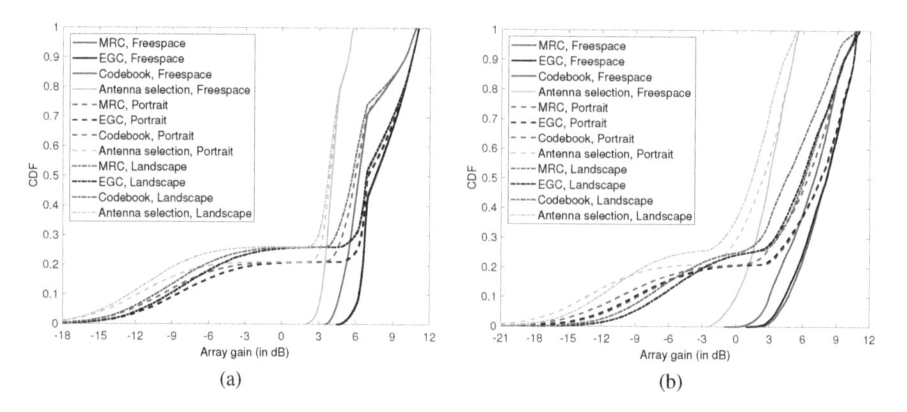

Figure 4.20 Beamforming performance with freespace and blockage model in portrait and landscape mode operations for the (a) face design and (b) edge design.

(labeled "Model 1") is captured by the center of the blocker (ϕ_1, θ_1), and the angular spread of the blocker (x_1, y_1) in azimuth and elevation with the blocking angles captured as $\phi \in \left[\phi_1 - \frac{x_1}{2}, \ \phi_1 + \frac{x_1}{2}\right]$ and $\theta \in \left[\theta_1 - \frac{y_1}{2}, \ \theta_1 + \frac{y_1}{2}\right]$ in azimuth and elevation, respectively. Over this spatial region, a simplistic flat 30 dB loss is assumed.

More recent studies in [168] show that this model is too pessimistic for form factor UE designs due to the use of horn antenna measurements (with smaller beamwidths) to model hand blockage loss. Thus, a modified blockage model is proposed in this work, which is labeled as "Model 2" in Table 4.3. Here, the spatial blockage region is retained from the 3GPP model and a log-normal blockage loss term is used. More realistic blockage models based on the observations in Chapter 2.2.4 can be considered, but the flavor of the conclusions remains the same.

We now study the spherical coverage CDFs with the blockage model in Table 4.3. Figures 4.20(a, b) plot the spherical coverage performance of different beamforming schemes with the blockage model in portrait and landscape mode for the face and edge designs. We now explain some of the observations in Figure 4.20. From Table 4.3, the blockage regions in portrait and landscape modes occupy the following fraction of physical or spatial angles:

$$\text{Physical angle loss}\Big|_{\text{Portrait}} = \frac{120^{\circ} \times 80^{\circ}}{360^{\circ} \times 180^{\circ}} = 14.81\% \tag{4.86}$$

$$\text{Physical angle loss}\Big|_{\text{Landscape}} = \frac{160^{\circ} \times 75^{\circ}}{360^{\circ} \times 180^{\circ}} = 18.52\%, \tag{4.87}$$

respectively. Since the spatial angles need to be weighted based on the Jacobian (see the discussion in Chapter 3.6.1), these blocked angles correspond to a CDF loss of

$$\text{CDF loss}\Big|_{\text{Portrait}} = \frac{1}{4\pi} \int_{\phi=\phi_{\text{p, l}}}^{\phi_{\text{p, u}}} \int_{\theta=\theta_{\text{p, l}}}^{\theta_{\text{p, u}}} \sin(\theta) \cdot d\theta d\phi = 21.07\% \tag{4.88}$$

$$\text{CDF loss}\Big|_{\text{Landscape}} = \frac{1}{4\pi} \int_{\phi=\phi_{\text{l, l}}}^{\phi_{\text{l, u}}} \int_{\theta=\theta_{\text{l, l}}}^{\theta_{\text{l, u}}} \sin(\theta) \cdot d\theta d\phi = 26.00\%, \tag{4.89}$$

where

$$\phi_{\text{p, l}} = 200^\circ \cdot \pi/180, \quad \phi_{\text{p, u}} = 320^\circ \cdot \pi/180, \tag{4.90}$$

$$\theta_{\text{p, l}} = 60^\circ \cdot \pi/180, \quad \theta_{\text{p, u}} = 140^\circ \cdot \pi/180, \tag{4.91}$$

$$\phi_{\text{l, l}} = -40^\circ \cdot \pi/180, \quad \phi_{\text{l, u}} = 120^\circ \cdot \pi/180, \tag{4.92}$$

$$\theta_{\text{l, l}} = 72.5^\circ \cdot \pi/180, \quad \theta_{\text{l, u}} = 147.5^\circ \cdot \pi/180. \tag{4.93}$$

The performance degradation in the tails of the portrait and landscape modes correspond to the CDF loss region estimates in (4.88) and (4.89), as expected. If a flat 30 dB loss is assumed (as with the 3GPP model), this blockage performance renders the tail region of spherical coverage completely irretrievable and the performance loss over this region is abrupt. On the other hand, with a log-normal loss model as assumed, this loss in performance is smoother allowing for some recovery over certain directions. Depending on the blockage model used, some performance can be retrieved in the tail with adaptive beam weights as illustrated in Chapter 4.2.4. In general, blockage can lead to a *bimodal* behavior of almost no loss over the unblocked region and minimally retrievable performance over the blocked region.

4.4.3 SPHERICAL COVERAGE COMPARISONS ACROSS DESIGNS

4.4.3.1 Face and Edge Designs

Since the face and edge designs, as described in Figure 3.8, are directly comparable with each other in a fair manner (due to the same codebook sizes), Figure 4.21(a) presents a head-to-head comparison of these designs in freespace and in portrait/landscape modes with blockage. From Figure 4.21(a), we observe that while both the designs are comparable in their respective top 20 percentile points of the sphere in portrait mode, the edge design appears to be better (by up to 1.5 dB) over the next 35 percentile points. The face design appears to be better over the remaining ≈ 20 percentile points before blockage effects kick in.

While both the face and edge designs are blocked over approximately 21% of the sphere in the portrait mode (see (4.88)), their crossovers can be explained by the following observations: Approximate array gain with the beamforming codebook at the 70-th percentile point for the face and edge designs are 7 and 8.5 dB, respectively. Similar numbers for the 50-th and 30-th percentile points are 6.5 vs. 7 dB and 5.5 vs. 5 dB, respectively. This tradeoff arises due to the structure of antenna arrays (2×2 planar arrays and 2×1 linear arrays in the face design vs. 4×1 linear arrays in the edge design). The better relative performance of the edge design over the face design in the middle 35 percentile points and its reversal in the next 20 percentile points is directly a result of the array gain tradeoffs.

On the other hand, the mismatch between the area blocked with the hand in the right-hand landscape mode (top short edge that is totally blocked in the edge design vs. the top front module that is only partially blocked in the face design) means that the face design appears to be uniformly better than the edge design (by up to 1.5 dB). From these observations, there does not appear to be an overwhelming advantage

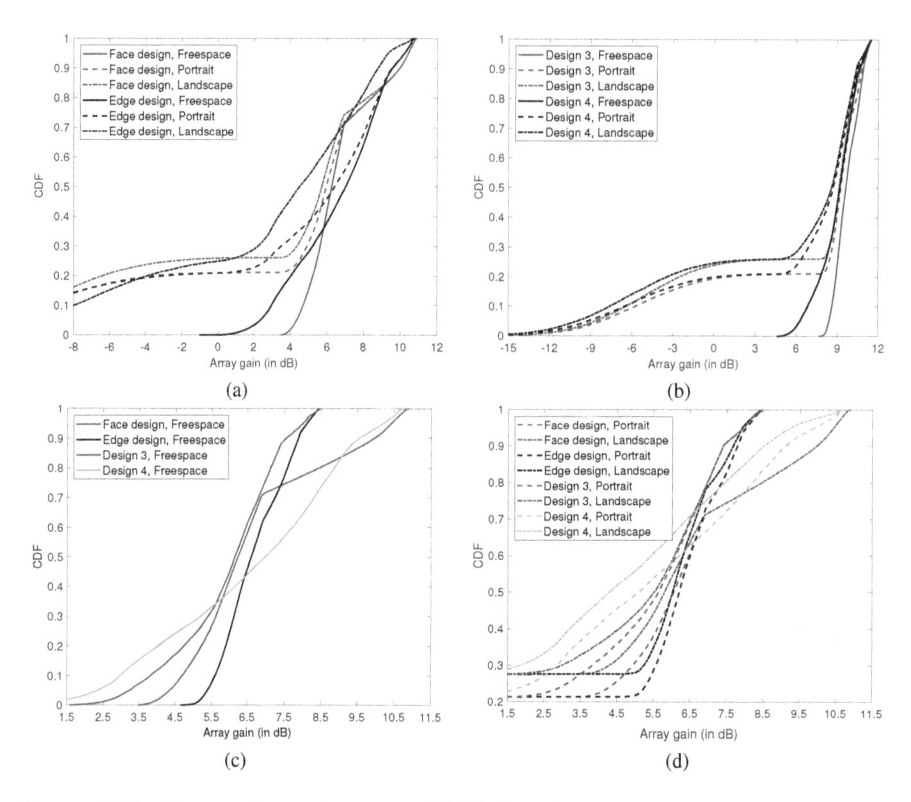

Figure 4.21 Comparative performance with blockage between (a) face and edge designs and (b) Designs 3 and 4. Comparative performance across the four designs in (c) freespace and (d) with blockage.

(defined as greater than 3 dB) for either design. This suggests that both designs are comparable in terms of performance and the choice between them should be based on implementation tradeoffs as described in Chapter 3.2.3.

4.4.3.2 Advanced Antenna Module Designs

We now perform a head-to-head comparison between more advanced designs (Designs 3 and 4) described in Chapter 3.2.3. Since both these designs have four antenna modules and have more subarrays (12 and 16, respectively) than either the face or edge designs (8 and 6, respectively), we use a beamforming codebook of larger size (size 48) than those used with the face and edge designs (size 24). Note that a smaller codebook size with Designs 3 and 4 can lead to coverage holes with poor spherical coverage tradeoffs. For Design 3, we use 4 beams for each polarization of the patch and dipole subarrays corresponding to 12 beams per antenna module for a codebook size of 48. For Design 4, we use 3 beams for each polarization of the patch corresponding to 12 beams per antenna module, also for a codebook size of 48.

While data is not presented here (see [76] for more detailed studies), similar to the face and edge designs, we observe that for both Designs 3 and 4, the beamforming codebooks are within 1–2 dB of the MRC/EGC performance suggesting the goodness of the codebook design principles. In particular, the worst-case points of the codebook's performance in freespace are, respectively, 3 and 6 dB away (which is better than the face and edge designs) from the peak gain for these designs. As intuitively expected from co-phasing with four antennas in either design, single antenna selection is approximately 5–6 dB worse than MRC/EGC. Blockage tradeoffs for both the designs are similar to those described earlier for the face and edge designs.

Since Designs 3 and 4 are directly comparable with each other in a fair manner, Figure 4.21(b) provides a comparison across these two designs. From this study, we observe that Design 3 has a universally (albeit slightly) better performance in freespace as well as with blockage over Design 4. This plot suggests that the use of dipoles over patches that scan the other side of the L can result in a better performance for diversity. Thus, the use of the appropriate/correct antenna modules is crucial for good performance in mmWave systems.

Figures 4.21(c, d) plot a comparative analysis across all the four designs in freespace and with blockage, respectively. Since the codebook size for Designs 3 and 4 are twice as much as the size for the face and edge designs, a 3 dB normalization penalty is imposed on Designs 3 and 4 in producing a fair comparison. From Figure 4.21(c), we observe that the face and edge designs are comparable for the top 20 percentile points of the sphere, whereas the 3 dB penalty hurts both Designs 3 and 4 over these points. Beyond this and over the next 40 percentile points, the edge design appears to have a smoother roll-off and degradation in performance, whereas the switch from a four to a two antenna element subarray hurts the face design more. Designs 3 and 4 have a steeper CDF curve, but the initial 3 dB penalty appears to continue to have an impact over this region resulting in the edge design being superior across all the four designs. The primary performance deterioration of the edge design appears to be in the tail (≈ 35) percentile points of the sphere—a region over which we would anyway expect performance degradation due to the hand position. By choosing to skip an antenna module at the bottom edge of the UE, which is most likely to steer beams toward the ground plane[14] in portrait mode, or is likely to be blocked due to the hand position in landscape mode, we can reduce the cost of antenna modules and still retain good performance overall.

While the above conclusions are made using freespace performance, Figure 4.21(d) presents the comparison across the four designs with portrait and landscape blockage. The conclusions with the portrait mode blockage appear to be similar to freespace. On the other hand, with landscape mode blockage, the edge design appears to have an up to 1 dB performance gap relative to the face design in the top 20 percentile points and remains fairly competitive across all the designs for the top 40 percentile points. The next 30 percentile points see a widening gap going all the way

---

[14]By construction, the edge design cannot capture ground bounces, if any, and this is the main performance tradeoff with this design.

up to 2.5 dB suggesting that the edge design can be used, but with some performance penalties in the lower coverage points in landscape mode. Nevertheless, the main advantage of the edge design is its reduced implementation complexity and reduction in cost as pointed out in Chapter 3.2.3. These two advantages of the edge design could help tolerate the performance loss over some parts of the sphere in landscape mode.

From our studies, we established the overhead of beam training as being the key determinant (and not the "theoretical" capabilities enabled with multiple antenna modules) for robust spherical coverage performance in practice. That is, it is not merely sufficient that the UE is packed with a large number of antenna modules, but that the subarrays in these modules are scanned with an appropriately designed beam codebook in a practical implementation. Further, the size of a good codebook has to scale with the number of antenna modules and can render the coverage gains unrealizable from a practical standpoint as the incurred latencies could be large. From this view, we established the goodness of the edge UE design that also has other advantages such as low cost and power consumption, implementation ease, and minimal exposure related challenges [169]. Also, note that while spherical coverage is important from an UE antenna module placement point-of-view, as Chapter 5 demonstrates, it is also important that there is a good coverage from multiple nodes in a practical network deployment.

4.4.4 SPECTRAL EFFICIENCY TRADEOFFS

Polarization diversity has historically been the focus of considerable interest in cellular system design [170, 171, 172, 173]. Beyond the rank-1 objectives captured by the spherical coverage metric, we are also interested in the polarization MIMO (or rank-2) performance of the face and edge designs. To understand this, we use the channel model described in (2.20) and study the performance with different analog beamforming schemes. The base station array is 8×4 with a progressive phase shift analog beamforming codebook of size 32. Four beamforming schemes are considered at the UE:

- Selection of single antenna elements emulating an initial acquisition phase
- Analog beamforming codebook of PPS beams of size 4
- Analog beamforming codebook of PPS beams with an increased granularity of beam scanning directions corresponding to size 8
- Per-tone SVD of $H(k)$ (that is, fully digital beamforming).

With these schemes, the achievable rate in the case of $L = 1$ and $L = 6$ clusters in the channel are presented in Figures 4.22(a, b). From these figures, we see that while the edge design is slightly inferior to the face design, the gap is minimal across all the schemes. On the other hand, while the per-tone SVD scheme provides an upper bound in terms of performance, this benchmark can be approached with different analog beamforming schemes. The gap in performance between the analog beamforming schemes and the upper bound is a function of many aspects:

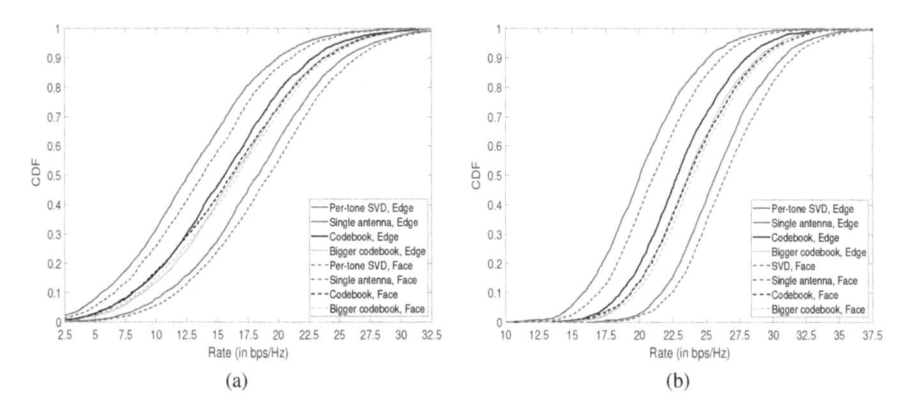

Figure 4.22 Spectral efficiency tradeoffs with polarization MIMO in the case of (a) $L = 1$ and (b) $L = 6$ clusters in the channel.

- Base station analog beamforming codebook's granularity/size
- UE's analog beamforming codebook's granularity/size
- Correlation at the base station between $F_{\mathrm{T}}^{\Theta}(\phi_{n,m}^{\mathrm{T}}, \theta_{n,m}^{\mathrm{T}})$ and $F_{\mathrm{T}}^{\Phi}(\phi_{n,m}^{\mathrm{T}}, \theta_{n,m}^{\mathrm{T}})$ vectors corresponding to the dominant cluster's AoD/ZoD angles
- Correlation at the UE between $F_{\mathrm{R}}^{\Theta}(\phi_{n,m}^{\mathrm{R}}, \theta_{n,m}^{\mathrm{R}})$ and $F_{\mathrm{R}}^{\Phi}(\phi_{n,m}^{\mathrm{R}}, \theta_{n,m}^{\mathrm{R}})$ vectors corresponding to the dominant cluster's AoA/ZoA angles.

Increased correlation between the polarization vectors at the base station and/or UE can be separated via the use of digital precoding codebooks as described in Chapter 4.6.4. This, in turn, is limited by the granularity of the digital beamforming codebook used for channel state feedback.

4.5 HYBRID BEAMFORMING TRANSMISSIONS

4.5.1 FOUR LAYER SU-MIMO SCHEMES

In many practical implementations, either the base station or the UE (or both devices) are equipped with 4 RF chains allowing hybrid beamforming over two spatial layers (assuming polarization MIMO over each spatial layer). At the base station, these RF chains can be deployed over independent sets of panels as illustrated in Figure 4.23 where the total power P is equally distributed across all the four RF chains. Each set of panels can then communicate with a spatial layer steered along the same direction corresponding to the use of a single transmission configuration indicator (TCI) state, or along different directions corresponding to the use of multiple TCI states for the implementation of the higher-rank scheme.

At a UE possibly equipped with multiple antenna modules, two flavors of higher-rank schemes are possible: intra- and inter-module schemes. With the former flavor, subsets of antenna elements within the same antenna module are grouped in different

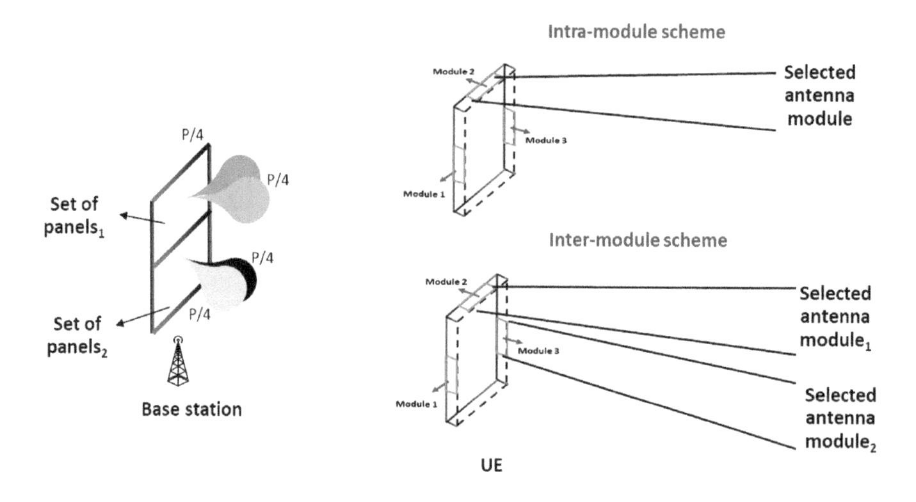

Figure 4.23 Intra- and inter-module UE side setups for the four layer schemes.

ways to generate two spatial layers. The main pros and cons of this approach are now discussed.

- Such an approach trades off the use of a single antenna module (associated with lower power consumption) with reduced number of antenna elements and hence reduced array gain per spatial layer
- Further, this approach leads to increased inter-layer interference due to increased correlation between the co-located sets of antenna elements assigned to the two spatial layers
- On the other hand, such an approach can be easily implemented with a single TCI state solution since the beam across the two spatial layers can be derived from a single quasi co-located beam [1] as reference.

With the latter flavor, antenna elements from different antenna modules are used for the two spatial layers. The pros and cons of this approach are as below:

- This approach leads to increased power consumption as multiple modules are turned on simultaneously
- On the other hand, it can minimize inter-layer interference as the modules are typically placed in the UE in different locations and therefore have different fields-of-view
- While all the antenna elements from each antenna module can be used over each spatial layer, poorer gains are anticipated over the second spatial layer as the link for this layer is established over a sub-dominant cluster in the channel
- Implementing this scheme typically requires a multiple TCI state solution requiring a higher control channel overhead.

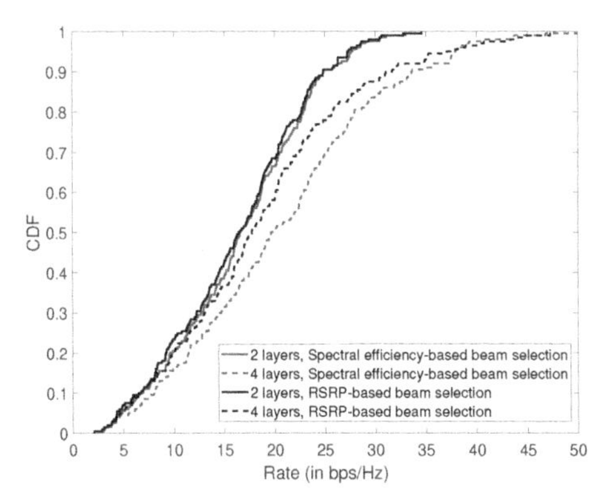

Figure 4.24 Comparison of rate CDFs with two and four layers schemes using a CDL-A channel structure.

For our studies, we assume a base station made of two 8×4 panels using an analog beamforming codebook of size 128 per panel (16 steerable beams in azimuth and 8 steerable beams in elevation). The UE consists of three antenna modules placed on the long edges and the top short edge as illustrated in Figure 4.23. Each antenna module consists of a 4×1 dual-polarized antenna array with an analog beamforming codebook of size 9. For the channel, we assume a CDL-A structure where the relative orientations of the clusters remain the same as defined in TR 38.901 [18], but with the main cluster's directions rotated randomly. Fading gains across the clusters follow a Rayleigh distribution. The achievable rate with two layers is given as

$$\text{Rate}\Big|_{\text{two layers}} = \log_2 \det\left(I_2 + \mathscr{H}\mathscr{H}^H\right) \tag{4.94}$$

$$\mathscr{H}(i,j) = g_i^H H_{ij} f_j, \; i,j = 1,2 \tag{4.95}$$

where $\mathscr{H}$ is the 2×2 post-analog beamformed matrix as seen at the baseband and H_{ij} is the port-to-port matrix as seen from the j-th port at the base station and the i-th port at the UE. Here, f_j and g_i are the beamforming vectors at the j-th and i-th ports at either end, respectively. Similarly, the achievable rate with four layers is given as

$$\text{Rate}\Big|_{\text{four layers}} = \log_2 \det\left(I_4 + \widehat{\mathscr{H}}\widehat{\mathscr{H}}^H\right) \tag{4.96}$$

$$\widehat{\mathscr{H}}(i,j) = g_i^H H_{ij} f_j, \; i,j = 1,\ldots,4 \tag{4.97}$$

with $\widehat{\mathscr{H}}$ denoting the 4×4 post-analog beamformed matrix as seen at the baseband.

In Figure 4.24, we plot the rate CDFs for two and four layer (inter-module) schemes, respectively. Spatial beam selection in the case of one/two spatial layers can be performed either by determining the beams that optimize a spectral efficiency estimate or those that optimize the RSRP as a metric. In the case of RSRP-based beam selection, the spatial beams only maximize the per-layer signal strengths and

Table 4.4

Relative rate improvement from two to four layers with different beam selection algorithms

Percentile point	Spectral efficiency-based	RSRP-based
90	35.4%	25.2%
50	18.7%	7.2%
20	16.7%	8.1%

need not optimize the spectral efficiency estimate. Thus, the realized performance with such a beam selection is necessarily sub-optimal. On the other hand, such an approach is easily amenable to 3GPP specifications as RSRP reporting is well-supported. From these plots, we observe that while spectral efficiency- and RSRP-based beam selection are comparable for two layer transmissions, there is a significant sub-optimality observed with four layer transmissions suggesting that more advanced beam selection mechanisms are useful for higher-rank schemes. Further, the relative rate improvement going from two to four layers at different operational points are presented in Table 4.4 showing that 15–35% improvement is possible with spectral efficiency-based beam selection. From Chapter 4.1, we note that four layer transmissions become relevant from a rate perspective as the SNR increases. This observation generally corresponds to increased four layer gains at higher percentile points in Table 4.4.

4.5.2 MU-MIMO TRANSMISSIONS

MU-MIMO is an important transmission scheme by which network-level spectral efficiency can be considerably improved [174–183]. Here, K users are simultaneously and opportunistically served by a base station, as illustrated in Figure 4.25. The base station and each user are assumed to be equipped with planar arrays of dimensions $N_{tx} \times N_{tz}$ antennas and $N_{rx} \times N_{rz}$ antennas, respectively. With $N_t = N_{tx} \cdot N_{tz}$ and $N_r = N_{rx} \cdot N_{rz}$, the base station and each user are assumed to have $M_t \leq N_t$ and $M_r \leq N_r$ RF chains, respectively.

Beam weights for multi-user transmissions can be obtained either via downlink training (CSI-RS based) or via uplink training (sounding reference signal or SRS based). There are a number of tradeoffs in terms of system level implementations between these two approaches.

- SRS is limited by the available link budget since the EIRP at the UE is typically much smaller than that at the base station
- CSI-RS resource allocation is more flexible than SRS in terms of density, location, etc.

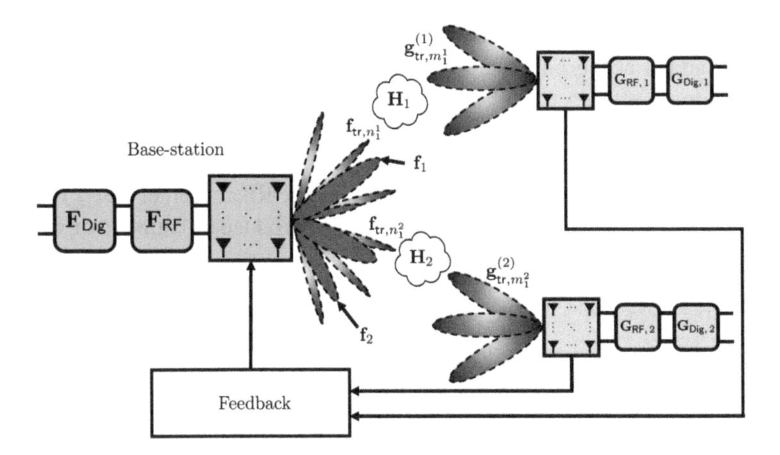

Figure 4.25 System model capturing a base station communicating with $K = 2$ users over an initial acquisition phase (beam patterns in dashed green) followed by multi-user beam design (illustrated in red) based on the feedback from each user.

1. For example, in 3GPP specifications [184], SRS can be 1, 2 or 4 consecutive symbols, but they are constrained to be allocated over the last 6 symbols of a slot
2. While CSI-RS is also constrained to be 1, 2 or 4 consecutive symbols, they can begin anywhere within a slot
3. CSI-RS can be periodic, semi-persistent or aperiodic

- While the SRS is UE-specific, CSI-RS can be a shared resource across UEs (even though it can also be UE-specific). Thus, the system level overhead may be larger with SRS than with CSI-RS
- At least at mmWave frequencies, MPE and blockage constraints can render SRS transmissions relatively not as useful for beam training as CSI-RS.

Thus, power and attendant thermal considerations as well as blockage and MPE favor CSI-RS based multi-user beam training over SRS based approaches. Nevertheless, since massive MIMO considerations at sub-7 GHz frequencies typically favor SRS based beam training [10], bootstrapping such approaches into mmWave frequencies could lead to the implementation of SRS based multi-user beam training in practice.

In the following illustration, we assume a CSI-RS based multi-user transmissions approach where the base station sends data along M_t RF chains to K users. In particular, the base station precodes r_m data-streams for the m-th user with the $r_m \times 1$ symbol vector s_m using the $M_t \times r_m$ digital/baseband precoder $F_{\mathsf{Dig},\,m}$ which is then up-converted to the carrier frequency by the use of the $N_t \times M_t$ RF precoder F_{RF}. This results in the following system equation at the k-th user:

$$y_k = \sqrt{\frac{\rho}{K}} H_k F_{\mathsf{RF}} \cdot \left[\sum_{m=1}^{K} F_{\mathsf{Dig},\,m} s_m \right] + n_k \tag{4.98}$$

where ρ is the pre-precoding SNR and $n_k \sim \mathscr{CN}(0, I_{N_r})$ is the $N_r \times 1$ white Gaussian noise vector added at the k-th user. We assume that s_m are i.i.d. standard complex Gaussian random vectors.

For the channel H_k between the base station and the k-th user, we assume a narrowband structure as described in (2.15) over L_k clusters. At the k-th user, we assume that y_k is down-converted with an $N_r \times M_r$ user-specific RF combiner $G_{\mathsf{RF},\,k}$ followed by a user-specific $M_r \times r_k$ digital combiner $G_{\mathsf{Dig},\,k}$ to produce an estimate of s_k as follows

$$\widehat{s}_k = G^H_{\mathsf{Dig},\,k} G^H_{\mathsf{RF},\,k} y_k \tag{4.99}$$

$$= \sqrt{\frac{\rho}{K}}\, G^H_{\mathsf{Dig},\,k} G^H_{\mathsf{RF},\,k} H_k F_{\mathsf{RF}} F_{\mathsf{Dig},\,k} s_k + \sqrt{\frac{\rho}{K}}\, G^H_{\mathsf{Dig},\,k} G^H_{\mathsf{RF},\,k} H_k F_{\mathsf{RF}} \sum_{m=1, m\neq k}^{K} F_{\mathsf{Dig},\,m} s_m + n_k. \tag{4.100}$$

The rate R_k realized at the k-th user by treating multi-user interference as noise with a mismatched decoder [185] usually serves as a lower bound to the achievable rate. This rate is given as

$$R_k = \log\det\left(I_{r_k} + \frac{\rho}{K} G^H_{\mathsf{Dig},\,k} G^H_{\mathsf{RF},\,k} H_k F_{\mathsf{RF}} F_{\mathsf{Dig},\,k} F^H_{\mathsf{Dig},\,k} F^H_{\mathsf{RF}} H^H_k G_{\mathsf{RF},\,k} G_{\mathsf{Dig},\,k} \cdot \mathbf{\Sigma}^{-1}_{\mathsf{intf}}\right) \tag{4.101}$$

where $\mathbf{\Sigma}_{\mathsf{intf}}$ denotes the covariance matrix of interference and noise

$$\mathbf{\Sigma}_{\mathsf{intf}} = I_{r_k} + \frac{\rho}{K} G^H_{\mathsf{Dig},\,k} G^H_{\mathsf{RF},\,k} H_k F_{\mathsf{RF}} \left(\sum_{m\neq k} F_{\mathsf{Dig},\,m} F^H_{\mathsf{Dig},\,m}\right) F^H_{\mathsf{RF}} H^H_k G_{\mathsf{RF},\,k} G_{\mathsf{Dig},\,k}. \tag{4.102}$$

We assume that each user has two RF chains for polarization MIMO transmissions and the base station transmits one spatial layer (over the polarizations) to each user. By not explicitly focusing on the polarization aspects and focusing only on the spatial domain, we have $M_r = r_k = 1$ (for all $k = 1, \ldots, K$) and $M_t = K \leq N_t$. The system model in (4.98) and (4.100) reduce to

$$\widehat{s}_k = G^H_{\mathsf{Dig},\,k} G^H_{\mathsf{RF},\,k} y_k \tag{4.103}$$

$$= \underbrace{G^H_{\mathsf{Dig},\,k}}_{1\times 1} \underbrace{G^H_{\mathsf{RF},\,k}}_{1\times N_r} \cdot \left(\sqrt{\frac{\rho}{K}} H_k \underbrace{F_{\mathsf{RF}}}_{N_t \times K} \cdot \underbrace{F_{\mathsf{Dig}}}_{K\times K} \cdot \underbrace{s}_{K\times 1} + n_k\right) \tag{4.104}$$

$$= \sqrt{\frac{\rho}{K}} \cdot g^H_k H_k \left[f_1 s_1, \ldots, f_K s_K\right] + g^H_k n_k \tag{4.105}$$

where

$$F_{\mathsf{Dig}} = \left[F_{\mathsf{Dig},1}, \ldots, F_{\mathsf{Dig},K}\right] \tag{4.106}$$

$$s = \left[s_1, \ldots, s_K\right]^T, \tag{4.107}$$

and the second equation follows assuming

$$f_k = F_{\mathsf{RF}} F_{\mathsf{Dig},k} \tag{4.108}$$
$$G_{\mathsf{RF},k} = g_k. \tag{4.109}$$

A simple realization of the hybrid precoding architecture is achieved by setting $F_{\mathsf{Dig}} = I_K$ and the desired f_k for the k-th user is set as the k-th column of F_{RF}. The desired f_k is such that $f_k^H f_k \leq 1$ and meets the phase and amplitude quantization requirements of analog precoding. In a practical implementation, F_{Dig} could primarily be used for sub-band precoding and in the narrowband context of this section, $F_{\mathsf{Dig}} = I_K$ would reflect such an implementation-driven model. The power constraint is equivalent to $\sum_{m=1}^{K} f_k^H f_k \leq K$ and R_k reduces to

$$R_k = \log\left(1 + \frac{\frac{\rho}{K} \cdot |g_k^H H_k f_k|^2}{1 + \frac{\rho}{K} \cdot \sum_{m \neq k} |g_k^H H_k f_m|^2}\right). \tag{4.110}$$

The focus now is to develop an advanced feedback mechanism and a systematic design of the multi-user beamforming structure based on a directional representation of the channel. This structure allows the base station to combat multi-user interference in simultaneous transmissions. For this, we assume that the base station is equipped with a size N codebook $\mathscr{F}_{\mathsf{tr}}$, defined as,

$$\mathscr{F}_{\mathsf{tr}} \triangleq \left\{ f_{\mathsf{tr},1}, \ldots, f_{\mathsf{tr},N} \right\}, \tag{4.111}$$

and the k-th user is equipped with a size M user-specific codebook $\mathscr{G}_{\mathsf{tr}}^k$, defined as,

$$\mathscr{G}_{\mathsf{tr}}^k \triangleq \left\{ g_{\mathsf{tr},1}^{(k)}, \ldots, g_{\mathsf{tr},M}^{(k)} \right\}. \tag{4.112}$$

In the initial acquisition phase, the top-P beam indices at the base station and each user that maximize an estimate of the received SNR are learned. In particular, the received SNR corresponding to the (m,n)-th beam index pair at the k-th user is given as

$$\mathsf{SNR}^{(k)}(m,n) = \left| \left(g_{\mathsf{tr},m}^{(k)} \right)^H H_k f_{\mathsf{tr},n} \right|^2. \tag{4.113}$$

Let the beam pair indices at the k-th user be arranged in non-increasing order of the received SNR and let the top-P beam pair indices be denoted as

$$\mathscr{M} = \left\{ \left(m_1^k, n_1^k \right), \ldots, \left(m_P^k, n_P^k \right) \right\}. \tag{4.114}$$

With the simplified notation of

$$\mathsf{SNR}_\ell^{(k)} \triangleq \mathsf{SNR}^{(k)}(m_\ell^k, n_\ell^k), \ \ell = 1, \ldots, P, \tag{4.115}$$

we have $\mathsf{SNR}_1^{(k)} \geq \ldots \geq \mathsf{SNR}_P^{(k)}$. With the initial acquisition methodology as described above, we now leverage the top-P beam information learned at the k-th user to estimate the channel matrix H_k and to design F_{RF} at the base station.

A typical use of the feedback information at the base station is to select the top beam indices for all the users and to leverage this information to construct a multi-user transmission scheme. Such an approach is adopted in [186], which proposes multi-user beam designs leveraging only the top beam pair index, (m_1^k, n_1^k), and intended to serve different objectives:

- Greedily (from each user's perspective) steering a beam to the best direction for that user (called the *beam steering* scheme)
- Using the information collected from different users to combat interference to other simultaneously scheduled users via a zeroforcing solution (called the *zero-forcing* scheme) and
- For leveraging both the beam steering and interference management objectives via a generalized eigenvector optimization (called the *generalized eigenvector* scheme).

If the beam pair (m_1^k, n_1^k) is blocked or fades, the k-th user requests the base station to switch to the beam index n_2^k and it switches to the beam with index m_2^k (and so on).

We can generalize the structures in [186] by leveraging *all* the top-P beam pair indices fed back from each user. For this, the base station intends to *reconstruct* or *estimate* a rank-P approximation of (a scaled version of) the channel matrix H_k corresponding to the k-th user as follows:

$$\widehat{H}_k = \sum_{\ell=1}^{P} \widehat{\alpha}_{k,\ell}\, \widehat{u}_{k,\ell}\, \widehat{v}_{k,\ell}^H \tag{4.116}$$

where $\widehat{u}_{k,\ell}$ and $\widehat{v}_{k,\ell}$ are defined as estimates of the array steering vectors $u_{k,\ell}$ and $v_{k,\ell}$, respectively. Given the channel model structure in (2.15), (4.116) is simplified by estimating $v_{k,\,\ell}$ and $|\alpha_{k,\,\ell}|$ by f_{tr,n_ℓ^k} and $\gamma_{k,\ell}$, respectively, where

$$\gamma_{k,\ell} \triangleq \sqrt{\mathcal{Q}_{\mathsf{amp}}\left(\mathsf{SNR}_\ell^{(k)}\right)} \tag{4.117}$$

for an appropriate choice of amplitude quantization function $\mathcal{Q}_{\mathsf{amp}}(\bullet)$. However, estimating $\widehat{H}_k$ as in (4.116) is not complete until we have an estimate for $\angle\alpha_{k,\ell}$ and $u_{k,\ell}$. The quantity $\angle\alpha_{k,\ell}$ can be estimated by the user with the same RS resource (or pilot symbol) transmitted during the initial acquisition phase with no additional training overhead. Therefore, we define $\varphi_{k,\ell}$ as an appropriate quantization of the phase of an estimate $\widehat{s}_{\mathsf{tr},k,\ell}$ of the pilot symbol $s_{\mathsf{tr},k,\ell}$

$$\varphi_{k,\ell} \triangleq \mathcal{Q}_{\mathsf{phase}}\left(\angle\widehat{s}_{\mathsf{tr},k,\ell}\right) \tag{4.118}$$

where

$$\widehat{s}_{\mathsf{tr},k,\ell} = \left(g_{\mathsf{tr},\,m_\ell^k}^{(k)}\right)^H \left[\sqrt{\rho}\, H_k f_{\mathsf{tr},\,n_\ell^k} s_{\mathsf{tr},k,\ell} + n_{k,\ell}\right]. \tag{4.119}$$

The noise term $n_{k,\ell}$ captures the additive noise in the initial acquisition process corresponding to the top-P beam pairs.

For $u_{k,\ell}$, we assume that the k-th user uses a multi-user reception beam g_k. In the simplest manifestation, g_k could be the best training beam learned in the initial acquisition phase, $g^{(k)}_{\mathsf{tr},m_1^k}$. We then note that the estimated SINR, defined as

$$\widehat{\mathsf{SINR}}_k \triangleq \frac{\frac{\rho}{K} \cdot |g_k^H \widehat{H}_k f_k|^2}{1 + \frac{\rho}{K} \cdot \sum_{m \neq k} |g_k^H \widehat{H}_k f_m|^2} \tag{4.120}$$

is only dependent on $\widehat{H}_k$ in the form of $g_k^H \widehat{H}_k$. Building on this fact, each user generates $\{\beta_{k,\ell}\}$, defined as,

$$\beta_{k,\ell} \triangleq g_k^H \widehat{u}_{k,\ell} \text{ where } \widehat{u}_{k,\ell} = g^{(k)}_{\mathsf{tr},\, m_\ell^k}. \tag{4.121}$$

It then quantizes the amplitude and phase of $\beta_{k,\ell}$ and feeds them back

$$\mu_{k,\ell} \triangleq \mathscr{Q}_{\mathsf{amp}}\left(|\beta_{k,\ell}|\right) \tag{4.122}$$

$$\nu_{k,\ell} \triangleq \mathscr{Q}_{\mathsf{phase}}\left(\angle \beta_{k,\ell}\right). \tag{4.123}$$

For both $\varphi_{k,\ell}$ and $\nu_{k,\ell}$, without loss in generality, relative phases with respect to $\varphi_{k,1}$ and $\nu_{k,1}$ (that is, $\varphi_{k,\ell} - \varphi_{k,1}$ and $\nu_{k,\ell} - \nu_{k,1}$) can be reported. While the feedback overhead increases linearly with P (the rank of the channel approximation), there are diminishing returns in terms of channel representation accuracy since the clusters captured in $\widehat{H}_k$ are sub-dominant as P increases and are eventually limited in number by L_k. Thus, it is useful to select P to tradeoff these two conflicting objectives.

Following the above discussion, the k-th user feeds back the $P \times 5$ matrix P_k, defined as

$$P_k \triangleq \begin{bmatrix} n_1^k & \gamma_{k,1} & 0 & \mu_{k,1} & 0 \\ n_2^k & \gamma_{k,2} & \varphi_{k,2} - \varphi_{k,1} & \mu_{k,2} & \nu_{k,2} - \nu_{k,1} \\ \vdots & \vdots & \vdots & \vdots & \vdots \\ n_P^k & \gamma_{k,P} & \varphi_{k,P} - \varphi_{k,1} & \mu_{k,P} & \nu_{k,P} - \nu_{k,1} \end{bmatrix} \tag{4.124}$$

and the base station approximates $g_k^H \widehat{H}_k$ as follows

$$g_k^H \widehat{H}_k = \sum_{\ell=1}^{P} \mu_{k,\ell}\, \gamma_{k,\ell} \cdot e^{j(\varphi_{k,\ell} + \nu_{k,\ell})} \cdot \left(f_{\mathsf{tr},n_\ell^k} \right)^H. \tag{4.125}$$

In other words, $g_k^H \widehat{H}_k$ is represented as a linear combination of the top-P beams as estimated from $\mathscr{F}_{\mathsf{tr}}$ in the initial acquisition phase. The weights in this linear combination correspond to the relative strengths of the clusters as distinguished by the codebook resolution (at both base station and UE). The base station uses the channel matrix constructed for each user based on its feedback information ($g_k^H \widehat{H}_k$) and generates a good beamformer structure.

We now present numerical studies to illustrate the advantages of the proposed beamforming solutions. The channel model corresponds to $L_k = 6$ clusters, AoDs uniformly distributed in a $120° \times 30°$ coverage area, and AoAs uniformly distributed in a $120° \times 120°$ coverage area for each of the $k = 1, \ldots, K$ users in the cell. We consider simultaneous transmissions from the base station to $K = 2$ users in the cell. For the initial acquisition codebooks, based on the beam broadening principles, we consider a size N codebook to cover the $120° \times 30°$ AoD space with a 16×4 planar array at the base station side where $N = 32$ and $N = 8$. For the codebooks at the UE, a refined beamforming codebook of size M where $M = 16$ is used over a 2×2 array. At this stage, it is worth noting that the performance of the proposed multi-user scheme is impacted by a number of system parameters such as:

- Granularity of $\mathscr{F}_{\text{tr}}$ and $\mathscr{G}_{\text{tr}}^k$ (initial acquisition codebook sizes)
- Coarseness of channel approximation (rank-P)
- Finite-rate feedback of channel reconstruction parameters and
- Quantization of the resulting multi-user beam weights.

A careful study of the impact of these different design parameters on the performance of the proposed multi-user scheme can be found in [159].

In Figure 4.26, we compare the performance of the proposed zeroforcing scheme with the beam steering scheme. We also benchmark/upper bound the performance with a fully digital system employing maximum ratio transmission (MRT)/MRC beams in the initial acquisition phase. In these plots, an $M = 16$ codebook is used at the UE. Figures 4.26(a, b) illustrate the trends with $N = 8$ and $N = 32$, respectively. We note that the proposed zeroforcing scheme improves the sum rate over a simple beam steering scheme especially as the rank approximation P increases. While the fully digital benchmark is significantly higher for small N, this gap narrows down as N increases suggesting that with a better codebook approximation, the proposed

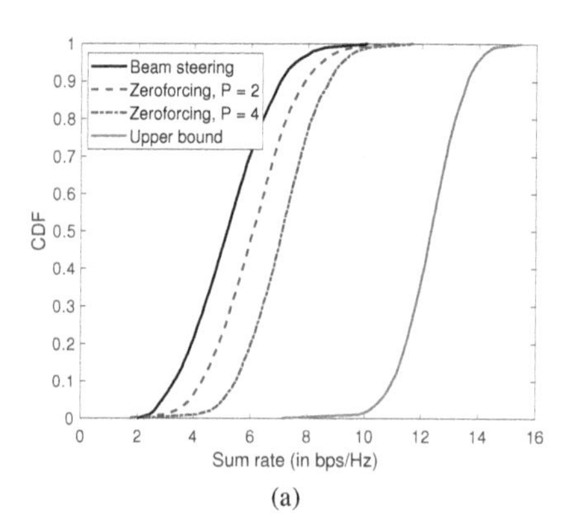

(a)

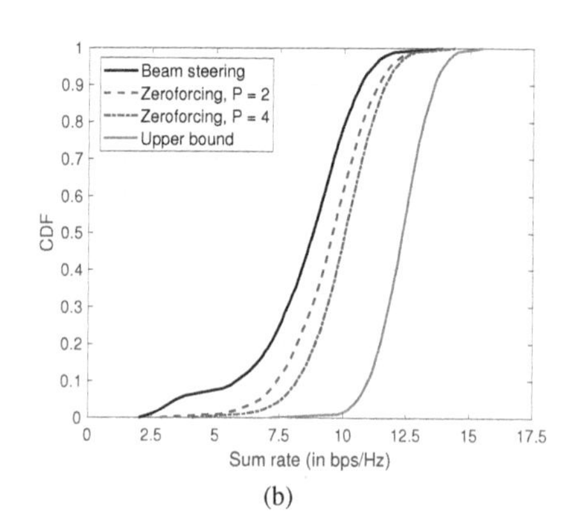

(b)

Figure 4.26 CDF of sum rates of the multi-user transmission schemes using a $M = 16$ codebook with (a) $N = 8$ and (b) $N = 32$.

channel reconstruction-based multi-user beamforming scheme can perform close to optimality.

MU-MIMO tradeoffs in a joint sub-7 GHz and mmWave deployment are considered in Chapter 5.5.3.

4.5.3 MULTI-TRP TRANSMISSIONS

Chapter 4.5.2 assumes that a UE is connected to a single base station at a given time. As Chapter 5 will showcase, in dense network deployments, the UE can be under the coverage of multiple neighboring base stations and can be connected to them if the connection to the serving base station drops (or if the signal strength deteriorates). While the UE is connected to the serving base station, the neighboring base stations are either serving other UEs and are potentially interfering with the UE of interest or are simply shut off. We now take the idea of dense network deployments further and leverage the techniques exposed in MU-MIMO studies by allowing neighboring base stations to cooperate. That is, we explore the idea of Coordinated Multipoint (COMP) transmissions at mmWave frequencies. In 3GPP parlance, this family of techniques as applicable to sub-7 GHz frequencies is also known as *multi-TRP transmissions* and is used as a mechanism for network-level coordination. The readers are referred to technical surveys in [187, 188, 189, 190, 191, 192, 193, 194] for a broader understanding of theoretical capabilities with different flavors of COMP and practical implementation challenges of COMP schemes such as CSI and synchronization challenges.

For mmWave frequencies with reliability considerations (due to blockage), [195] has proposed an iterative optimization solution using second-order cone programming solutions for weighted sum rate maximization in coordinated transmissions. Given the high energy usage at mmWave frequencies, energy-efficient cooperative precoding across TRPs is studied in [196] where the sum power consumption across the TRPs is minimized subject to per-user spectral efficiency and per-TRP peak power constraints. Stochastic geometric tools are used to show that cooperation from randomly located TRPs decreases the probability of outage due to blockage and increases the coverage probability in [197]. Stochastic geometric tools are also used in studying coverage improvement with dynamic TRP selection policies in [198]. There have also been developments in 3GPP specifications to support transmissions and receptions from multiple panels (or antenna modules) and multiple TRPs [199].

To put the ideas in context, consider transmissions of a common data symbol $\mathsf{s} \sim \mathscr{CN}(0,1)$ from K TRPs to a UE as illustrated in Figure 4.27. Let H_k denote the $N_r \times N_{t,k}$ channel matrix from TRP$_k$ to the UE where N_r and $N_{t,k}$ correspond to the number of antenna elements at the UE and TRP$_k$, respectively. For the channel H_k, we assume the narrowband structure as in (2.15). If TRP$_k$ uses an $N_{t,k} \times 1$ unit-norm beamforming vector f_k with a transmit power ρ_k and the UE uses a unit-norm beamforming vector g, the system model in terms of the decoded symbol $\widehat{\mathsf{s}}$ is given

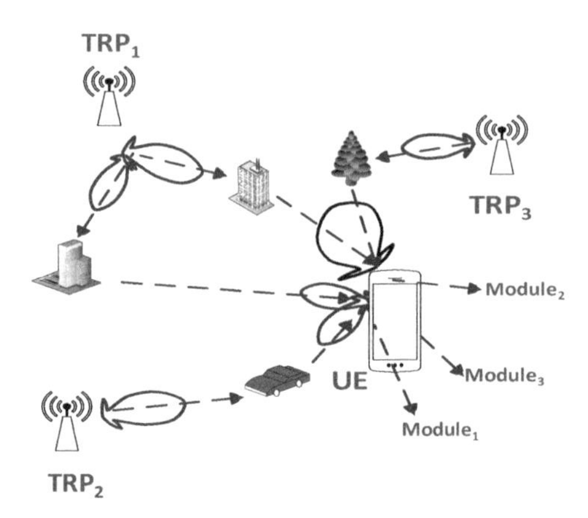

Figure 4.27 Typical small cell setting with cooperative transmissions from multiple TRPs.

as

$$\widehat{\mathsf{s}} = g^H \cdot \left(\sum_{k=1}^{K} \sqrt{\rho_k} H_k f_k \mathsf{s} + n \right) \tag{4.126}$$

where n denotes the $N_r \times 1$ complex white Gaussian noise added at the UE end with $n \sim \mathscr{CN}(0, I_{N_r})$ and independent of s. The received SNR in this operation can be easily computed as

$$\mathsf{SNR}\Big|_{\mathsf{multi-TRP}} = \left| \sqrt{\rho_1} \cdot g^H H_1 f_1 + \ldots + \sqrt{\rho_K} \cdot g^H H_K f_K \right|^2. \tag{4.127}$$

We are interested in the choice of $\{f_1, \ldots, f_K, g\}$ to maximize SNR.

For any given choice of $\{f_1, \ldots, f_K\}$, it is straightforward to identify g_{opt} that maximizes SNR for the multi-TRP case as

$$g_{\mathsf{opt}} = \frac{\sqrt{\rho_1} H_1 f_1 + \cdots + \sqrt{\rho_K} H_K f_K}{\left\| \sqrt{\rho_1} H_1 f_1 + \cdots + \sqrt{\rho_K} H_K f_K \right\|} \tag{4.128}$$

where $\| \cdot \|$ denotes the two-norm operation of a vector leading to

$$\mathsf{SNR}\Big|_{\mathsf{multi-TRP}} = \left\| \sqrt{\rho_1} H_1 f_1 + \cdots + \sqrt{\rho_K} H_K f_K \right\|^2. \tag{4.129}$$

To compare the performance of any constructive scheme, the following upper bound to SNR can be easily established [200]:

$$\mathsf{SNR} \leq \left(\sum_{k=1}^{K} \sqrt{\rho_k \cdot \lambda_1 (H_k^H H_k)} \right)^2. \tag{4.130}$$

The above upper bound implicitly assumes that the dominant eigenmode can be excited from every TRP and co-phasing can be performed across TRPs to mitigate inter-TRP interference. Leveraging the sparse structure of (2.15), we propose a constructive coordinated beamforming approach for the $K = 2$ case. Generalization to the $K > 2$ case is straightforward and is not provided here. The proposed approach consists of two steps: Single-user beam training followed by enhanced feedback of co-phasing factors to enable coordinated transmissions.

Step 1 (Single-User Beam Training): Motivated by the hierarchical beam training protocol [147, 201], we assume that each TRP beam trains the UE separately and unhindered by other TRPs' interference or simultaneous transmissions. For beam training, we assume that each TRP uses an analog beamforming codebook of narrow beamwidth beams. To be precise, let $\mathscr{F}_1$ and $\mathscr{F}_2$ denote the two codebooks used at the two TRPs corresponding to sizes M and N, respectively. Let the beamforming vectors in the two codebooks be denoted as

$$\mathscr{F}_1 \triangleq \{c_1, \ldots, c_M\} \text{ and } \mathscr{F}_2 \triangleq \{d_1, \ldots, d_N\} \tag{4.131}$$

where $\|c_i\| = \|d_j\| = 1,\ i = 1, \ldots, M, j = 1, \ldots, N$. Over the beam training period, let the UE use a codebook of size P denoted as

$$\mathscr{G} \triangleq \{e_1, \ldots, e_P\} \tag{4.132}$$

where $\|e_i\| = 1, i = 1, \ldots, P$.

In beam training, the UE estimates the top-J beam indices to be used at each TRP (as well as the respective beams to be used at the UE). The UE feeds back the top-J beam indices along with the RSRPs seen from each TRP via the agreed 5G-NR beam management procedure (e.g., TCI state feedback for the beam indices) [202]. In the special case of $J = 1$ (reporting of only the best beam), let $f_1 = c_i$ and $g_1 = e_k$ denote the best choices of beams to be used at the two ends of the first link (for some choices of i and k). Similarly, let $f_2 = d_j$ and $g_2 = e_\ell$ denote the best choices of beams to be used at the two ends of the second link (for some choices of j and ℓ). Intuitively speaking, these beam pairs (c_i, e_k) and (d_j, e_ℓ) can be used to construct a rank-1 approximation of H_1 and H_2, respectively. Let the associated RSRPs with these two best beam pairs be indicated as $\mathsf{RSRP}_{ki}^{(1)}$ and $\mathsf{RSRP}_{\ell j}^{(2)}$, respectively.

Step 2 (Co-phasing Factor Estimation): In addition to the beam indices to be used at each TRP, we assume that the UE can provide enhanced feedback as described next. Here, the UE feeds back co-phasing information to be used across the beams at each TRP. For this, the UE correlates the estimated post-beamformed signal/symbol learned with the best beam pair from the first TRP with the estimated post-beamformed signal/symbol learned with the best beam pair from the second TRP. That is, it obtains

$$\phi = \angle \widehat{s_1} \cdot \widehat{s_2}^{\star} \tag{4.133}$$

where

$$\widehat{s_1} = e_k^H \cdot (\sqrt{\rho_1} H_1 c_i s + n_1) \quad (4.134)$$
$$\widehat{s_2} = e_\ell^H \cdot (\sqrt{\rho_2} H_2 d_j s + n_2). \quad (4.135)$$

At the receive side, the UE uses the (matched filtering-based) combining beam g, defined as,

$$g = \frac{\sqrt{\mathsf{RSRP}_{ki}^{(1)}} \cdot e_k + e^{j\phi} \cdot \sqrt{\mathsf{RSRP}_{\ell j}^{(2)}} \cdot e_\ell}{\left\| \sqrt{\mathsf{RSRP}_{ki}^{(1)}} \cdot e_k + e^{j\phi} \cdot \sqrt{\mathsf{RSRP}_{\ell j}^{(2)}} \cdot e_\ell \right\|} \quad (4.136)$$

for coordinated reception where the RSRPs are as defined in Step 1.

Note that the co-phasing factor estimation step implicitly assumes phase noise and carrier frequency offset (CFO) coherence. A simple approach to minimize the impact of non-coherence is to use a contiguous set of symbols/sub-symbols for beam training from both TRPs. It is also important to note that the above coordinated scheme at the TRPs and the UE side assume a rank-1 "effective channel" approximation of H_1 and H_2. This approach is useful and near-optimal if the dominant cluster of H_1 and H_2 dominate the other clusters.

In practice, multiple clusters could dominate the channel structure in (2.15) with comparable powers. In this context, similar to the MU-MIMO approach, we now propose a natural higher-rank "effective channel" approximation. We illustrate the proposal with a rank-2 "effective channel" approximation scheme. Here, in addition to Step 1, let (c_m, e_o) and (d_n, e_p) denote the second-best beam pairs from the two TRPs to the UE with associated RSRPs given as $\mathsf{RSRP}_{om}^{(1)}$ and $\mathsf{RSRP}_{pn}^{(2)}$, respectively. The UE feeds back beam indices i and m with their respective RSRPs, $\mathsf{RSRP}_{ki}^{(1)}$ and $\mathsf{RSRP}_{om}^{(1)}$, to the first TRP along with beam indices j and n with their respective RSRPs, $\mathsf{RSRP}_{\ell j}^{(2)}$ and $\mathsf{RSRP}_{pn}^{(2)}$, to the second TRP. As indicated in Step 2, the UE estimates a co-phasing factor to use across the beams and feeds this information back to each TRP. The beams used in the coordinated transmission at the two TRPs are:

$$f_1 = \frac{e^{j\phi_1}\sqrt{\mathsf{RSRP}_{ki}^{(1)}} \cdot c_i + e^{j\phi_2}\sqrt{\mathsf{RSRP}_{om}^{(1)}} \cdot c_m}{\left\| e^{j\phi_1}\sqrt{\mathsf{RSRP}_{ki}^{(1)}} \cdot c_i + e^{j\phi_2}\sqrt{\mathsf{RSRP}_{om}^{(1)}} \cdot c_m \right\|} \quad (4.137)$$

$$f_2 = \frac{e^{j\phi_3}\sqrt{\mathsf{RSRP}_{\ell j}^{(2)}} \cdot d_j + e^{j\phi_4}\sqrt{\mathsf{RSRP}_{pn}^{(2)}} \cdot d_n}{\left\| e^{j\phi_3}\sqrt{\mathsf{RSRP}_{\ell j}^{(2)}} \cdot d_j + e^{j\phi_4}\sqrt{\mathsf{RSRP}_{pn}^{(2)}} \cdot d_n \right\|}. \quad (4.138)$$

The four co-phasing factors are $\phi_1, \ldots, \phi_4$. Without loss in generality, we can set ϕ_1 to zero. At the UE side, while an analog to (4.136) can be constructed, due to the use of smaller array dimensions (denoted as N_r), we propose a simpler alternative approach. Here, both the TRPs *jointly* train the UE with beams f_1 and f_2 over N_r

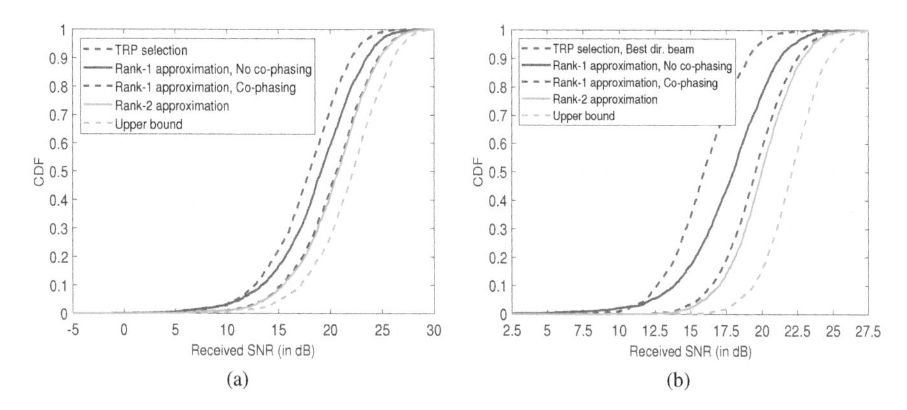

Figure 4.28 Performance of different multi-TRP schemes with the number of clusters in the channel (L_k) corresponding to (a) $L_k = 2$ and (b) $L_k = 6$.

consecutive CSI-RS symbol opportunities with a common training symbol s. The UE estimates the effective vector channel seen as

$$h = \sqrt{\rho_1} H_1 f_1 \mathsf{s} + \sqrt{\rho_2} H_2 f_2 \mathsf{s} + n \tag{4.139}$$

and uses $g = \frac{h}{\|h\|}$ for multi-TRP reception.

We now illustrate the performance improvement seen with the proposed scheme for coordinated transmissions in the multi-TRP setup with two numerical studies. For this, we consider the scenario with $K = 2$ TRPs and where both TRPs are equipped with $N_{t,1} = N_{t,2} = 16$ antenna elements in a ULA structure with $\lambda/2$ inter-antenna element spacings. We assume that the UE is equipped with an $N_r = 4$ antenna element ULA, also with $\lambda/2$ spacing. We assume infinite precision codebook sizes for analog beamforming at both the TRPs and the UE (that is, $M = N = P \to \infty$). Such high-precision codebooks are expected to be practical if the P-1, P-2, P-3 phases are completed [203] and the channel remains sufficiently stationary to allow the use of coordinated transmissions across TRPs. We consider two scenarios where $L_k = 2$ and $L_k = 6$ for both $k = 1, 2$. The former scenario corresponds to lesser number of clusters, typical of outdoor/suburban deployments and the latter scenario corresponds to more clusters, typical of indoor/downtown deployments.

Figures 4.28(a, b) illustrate the performance of different coordinated transmission schemes in the two scenarios: $L_k = 2$ and $L_k = 6$, respectively. The schemes considered here include selection of the best TRP from amongst the two TRPs, coordinated transmission with a rank-1 "effective channel" approximation as well as a rank-2 "effective channel" approximation. The performance of these schemes are benchmarked against the generally unrealizable upper bound as described in (4.130). At this stage, recall that the upper bound is *not realizable even with full CSI* and requires CSI of the first TRP-to-UE link at the second TRP and *vice versa*. Thus, a gap to the upper bound is not indicative of the poorness of the proposed schemes. On the other hand, a significant performance improvement over TRP selection (the state-of-the-art approach considered in implementations today) is indicative of the goodness of the proposed scheme and the promise of coordination [191].

From this perspective, we observe that while the proposed scheme leads to substantial performance improvement over a naïve TRP selection strategy, much of the gains can be attributed to the appropriate co-phasing factor estimation in the rank-1 approximation. Further, as we increase the granularity in terms of approximating the effective channel, the performance improves as seen from the rank-2 approximation curves. Nevertheless, the gap to the upper bound indicates that while a higher-rank approximation of the channel can perhaps improve performance, the feedback overhead associated with co-phasing factors need to also be accounted for in any such study.

4.6 APPENDIX

4.6.1 MIMO RATE COMPUTATION

Let $H(k)$ denote the $N_r \times N_t$ channel matrix between a transmitter with N_t transmit antennas and a receiver with N_r receive antennas over the k-th subcarrier. With the system model as given in (4.1) and with a linear processing as described in (4.2), the achievable spectral efficiency R (in bits per channel use) is given as

$$R = I(s;\widehat{s}) = h(\widehat{s}) - h(\widehat{s}|s) \tag{4.140}$$

where $h(\cdot)$ and $h(\cdot|\cdot)$ denote the differential and conditional differential entropies of the underlying random vectors [204]. Assuming that s comes from a proper complex Gaussian codebook/constellation, we have [204, 39]

$$h(\widehat{s}) = \log_2 \det(\pi e \cdot \mathbf{\Sigma}_{\widehat{s}}) \tag{4.141}$$

where $\mathbf{\Sigma}_{\widehat{s}}$ denotes the covariance matrix of $\widehat{s}$. With this notation, R is given as

$$R = \log_2 \det\left((\mathbf{\Sigma}_{G^H n})^{-1} \mathbf{\Sigma}_{G^H HFs + G^H n} \right). \tag{4.142}$$

Further, assuming that s and n are statistically independent of each other, we have

$$\mathbf{\Sigma}_{G^H n} = \sigma_{\mathsf{n}}^2 \cdot G^H G \tag{4.143}$$

$$\mathbf{\Sigma}_{G^H HFs + G^H n} = \sigma_{\mathsf{s}}^2 \cdot G^H HFF^H H^H G + \sigma_{\mathsf{n}}^2 \cdot G^H G, \tag{4.144}$$

which leads to the simplification for R as given in (4.3) and is reproduced below:

$$R = \log_2 \det\left(I_r + \frac{\sigma_{\mathsf{s}}^2}{\sigma_{\mathsf{n}}^2} \cdot (G^H G)^{-1} G^H HFF^H H^H G \right). \tag{4.145}$$

Note that the rate in (4.145) is invariant to the norm of G. In other words, any choice of norm changes the symbol as well as noise strength in proportionate measure rendering the SNR and the achievable rate invariant to the norm. Thus, it is typical to use G normalized to an appropriate norm (typically set to 1).

We are interested in selecting the matrix pair (F, G) to maximize R as quantified in (4.145). In general, this optimization appears to be a hard problem. In this direction, we first assume that F is fixed and is known at the receiver, and then consider the optimization of G for a given F. We can then optimize the intermediate functional R over F. Even with this decomposition, maximizing R appears to be difficult. Thus, we assume that F is fixed and study the structure of G to minimize the mean squared error (MSE) between s and $\widehat{s}$. This LMMSE structure is given as [205, 206, 207, 208]

$$G_{\mathsf{opt}} = \sigma_{\mathsf{s}}^2 \cdot \left(\sigma_{\mathsf{n}}^2 \cdot I_{N_r} + \sigma_{\mathsf{s}}^2 \cdot HFF^H H^H\right)^{-1} \cdot HF \quad (4.146)$$

$$= \frac{\sigma_{\mathsf{s}}^2}{\sigma_{\mathsf{n}}^2} \cdot HF \cdot \left(I_r + \frac{\sigma_{\mathsf{s}}^2}{\sigma_{\mathsf{n}}^2} \cdot F^H H^H HF\right)^{-1}, \quad (4.147)$$

where (4.147) follows from the matrix inversion lemma (see Appendix 4.6.3). With the choice of G_{opt} as in (4.147), it can be seen that

$$R = \log_2 \det\left(I_r + \frac{\sigma_{\mathsf{s}}^2}{\sigma_{\mathsf{n}}^2} \cdot F^H H^H HF\right). \quad (4.148)$$

For the optimization of R in (4.148) over F, note that the average transmit power P translates to $P = \mathsf{E}[s^H s] = \sigma_{\mathsf{s}}^2 \cdot r$ leading to $\sigma_{\mathsf{s}}^2 = \frac{P}{r}$. We also have

$$\begin{aligned}
\sigma_{\mathsf{s}}^2 \cdot r &= \mathsf{E}[s^H s] && (4.149)\\
&= \mathsf{E}\left[(Fs)^H (Fs)\right] && (4.150)\\
&= \mathsf{E}\left[\mathsf{Tr}\left(s^H F^H Fs\right)\right] && (4.151)\\
&= \mathsf{E}\left[\mathsf{Tr}\left(Fss^H F^H\right)\right] && (4.152)\\
&= \mathsf{Tr}\left(F\mathsf{E}\left[ss^H\right] F^H\right) && (4.153)\\
&= \mathsf{Tr}\left(F \cdot \sigma_{\mathsf{s}}^2 I_r \cdot F^H\right) && (4.154)\\
&= \sigma_{\mathsf{s}}^2 \cdot \mathsf{Tr}\left(F^H F\right) && (4.155)
\end{aligned}$$

leading to a Frobenius norm constraint on F, namely $\mathsf{Tr}\left(F^H F\right) = r$. In the above series of equalities, (4.151) follows from the fact that $s^H F^H Fs$ is a scalar, (4.152) from the fact that $\mathsf{Tr}(AB) = \mathsf{Tr}(BA)$, and (4.153) from the linearity of expectation and trace operations. Thus, the optimization problem of interest can be rewritten as

$$F_{\mathsf{opt}} = \arg \max_{F \,:\, \mathsf{Tr}\left(F^H F\right) = r} \log_2 \det\left(I_r + \frac{P}{r\sigma_{\mathsf{n}}^2} \cdot F^H H^H HF\right). \quad (4.156)$$

4.6.2 OPTIMAL SCHEME ASSUMING PERFECT CSI

To understand the structure of F_{opt} in (4.156), we provide some background into matrix algebra in Appendix 4.6.3.

Let the singular value decomposition (SVD) of $H^H H$ and F be given as $H^H H = UDU^H$ and $F = U_F D_F V_F^H$, respectively, where U and U_F are $N_t \times N_t$ unitary matrices, V_F is an $r \times r$ unitary matrix, D is $N_t \times N_t$ diagonal and D_F is $N_t \times r$ diagonal. Note that if the SVD of H is written as $U_H D_H V_H^H$, then we have $U = V_H$ and

$D = D_H^H D_H = \text{diag}(\lambda_i(H^H H))$. Without loss in generality, let the diagonal entries of D be assumed to be arranged in non-increasing order.

With this background, we can write R as

$$\begin{aligned} R &= \log_2 \det\left(I_r + \frac{P}{r\sigma_n^2} \cdot F^H H^H H F\right) && (4.157) \\ &= \log_2 \det\left(I_r + \frac{P}{r\sigma_n^2} \cdot V_F D_F^H U_F^H U D U^H U_F D_F V_F^H\right) && (4.158) \\ &= \log_2 \det\left(I_r + \frac{P}{r\sigma_n^2} \cdot D_F^H \underbrace{\widetilde{U} D \widetilde{U}^H}_{=Z} D_F\right) && (4.159) \end{aligned}$$

where $\widetilde{U} = U_F^H U$ and we use the Sylvester's identity (see Chapter 4.6.3) to establish the invariance of R to V_F. That is, without loss in generality, we can choose $V_F = I_r$. Applying the Hadamard inequality (also, see Chapter 4.6.3), we have

$$\begin{aligned} R &\leq \sum_{i=1}^{r} \log_2\left(1 + \frac{P}{r\sigma_n^2} \cdot \left(D_F^H Z D_F\right)_{ii}\right) && (4.160) \\ &= \sum_{i=1}^{r} \log_2\left(1 + \frac{P}{r\sigma_n^2} \cdot Z_{ii} \left(D_{F,ii}\right)^2\right). && (4.161) \end{aligned}$$

Note that the above upper bound can be met with equality if $Z = \widetilde{U} D \widetilde{U}^H$ is a diagonal matrix. This is possible if and only if $\widetilde{U}$ is a permutation matrix P resulting in $U_F = UP^T$. Let P be such that

$$Z = PDP^T = \text{diag}\left[D_{\pi_1}, \ldots, D_{\pi_{N_t}}\right] \quad (4.162)$$

and we have

$$R \leq \sum_{i=1}^{r} \log_2\left(1 + \frac{P}{r\sigma_n^2} \cdot \left(D_{F,ii}\right)^2 D_{\pi_i}\right). \quad (4.163)$$

We intend to maximize the above expression over D_F. However, since

$$r = \text{Tr}(F^H F) = \text{Tr}(D_F^H D_F) = \sum_{i=1}^{r} \left(D_{F,ii}\right)^2, \quad (4.164)$$

the optimization over D_F in (4.163) can be made invariant to P. Thus, our goal is the maximization of the objective function for the constrained optimization with Lagrangian parameter γ:

$$\mathscr{L} = \sum_{i=1}^{r} \log_2\left(1 + \frac{P}{r\sigma_n^2} \cdot \left(D_{F,ii}\right)^2 D_i\right) + \gamma \cdot \left(\sum_{i=1}^{r} \left(D_{F,ii}\right)^2 - r\right). \quad (4.165)$$

Without loss in generality, we can assume that $D_i > 0$ or else, we can set $D_{F,ii} = 0$. By observing the critical points of $\mathscr{L}$, we have the *waterfilling* solution:

$$(D_{F,ii})^2 = \frac{r\sigma_n^2}{P} \cdot \left(\mu - \frac{1}{D_i}\right)^+ \tag{4.166}$$

where $x^+ = \max(x, 0)$ and μ satisfies the constraint $\sum_{i=1}^{r} \left(\mu - \frac{1}{D_i}\right)^+ = \frac{P}{\sigma_n^2}$.

In general, the number of eigenmodes excited by the waterfilling solution is non-decreasing as $\frac{P}{\sigma_n^2}$ increases. It can be shown that switching from rank-1 to rank-2 transmissions is optimal when

$$\frac{P}{\sigma_n^2} \geq \frac{\lambda_1(H^H H) - \lambda_2(H^H H)}{\lambda_1(H^H H) \cdot \lambda_2(H^H H)} \tag{4.167}$$

and switching from rank-$(k-1)$ to rank-k transmissions is optimal when $\frac{P}{\sigma_n^2}$ is approximately larger than $\frac{k}{\lambda_k(H^H H)}$ [209].

4.6.3 COMMONLY USED FACTS FROM MATRIX ALGEBRA

The readers are referred to classic texts such as [210, 211, 212, 213] for a solid background on matrix algebra facts.

Matrix Decompositions: Consider an $n \times n$ square matrix A with eigen-pairs (x_i, λ_i) for $i = 1, \ldots, n$. That is, $Ax_i = \lambda_i x_i$ for each i. Writing the above in the form of a matrix relationship, we have

$$A \cdot \underbrace{[x_1, \ldots, x_n]}_{=X} = [Ax_1, \ldots, Ax_n] = [\lambda_1 x_1, \ldots, \lambda_n x_n] \tag{4.168}$$

$$= [x_1, \ldots, x_n] \cdot \underbrace{\begin{bmatrix} \lambda_1 & 0 & \cdots & 0 \\ 0 & \lambda_2 & \cdots & 0 \\ \vdots & \vdots & \ddots & \vdots \\ 0 & \cdots & 0 & \lambda_n \end{bmatrix}}_{=\Lambda}. \tag{4.169}$$

Thus, in block matrix formulation, we have $AX = X\Lambda$. If the eigenvectors of A are linearly independent, then $\det(X) \neq 0$ (which implies that X is invertible) and we have $A = X\Lambda X^{-1}$, which is a *decomposition* of A along its eigenvectors and is called the *eigen-decomposition* of A. Some conditions under which the eigenvectors of A are linearly independent are:

- If the eigenvalues of A are unique
- If A is a normal matrix satisfying $AA^H = A^H A$. Common examples of normal matrices are unitary and Hermitian matrices. In the normal matrix case, at least one set of eigenvectors X can be made orthonormal satisfying $XX^H = I$ or $(X)^{-1} = X^H$. With this set of eigenvectors, we have $A = X\Lambda X^H$. If A is real symmetric, in addition, X can be made real and we thus have $A = X\Lambda X^T$.

Linear independence of eigenvectors is the critical condition needed for a possible eigen-decomposition of a matrix A. For example, consider the 2×2 matrix

$$A = \begin{bmatrix} 1 & 2 \\ 0 & 1 \end{bmatrix}. \tag{4.170}$$

Being an upper triangular matrix, it is clear that the two eigenvalues of A are 1 and 1. Also, solving for the eigenvector equation, we see that both eigenvectors should be of the form $[x,\ 0]^T$ illustrating that the two eigenvectors are linearly dependent and that there is no way to diagonalize A.

While the focus of eigen-decomposition is on square matrices, there are many other ways in which we can decompose a (general) rectangular $n \times k$ matrix A. Some of these decompositions include:

- LU decomposition which decomposes a square matrix A ($n = k$) as $A = LU$ with L being a lower triangular matrix and U being an upper triangular matrix.
- Cholesky decomposition is a special case of LU decomposition for a Hermitian positive semidefinite matrix A which can be decomposed as $A = LL^H$ where $L^H = U$.
- QR decomposition which decomposes an $n \times k$ matrix A as $A = QR$ with Q being an $n \times n$ unitary matrix and R being an $n \times k$ upper triangular matrix.
- Schur decomposition or triangularization which decomposes a square matrix A as $A = UTU^H$ where U is unitary and T is upper triangular. If A is normal, then T is diagonal and Schur decomposition reduces to the eigen-decomposition.
- Most important from a MIMO point-of-view is the SVD where a rectangular matrix A is decomposed as $A = UDV^H$ with U being $n \times n$ unitary, V being $k \times k$ unitary and D being $n \times k$ and having non-negative entries on the main diagonal and zeros everywhere else. The columns of U and V are eigenvectors of AA^H and A^HA, respectively. The diagonal entries of D are said to be the singular values of A and are the square root of eigenvalues of AA^H (which are the same as the non-trivial eigenvalues of A^HA if $n \leq k$).

We now present some other matrix algebra results of relevance in MIMO analysis.

Matrix Inversion Lemma (Used in (4.25) and (4.147)): Let A and C be invertible $n \times n$ and $m \times m$ matrices with U and V being $n \times m$ and $m \times n$, respectively. Also, assume that $A + UCV$ is invertible. Then, the matrix inversion lemma (sometimes called the Woodbury identity) is given as

$$(A+UCV)^{-1} = A^{-1} - A^{-1}U\left(C^{-1} + VA^{-1}U\right)^{-1}VA^{-1}. \tag{4.171}$$

Sylvester Identity (Used in (4.159)): Let A and B be $n \times m$ and $m \times n$ matrices. The identity due to Weinstein-Aronszajn (sometimes attributed to Sylvester) states that

$$\det(I_n + AB) = \det(I_m + BA). \tag{4.172}$$

Hadamard Inequality (Used in (4.160)): The Hadamard inequality states that for an $n \times n$ matrix A with column vectors $a_i,\ i = 1, \ldots, n$, we have

$$|\mathsf{det}(A)| \leq \prod_{i=1}^{n} \|a_i\| \tag{4.173}$$

with $\|\cdot\|$ denoting the two-norm of a vector and with equality if and only if the vectors are orthogonal. This can be applied to a positive semidefinite matrix B to show that

$$\mathsf{det}(B) \leq \prod_{i=1}^{n} B_{ii} \tag{4.174}$$

with equality if and only if B is diagonal. The proof of this claim is straightforward by using the Cholesky decomposition to write B as $B = U^H U$ for some upper triangular matrix U with column vectors u_i. We have

$$\mathsf{det}(B) = \mathsf{det}\left(U^H U\right) = |\mathsf{det}(U)|^2 \leq \left(\prod_{i=1}^{n} \|u_i\|\right)^2 = \prod_{i=1}^{n} B_{ii} \tag{4.175}$$

with the last step following since

$$B_{ii} = \left(U^H U\right)_{ii} = \prod_{i=1}^{n} u_i^H u_i = \prod_{i=1}^{n} \|u_i\|^2. \tag{4.176}$$

For equality since U is an upper triangular matrix, we have

$$|\mathsf{det}(U)| = \prod_{i=1}^{n} |U_{ii}|. \tag{4.177}$$

We have $|U_{ii}| = \|u_i\|$ for all i and equality in Hadamard inequality if and only if U is diagonal.

Matrix Perturbation Theory: We start with an illustrative example of matrix perturbation theory. Consider the 2×2 identity matrix A with eigenvalues being 1 and 1. One set of unit-norm eigenvectors are $\begin{bmatrix} 1 \\ 0 \end{bmatrix}$ and $\begin{bmatrix} 0 \\ 1 \end{bmatrix}$, respectively. Now consider a *structured* perturbation around A that leads to an erroneous representation of A of the form:

$$\widehat{A}(\varepsilon) = \begin{bmatrix} 1 & \varepsilon \\ \varepsilon & 1 \end{bmatrix}. \tag{4.178}$$

It is straightforward to check that the eigenvalues of $\widehat{A}(\varepsilon)$ are $1+\varepsilon$ and $1-\varepsilon$ with unit-norm eigenvectors being $\frac{1}{\sqrt{2}} \cdot \begin{bmatrix} 1 \\ 1 \end{bmatrix}$ and $\frac{1}{\sqrt{2}} \cdot \begin{bmatrix} 1 \\ -1 \end{bmatrix}$, respectively. Clearly, this example illustrates the case where a small perturbation in A leads to a small error in its eigenvalues, but a large deviation[15] in the eigenvectors (45° away from the above choice of eigenvectors of A).

The critical part in the above observation is the *eigen-gap* between the eigenvalues of A with a smaller eigen-gap leading to bigger deviation in the distance between the eigenvectors of the unperturbed and perturbed matrices. The Davis-Kahan $\sin(\theta)$ theorem [211] captures the criticality of eigen-gap in perturbation bounds. The following variant of this theorem [214] says the following.

Let $\mathbf{\Lambda} = \mathsf{diag}(\lambda_j)$ be an $n \times n$ positive definite diagonal matrix with distinct eigenvalues. Let X be a Hermitian matrix and consider the perturbed matrix $S(\varepsilon) = \mathbf{\Lambda} + \varepsilon X$. Then, for sufficiently small ε, we have $\lambda_j(\varepsilon) \triangleq \lambda_j(S(\varepsilon))$ being distinct and we can choose $\widehat{u}(\varepsilon)$ to be an eigenvector of $S(\varepsilon)$ corresponding to the eigenvalue $\lambda_j(\varepsilon)$ such that

$$\begin{aligned} \left(\widehat{u}(\varepsilon)\right)_j &= 1 + \mathcal{O}(\varepsilon^2) && (4.179) \\ |(\widehat{u}(\varepsilon))_i| &\leq \varepsilon \cdot \frac{|X_{ij}|}{|\lambda_i - \lambda_j|} + \mathcal{O}(\varepsilon^2). && (4.180) \end{aligned}$$

Here, $\mathcal{O}(\bullet)$ denotes the order notation:

$$f(x) = \mathcal{O}\Big(g(x)\Big) \text{ as } x \to 0 \text{ if } \left| \lim_{x \to 0} \frac{f(x)}{g(x)} \right| \leq M \text{ for some } M > 0. \quad (4.181)$$

Since $\mathbf{\Lambda}$ is a diagonal matrix, it is already diagonalized with the eigenvectors being the columns of the identity matrix. Thus, the above result states that the diagonal entries of the eigenvector matrix of $S(\varepsilon)$ are centered around 1, but the perturbation in the off-diagonal entries are limited only as a function of the eigen-gap of $\mathbf{\Lambda}$.

4.6.4 A BRIEF GLIMPSE OF CHANNEL STATE FEEDBACK SCHEMES IN 3GPP

The readers are referred to [24, 202, 215, 216] for a more careful study of channel state feedback schemes in theory and practice.

LTE Rel. 8 considers codebook-based precoding for the cellular downlink via cell-specific reference signals (RSs) known as Common Reference Signals (CRSs) for the use of up to $N = 4$ RF chains (with the number of layers, or rank being less than the number of RF chains) where no UE-specific processing is applied. From

[15] Note that a different set of eigenvectors could have been chosen for A and the deviation of the eigenvectors of $\widehat{A}(\varepsilon)$ with respect to the choice of eigenvectors of A would have been different. Nevertheless, the sensitivity of the eigenvector to perturbations is the important effect to consider here.

a precoder perspective, for signaling over N RF chains, we define an $N \times N$ DFT matrix precoder parameterized by a shift parameter L as follows [217]:

$$\mathscr{W}_\ell(m,n) = \frac{1}{\sqrt{2}} \cdot e^{\frac{j2\pi m}{N} \cdot \left(n+\frac{\ell}{L}\right)}, \quad 0 \leq m,n \leq N-1, \ 0 \leq \ell \leq L-1. \qquad (4.182)$$

For the special case of $N = 2$ and $L = 4$, from the above definitions, we have

$$\mathscr{W}_0 = \frac{1}{\sqrt{2}} \cdot \begin{bmatrix} 1 & 1 \\ 1 & -1 \end{bmatrix}, \qquad \mathscr{W}_1 = \frac{1}{\sqrt{2}} \cdot \begin{bmatrix} 1 & 1 \\ \frac{1+j}{\sqrt{2}} & \frac{-1-j}{\sqrt{2}} \end{bmatrix}, \qquad (4.183)$$

$$\mathscr{W}_2 = \frac{1}{\sqrt{2}} \cdot \begin{bmatrix} 1 & 1 \\ j & -j \end{bmatrix}, \qquad \mathscr{W}_3 = \frac{1}{\sqrt{2}} \cdot \begin{bmatrix} 1 & 1 \\ \frac{-1+j}{\sqrt{2}} & \frac{1-j}{\sqrt{2}} \end{bmatrix}. \qquad (4.184)$$

The LTE codebook $\mathscr{C}_2^1$ for rank-1 with $N = 2$ RF chains (the subscript denotes the number of RF chains and the superscript denotes the rank) consists of the individual columns of $\mathscr{W}_0$ and $\mathscr{W}_2$:

$$\mathscr{C}_2^1 = \left\{ \frac{1}{\sqrt{2}} \begin{bmatrix} 1 \\ 1 \end{bmatrix}, \frac{1}{\sqrt{2}} \begin{bmatrix} 1 \\ -1 \end{bmatrix}, \frac{1}{\sqrt{2}} \begin{bmatrix} 1 \\ j \end{bmatrix}, \frac{1}{\sqrt{2}} \begin{bmatrix} 1 \\ -j \end{bmatrix} \right\}. \qquad (4.185)$$

The codebook $\mathscr{C}_2^2$ for rank-2 with $N = 2$ RF chains consists of $\mathscr{W}_0$, $\mathscr{W}_2$ and the scheme that performs RF chain/port selection and is given as

$$\mathscr{C}_2^2 = \left\{ \frac{1}{2} \begin{bmatrix} 1 & 1 \\ 1 & -1 \end{bmatrix}, \frac{1}{2} \begin{bmatrix} 1 & 1 \\ j & -j \end{bmatrix}, \frac{1}{\sqrt{2}} \begin{bmatrix} 1 & 0 \\ 0 & 1 \end{bmatrix} \right\}. \qquad (4.186)$$

The third precoder matrix in rank-2 transmissions corresponds to the use of two layers over two RF chains in open loop mode without any feedback. Note that independent of the rank, the codebooks are such that the Frobenius norm of the precoding vector/matrix is normalized to one; see, [218, Table 6.3.4.2.3-1] for details. These codebooks are also presented in [218, Table 6.3.4.2.3-1] and [202, Type-I single panel codebook].

Some important properties considered in the design of $\mathscr{C}_2^1$ and $\mathscr{C}_2^2$ include:

- *M-ary PSK alphabet*: The codebook entries for both rank-1 and rank-2 come from $\{\pm 1, \pm j\}$ which form the QPSK alphabet. Similarly, all the entries of $\mathscr{W}_i, 0 \leq i \leq 3$ come from a 8-PSK alphabet. Such a limited choice of alphabet for the codebook entries leads to reduction in UE complexity whcrein the channel quality indicator (CQI) can be computed with low-complexity without any recourse to complicated vector multiplication operations. For example, the choice of a QPSK alphabet reduces multiplication operations to addition operations in CQI computation. In general, the codebook entries could be limited to M-ary PSK.
- *Nesting*: The precoders used for rank-1 are column vectors of the precoders used for rank-2. In general, lower rank precoders forming a subset of higher-rank

precoders can simplify CQI computation across all the ranks since CQI for lower rank can be derived based on information obtained with higher rank.

- *Constant modulus codebook entries*: All the codebook entries (except the third rank-2 open loop precoder) have constant modulus or equal amplitude to allow balanced power on the different ports. Thus, the PAs that drive each port are equally excited.

The DFT precoder codebook design in (4.182) can be used for the $N = 4$ RF chains case with different values of L [217]. An alternate proposal[16] for an $N \times N$ precoder matrix is to use an $N \times 1$ generating vector $\mathscr{U}$ of unit-norm and consider the Householder transformation induced by $\mathscr{U}$, which is defined as,

$$\mathscr{W} \triangleq I_N - 2\,\mathscr{U}\,\mathscr{U}^H. \tag{4.187}$$

The Householder transformation is a linear transformation that describes the reflection of $\mathscr{U}$ about a plane containing the origin. That is,

$$\mathscr{W}\mathscr{U} = \left(I_N - 2\,\mathscr{U}\,\mathscr{U}^H\right)\cdot \mathscr{U} = -\mathscr{U}. \tag{4.188}$$

Note that $\mathscr{W}$ is a unitary Hermitian matrix with entries given as

$$\mathscr{W}(m,m) \;=\; 1 - 2\,|\mathscr{U}(m)|^2 \tag{4.189}$$
$$\mathscr{W}(m,n) \;=\; -2\,\mathscr{U}(m)\,\mathscr{U}(n)^\star, \;\; m \neq n. \tag{4.190}$$

If $|\mathscr{U}(m)| = \frac{1}{\sqrt{N}}$, we have $\mathscr{W}(m,m) = 1 - \frac{2}{N}$ and $|\mathscr{W}(m,n)| = \frac{2}{N}$ if $m \neq n$. With $N = 4$, this means that the entries of $\mathscr{W}$ have equal amplitude, $|\mathscr{W}(m,n)| = \frac{1}{2}$ for all (m,n) and thus, the signals seen across all the PAs are balanced.

While a codebook of precoders derived from either the DFT transformation in (4.182) or the Householder transformation in (4.187) lead to comparable performance, the Householder construction reduces the UE complexity in CQI computation in terms of the number of matrix inversions required. The LTE codebook for rank-4 with $N = 4$ RF chains corresponds to 16 Householder matrices (of size 4×4). These are induced by 16 unit-norm $\mathscr{U}$'s with equal amplitude and chosen appropriately from a QPSK or an 8-PSK alphabet which reduces UE complexity; see, [218, Table 6.3.4.2.3-2] for details. From these 16 Householder matrices, the precoders for reduced rank are derived based on column subset selection (selected appropriately and a static choice as described in [218]) of the 4×4 matrices thereby guaranteeing a nested property. The choice of Householder matrices also reduces the feedback overhead for rank-1 transmissions with $N = 4$ (relative to a DFT structure).

LTE Rel. 9 extends the above codebook-based precoding approaches to non-codebook based precoding for up to $N = 2$ RF chains (only $N = 1$ RF chain is supported in Rel. 8) using precoded UE-specific RSs for data demodulation, also known

---

[16]Given that Householder reflections and Givens rotations are canonical operations on matrix spaces, a Givens codebook has also been considered for the $N = 4$ case, but was dropped due to its lack of generalizability potential [215].

as DMRSs. Since the UE receives the known RS which is precoded, it does not need to know the precoder in advance, thus enabling precoding beyond the codebook.

Since DMRSs can be allocated to a UE only when it is scheduled, in Rel. 10 to Rel. 12 (also known as LTE-Advanced), up to $N = 8$ RF chains are supported by the introduction and development of cell-specific CSI-RSs which allow CSI measurements without the UE being scheduled for data transmissions. The same codebook structure as in Rel. 8 for $N = 2$ and $N = 4$ RF chains are assumed in Rel. 10. However, for the $N = 8$ case, a *dual codebook* structure is proposed [219]. The justification behind this choice is that $N = 8$ RF chains are typically used over a 4×1 dual-polarized antenna array design (typically a ULA with $\lambda/2$ inter-antenna element spacings) in practical deployments. The relatively close spacings of the four co-polarized antenna elements leads to high (long-term) correlation with a small angular spread (typical of urban-macro deployments), which can be used to reduce the CSI overhead by leveraging the correlation. On the other hand, the low correlation between cross-polarized antenna elements improves diversity and leveraging this requires a feedback codebook as described earlier.

Combining these two disparate ideas, the precoder W to be used over the eight antenna elements is given as $W = W_1 W_2$ with W_1 typically corresponding to DFT beam weights[17] from a *grid-of-beams* that capture the channel's cluster structure and W_2 corresponding to selecting the best beam from the grid and *co-phasing* to combine the energy across polarizations. To reduce feedback overhead, a single value is reported for W_1 over the entire system bandwidth leveraging the observation that spatial covariance information shows little variation across frequency [228], whereas W_2 allows sub-band reporting. A typical 1- and 2-layer precoding matrix are of the form (see [218, Sec. 6.3.4.2.3] and [24, Sec. 7.2.4]):

$$\mathcal{C}_8^1(m,n) = \begin{bmatrix} 1 \\ e^{\frac{j2\pi m}{32}} \\ e^{\frac{j4\pi m}{32}} \\ e^{\frac{j6\pi m}{32}} \\ e^{\frac{j\pi n}{2}} \\ e^{\frac{j\pi n}{2}+\frac{j2\pi m}{32}} \\ e^{\frac{j\pi n}{2}+\frac{j4\pi m}{32}} \\ e^{\frac{j\pi n}{2}+\frac{j6\pi m}{32}} \end{bmatrix} = \underbrace{\begin{bmatrix} 1 & 0 \\ e^{\frac{j2\pi m}{32}} & 0 \\ e^{\frac{j4\pi m}{32}} & 0 \\ e^{\frac{j6\pi m}{32}} & 0 \\ 0 & 1 \\ 0 & e^{\frac{j2\pi m}{32}} \\ 0 & e^{\frac{j4\pi m}{32}} \\ 0 & e^{\frac{j6\pi m}{32}} \end{bmatrix}}_{W_1} \cdot \underbrace{\begin{bmatrix} 1 \\ e^{\frac{j\pi n}{2}} \end{bmatrix}}_{W_2} \tag{4.191}$$

[17] In general, W_1 captures the spatial covariance matrix of a dual-polarized antenna setup. The DFT structure captures this covariance matrix in an ideal scenario. Dual codebook structures beyond the DFT structure that take into account the true spatial covariance matrix of the antenna array under practical impairments have been explored in [220, 221, 222, 223, 224, 225, 226, 227].

$$\mathscr{C}_8^2(m,m',n) = \begin{bmatrix} 1 & 1 \\ e^{\frac{j2\pi m}{32}} & e^{\frac{j2\pi m'}{32}} \\ e^{\frac{j4\pi m}{32}} & e^{\frac{j4\pi m'}{32}} \\ e^{\frac{j6\pi m}{32}} & e^{\frac{j6\pi m'}{32}} \\ e^{\frac{j\pi n}{2}} & -e^{\frac{j\pi n}{2}} \\ e^{\frac{j\pi n}{2}+\frac{j2\pi m}{32}} & -e^{\frac{j\pi n}{2}+\frac{j2\pi m'}{32}} \\ e^{\frac{j\pi n}{2}+\frac{j4\pi m}{32}} & -e^{\frac{j\pi n}{2}+\frac{j4\pi m'}{32}} \\ e^{\frac{j\pi n}{2}+\frac{j6\pi m}{32}} & -e^{\frac{j\pi n}{2}+\frac{j6\pi m'}{32}} \end{bmatrix} \tag{4.192}$$

$$= \underbrace{\begin{bmatrix} 1 & 0 & 1 & 0 \\ e^{\frac{j2\pi m}{32}} & 0 & e^{\frac{j2\pi m'}{32}} & 0 \\ e^{\frac{j4\pi m}{32}} & 0 & e^{\frac{j4\pi m'}{32}} & 0 \\ e^{\frac{j6\pi m}{32}} & 0 & e^{\frac{j6\pi m'}{32}} & 0 \\ 0 & 1 & 0 & 1 \\ 0 & e^{\frac{j2\pi m}{32}} & 0 & e^{\frac{j2\pi m'}{32}} \\ 0 & e^{\frac{j4\pi m}{32}} & 0 & e^{\frac{j4\pi m'}{32}} \\ 0 & e^{\frac{j6\pi m}{32}} & 0 & e^{\frac{j6\pi m'}{32}} \end{bmatrix}}_{W_1} \cdot \underbrace{\begin{bmatrix} 1 & 0 \\ e^{\frac{j\pi n}{2}} & 0 \\ 0 & 1 \\ 0 & -e^{\frac{j\pi n}{2}} \end{bmatrix}}_{W_2} \tag{4.193}$$

for different choices of m, m' and n.

Note that the first four entries of $\mathscr{C}_8^1(m,n)$ can be written as

$$\begin{bmatrix} 1 \\ e^{\frac{j2\pi m}{32}} \\ e^{\frac{j4\pi m}{32}} \\ e^{\frac{j6\pi m}{32}} \end{bmatrix} = \begin{bmatrix} 1 \\ e^{\frac{j2\pi}{4}\cdot\frac{m}{8}} \\ e^{\frac{j2\pi}{4}\cdot\frac{2m}{8}} \\ e^{\frac{j2\pi}{4}\cdot\frac{3m}{8}} \end{bmatrix} \tag{4.194}$$

and can be interpreted to correspond to DFT beam weights (steered toward $\cos^{-1}(2m/32)^{\circ}$) for a 4×1 array, but oversampled by a factor of 8 (that is, 8 equi-spaced beams occupy the beamwidth of a single beam indexed by m with a 4×1 array) and selected from this oversampled set. The beam patterns corresponding to this set of 32 beam weights are plotted in Figure 4.29. This figure shows that all the beams lead to a peak array gain of $10 \cdot \log_{10}(4) \approx 6$ dB. However, while the beam indices corresponding to $m = 0, 8, 16$ and 24 (in solid line types) are orthogonal to each other and coarsely sample the beamspace, the intermediate values of m lead to beams that finely sample the beamspace. Note that the beam patterns in Figure 4.29 would be similar to what a Butler matrix architecture would produce with variable phase shifts.

The precoder for the first four antenna elements for the two layers in (4.193) are indexed by m and m' with $0 \leq \{m, m'\} \leq 31$. With this approach, m or m' can be from

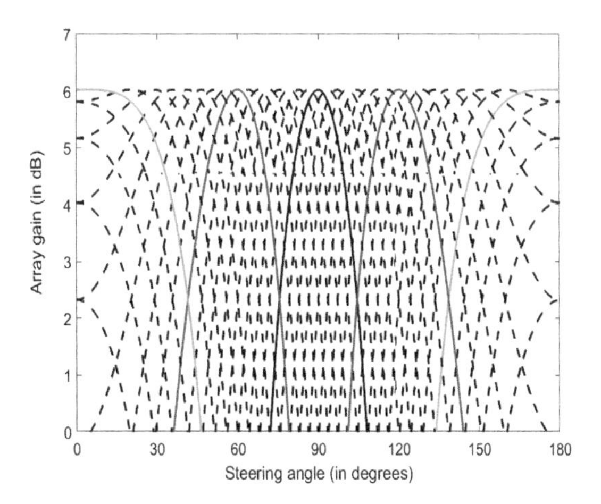

Figure 4.29 Array gain patterns of oversampled DFT beams with $N = 4$ and an oversampling factor of 8. Solid line types indicate orthogonal beams and dashed line types indicate non-orthogonal/oversampled beams.

one of 32 choices for 1 or 2 layers, 16 for 3 or 4 layers, 4 for 5–7 layers and 1 for 8 layers. The $\pm e^{j\pi n/2}$ factor used over the next four antenna elements is a choice from the set $\{1, j, -1, -j\}$ (for $n = 0, 1, 2, 3$) that allows combining of energy from the 8 antenna elements. In this sense, this approach is similar to $\mathscr{C}_2^1$ in (4.185) where the QPSK alphabet is used for co-phasing uncorrelated antennas (as the cross-polarized antenna elements show low correlation). In view of the dual codebook structure, the Rel. 8/9 codebooks for less than 8 RF chains can be seen as a special case of the dual codebook structure with W_1 set to I_N. Nevertheless, in Rel. 12, the dual codebook structure is extended to cover the $N = 4$ RF chains scenario with control signaling determining whether the old (Rel. 8/9) codebook or the dual codebook is used.

In this context, a *Type-I* codebook consists of generalizing the dual codebook structure in the $N = 8$ RF chain case from Rel. 10. Here, a typical rank-1 precoder at the TXRU level parameterized by a triplet (m, ℓ, n) is of the form:

$$\mathscr{C}^1_{2N_1N_2}(m,\ell,n) = \begin{bmatrix} u_m \\ e^{\frac{j2\pi\ell}{O_1N_1}} \cdot u_m \\ \vdots \\ e^{\frac{j2\pi\ell(N_1-1)}{O_1N_1}} \cdot u_m \\ e^{\frac{j\pi n}{2}} \cdot u_m \\ e^{\frac{j\pi n}{2}} \cdot e^{\frac{j2\pi\ell}{O_1N_1}} \cdot u_m \\ \vdots \\ e^{\frac{j\pi n}{2}} \cdot e^{\frac{j2\pi\ell(N_1-1)}{O_1N_1}} \cdot u_m \end{bmatrix}, \; u_m\Big|_{N_2\times 1} = \begin{bmatrix} 1 \\ e^{\frac{j2\pi m}{O_2N_2}} \\ \vdots \\ e^{\frac{j2\pi m(N_2-1)}{O_2N_2}} \end{bmatrix}. \tag{4.195}$$

In (4.195), u_m corresponds to a DFT beam weight vector used over N_2 dimensions (in azimuth) with an oversampling factor of O_2 for a fine-grained grid of beams from which the m-th beam is selected. Another beam that corresponds to DFT over N_1 dimensions with an oversampling factor of O_1 is produced to target the elevation domain and a Kronecker product of these two beams constitutes $\mathscr{C}^1_{2N_1N_2}(m, \ell, n)$.

Given the focus on elevation steering in FD-MIMO, $N_1 \geq N_2$ with $O_1 = 4$ for all choices of N_1 (since $N_1 > 1$) and $O_2 = 4$ if $N_2 > 1$ (and $O_2 = 1$ otherwise). Rel. 13 supports 4, 8, 12 and 16 RF chains, whereas Rel. 14 extends support for 24 and 32 RF chains. In particular, the 4 RF chain case supports $N_1 = 2, N_2 = 1$ (denoted for simplicity as the $(2,1)$ configuration), whereas the 8 RF chain case supports the $(2,2)$ and $(4,1)$ configurations. For the 12 and 16 RF chain cases, $(3,2)$ and $(6,1)$ configurations and $(4,2)$ and $(8,1)$ configurations are supported, respectively. In the 24 and 32 RF chain cases, the following are supported: $(4,3)$, $(6,2)$ and $(12,1)$ configurations and $(4,4)$, $(8,2)$ and $(16,1)$ configurations, respectively. Generalizations of the precoder structure as described above to different number of layers, choices of N_1 and N_2 and multiple panels are described in [202, Sec. 5.2.2.2].

The Type-I codebook is typically meant for SU-MIMO transmissions from a single panel or from multiple panels. However, due to the high fidelity of CSI needed in MU-MIMO transmissions, the Type-I codebook leads to increased inter-user interference. In these scenarios, a *Type-II* codebook is provided in the standard specifications where, mimicking the SVD structure, the precoder is described as a linear combination of L beams. In the 4 port case, the linear combination is limited to two beams, whereas for the case where there are more than 4 ports, the linear combination is limited to four beams. The amplitudes and phases in the linear combination can be sub-band based and the phase can come from a QPSK/8-PSK alphabet. Within the Type-II class, a port selection variant, an enhanced version and an enhanced port selection variant can also be configured. We refer the readers to [202] for more details.

5 System Level Tradeoffs and Deployment Aspects

The focus of this chapter is on core system level performance aspects of mmWave systems. The chapter covers three broad topics:

- Signal quality and management
- Interference quality and management and
- Deployment considerations.

The chapter begins with baseline performance characterization of mmWave systems under idealized assumptions. Due to the more challenging propagation conditions as described in the earlier chapters, it is recognized that such systems will need a greater degree of infrastructure densification. Since such systems are more coverage-limited as opposed to being capacity-limited, the densification strategy can consider a variety of in-band relay options such as different flavors of repeaters. Performance vs. cost tradeoffs for different relay options are outlined along with a brief overview of protocol support needed for operation. It is to be noted that dynamic beamforming and beam tracking on each link are important components of signal quality management in a mobile mmWave system.

The second part of the chapter focuses on interference in mmWave systems. Due to the narrow beamwidth of the beams, average operational interference at typical network densities is low compared with that observed in sub-7 GHz systems. This allows for a high-degree of spatial reuse for mmWave frequencies especially when deployed in dense "hotspot" areas (both indoor and outdoor). If needed, simple interference management schemes may also be employed to overcome the bursty interference that might occur sporadically. Finally, the chapter focuses on network planning for the deployment of mmWave systems. Compared with lower frequency bands, mmWave networks tend to be more sensitive to blockage due to clutter in the deployed environment.

A network planning approach that exploits the significant advances in object detection via machine learning (ML) is outlined. An optimization framework that accounts for specific environmental clutter and the use of multiple nodes of different types is showcased. The chapter concludes with a number of important deployment results ranging from foliage sensitivity, cost benefits of using relays, etc. to the important topic of how sub-7 GHz and mmWave systems can be jointly designed and how they complement each other extremely well.

DOI: 10.1201/9781032703756-5

Table 5.1
System simulation comparisons between sub-7 GHz and mmWave systems

Design parameter	Sub-7 GHz	Millimeter wave
EIRP	57 dBm	59 dBm
Number of ports	64	4
Number of antennas	96	192
Maximum number of layers (SU-MIMO)	4	2
Bandwidth	100 MHz	800 MHz
Subcarrier spacing	30 kHz	120 kHz
Modulation	QPSK to 256-QAM	QPSK to 64-QAM
Maximum spectral efficiency	29.6 bps/Hz	11 bps/Hz
Peak downlink rate	1.9 Gbps	4.4 Gbps

5.1 COVERAGE OF MMWAVE DEPLOYMENTS

5.1.1 BASELINE PERFORMANCE CHARACTERIZATION

To illustrate the baseline mmWave performance, we consider a Manchester, UK geography with the following deployment considerations:

- 54 sub-7 GHz macro cells
- 77 mmWave cells
- 54 sub-7 GHz macro cells + 14 mmWave small cells
- 54 sub-7 GHz macro cells + 38 mmWave small cells
- 54 sub-7 GHz macro cells + 77 mmWave small cells.

The system simulations for sub-7 GHz and mmWave make the assumptions as described in Table 5.1. For the different deployment scenarios, sub-7 GHz macro cells are deployed exclusively on the rooftops of buildings. Millimeter wave small cells are exclusively deployed on the poles at the street level since this location leads to better channel conditions (a higher proportion of LOS links and smaller distances from the base station to the UEs) than if the mmWave small cells are deployed on a rooftop mounting. In this study, we assume a user concentration that is 10 times larger at hotspot areas (relative to non-hotspot areas).

Figure 5.1 plots the CDF of the throughput on the downlink with a SU-MIMO solution constrained to 4 layers in sub-7 GHz and 2 layers in mmWave. We observe the following:

- Relative to sub-7 GHz systems with better propagation conditions, mmWave deployments require more base stations for similar coverage requirements. In the case of 54 sub-7 GHz macros vs. 77 mmWave small cells, we see a median rate increase by a factor of ≈ 5.5 with mmWave taking advantage of the increased

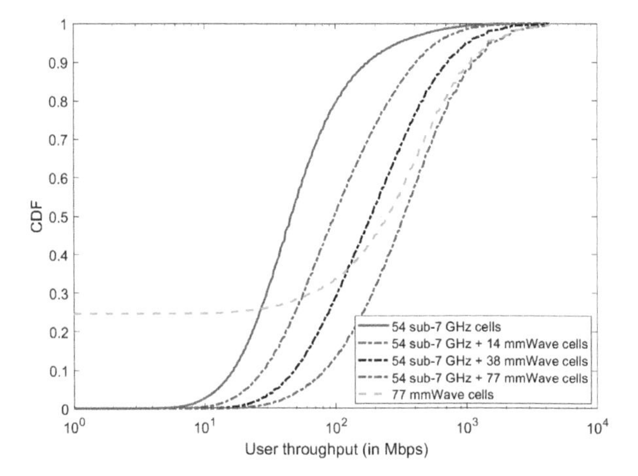

Figure 5.1 Throughput with different sub-7 GHz (rooftop mounted) and mmWave (pole mounted) deployments.

available bandwidth at these frequencies. This observation illustrates the enormous promise of mmWave in capacity building in wireless networks.

- However, with a mmWave-only solution, we observe a severe tail degradation in performance due to fading and blockage. On the other hand, a sub-7 GHz-only solution can lead to improved tail performance.
- Traversing between these two extremes, a non-standalone solution of sub-7 GHz and mmWave with increasing density of mmWave cells leading to a densified network can improve the throughput performance *significantly*. In particular, a joint deployment of 54 sub-7 GHz macros and 77 mmWave small cells leads to a median rate improvement by a factor of $\approx$7.5. The tail behavior in such a deployment also shows improvement. In particular, at the 90-th percentile, we see $\approx$5 times rate increase.

The focus of the rest of the chapter is on different densification options.

From a temporal standpoint, Chapter 2.3.2 showed that the time-scales at which hand/body blockage can disrupt mmWave signals are on the order of a few hundreds of milliseconds. Similarly, the time-scales at which fading leads to the need for beam switching are also on the order of a few tens to a few hundreds of milliseconds depending on the mobility, channel structure, operational bandwidth, etc. These signal deteriorations can be addressed in multiple ways. One possibility is antenna module densification at UE which allows many viable clusters in the channel to be observed at the UE with significant signal strengths. This approach is explored in Chapter 3.2.3 where the tradeoffs between the use of different types of geometrically structured antenna modules is considered. A necessary prerequisite for the viability of multiple clusters at the UE is network densification which leads to a reduction in the inter-site distances (ISDs). This is the subject of the next discussion. Further, taking advantage of multiple clusters requires a better quality of RS scheduling, which is studied in Chapter 5.6.1.

5.1.2 ULTRA DENSIFICATION

The focus of this section is on network densification. As the level of network densification increases, a large fraction of devices are under coverage from multiple base stations, many of which can establish LOS links to the devices. Network densification improves the robustness of coverage and also leads to interference that has to be properly managed. It is common to assume that every base station serves at least one device. Taking this step further leads to a regime of *ultra densification* where the number of base stations on average exceeds the number of active devices over a set of resource blocks of interest. One utility of ultra densification is to overcome dynamic blockage such as hand/body blockage studied in Section 2.2.4, or blockages associated with mobility. The impact of dynamic blockage on performance heavily depends on the local geometry and device orientation with respect to the base stations. Occurrence of dynamic blockage is therefore hard to predict. Network deployment has to be much denser to reduce the likelihood of outage caused by dynamic blockage to an acceptable level. As a result, many base stations are deployed for the purpose of redundancy (or diversity) rather than for capacity purposes and only need to be turned on when dynamic blockage occurs. Base stations that have no devices to serve can be temporarily turned off to avoid generating interference and to yield network energy savings.

Figure 5.2 illustrates the three levels of network densification. In this figure, an oval simplistically represents a directional link/beam when a base station transmits to a device. In a normal density deployment illustrated in Figure 5.2(a), most devices are served by one base station. In a dense deployment illustrated in Figure 5.2(b), many devices are under coverage from multiple base stations. In an ultra dense deployment illustrated in Figure 5.2(c), most (if not all) devices are under coverage from multiple base stations and some base stations have at most one device in their coverage. In general, the SINR increases from a normal density to dense deployment because both signal and interference powers increase proportionally while the noise power remains the same. As the network becomes denser, the SINR can drop because a device is under the coverage of more base stations with LOS links. If the device is served by one of these base stations, the remaining ones serve other devices and can cause severe interference to the device. Thus, when the network grows ultra dense while keeping the device density fixed, many of the network nodes are turned off. Relative to the serving base station, the interfering base stations are in effect pushed farther away. As a result, the SINR increases.

5.1.3 JOINT DENSIFICATION WITH SUB-7 GHZ AND MMWAVE NETWORKS

An important aspect of mmWave deployments is to consider the role of sub-7 GHz networks in the deployments. While many options exist for densifying with mmWave

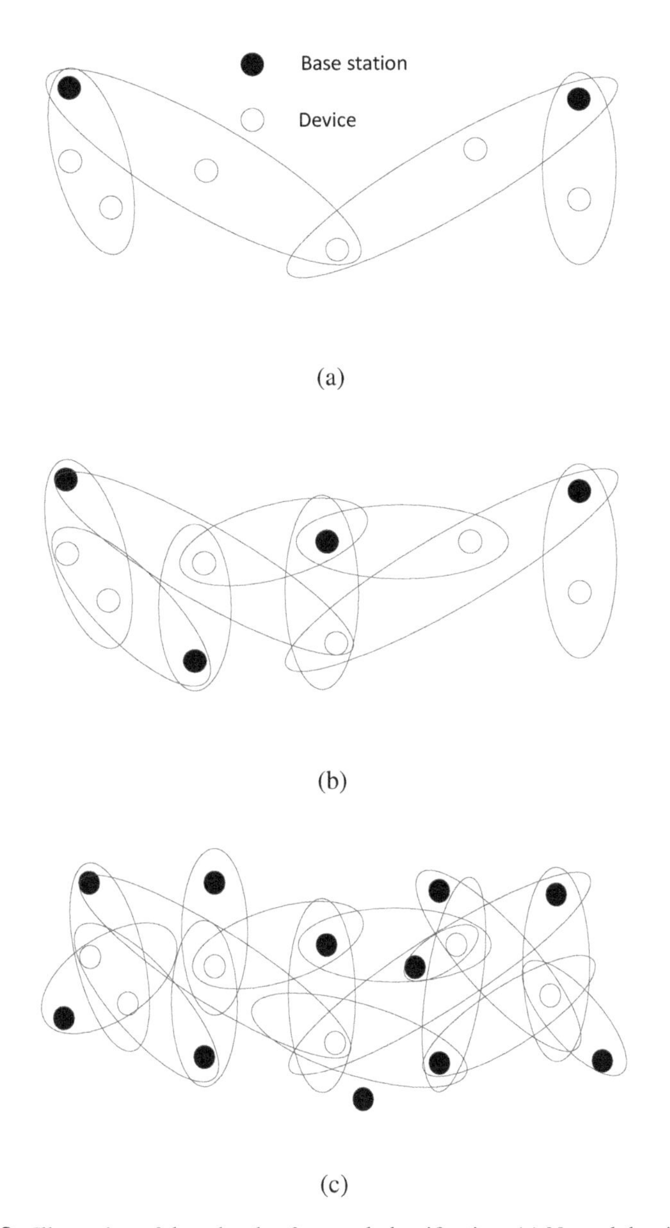

Figure 5.2 Illustration of three levels of network densification: (a) Normal density deployment. (b) Dense deployment. (c) Ultra dense deployment. Each ellipse corresponds to a directional link between the two nodes.

alone, in practice, deployments could benefit from *jointly* considering sub-7 GHz and mmWave band options. The sub-7 GHz deployments could be existing deployments being leveraged, new deployments, or a mixture of the two. Considering the sub-7 GHz deployment while planning the mmWave deployment can alleviate some of the burden on deploying mmWave nodes and allow the deployment to focus on regions where mmWave frequencies can provide the maximum capacity benefit. The two networks, working in concert with each other, can provide the most optimal deployment. Here are some considerations that make this option attractive.

- Sub-7 GHz coverage can be good and thereby lighten the burden on mmWave coverage requirements, especially in regions of deployment which are not capacity-constrained.
- Most mmWave bands of interest have 4–8 times the bandwidth over that available at typical sub-7 GHz bands. As will be shown later in this chapter, higher-order MU-MIMO as a multiplexing scheme at sub-7 GHz, while feasible, cannot fully counter the bandwidth advantage of mmWave frequencies. This is because it requires sufficient orthogonalization across the multiplexed users (that is, channel decorrelation), power splitting on the downlink, and increased interference on the uplink.
- For practical physical aperture sizes, mmWave beams have relatively narrower beamwidths than beams at sub-7 GHz frequencies. In combination with the poorer propagation, this provides significant inter-cell interference resilience at mmWave frequencies. Therefore, mmWave systems end up being a far better option for densifying in hotspot areas.

In a later section, these aspects of sub-7 GHz and mmWave joint designs for deployment will be described in detail along with numerical studies.

5.2 NETWORK DENSIFICATION OPTIONS

It is important to consider cost-effective ways to achieve densification of mmWave networks. The main challenges in network densification are site acquisition and fiber deployment costs. If fiber deployment costs can be reduced via relay nodes or eliminated, site acquisition can be made cheaper and hence easier. This is especially the case where multiple site candidates can be found. Therefore, using wireless relay nodes to extend coverage for network densification is attractive for 5G networks.

In recent years, 3GPP has defined different types of relay nodes for 5G networks. Some of these include:

- Integrated access and backhauling (IAB) feature using Layer 2-based[1] decode-and-forward relay nodes has been introduced in Rel. 16 [229]. Further enhancements for IAB have been added in Rel. 17 and 18. The standard specifications

[1]A Layer 1, Layer 2 or Layer 3 control signal originates from the physical, MAC or network layers, respectively.

seamlessly support IAB operation where the backhaul and access links are in the same band or in different bands.

In Rel. 19, wireless access and backhaul (WAB) systems are expected to be standardized. Here, the backhaul is via a full-stack UE instead of a Layer 2-based node as in IAB. The advantage of such an approach is to enable the reuse of the existing implementations of UEs and small cells so that cascading them with a simple interface would enable WAB. Note that the simplest WAB can reuse the existing specifications before Rel. 18, even as further optimization is being pursued in Rel. 19.

- Conventional or simple repeaters which simply amplify-and-forward any signal that they receive have been studied at 3GPP for Rel. 17 with the corresponding RF requirements specified in [230] targeting both sub-7 GHz and mmWave frequencies.
- Smart repeaters (also called network-controlled repeaters) that perform amplify-and-forward operation with the capability to receive and process side control information from the network have been studied in Rel. 18 [231]. The side control information can include beam information for the access link, on-off information, and TDD downlink or uplink configuration information.
- The UE-to-Network Layer 2/Layer 3-based relay via sidelink has been introduced in Rel. 17 [232] for proximity-based services (ProSe) in 5G systems.

We will now outline the performance and complexity tradeoffs of the different types of relays stated above (except for the sidelink relay which is not treated here). A brief comparison across these three relay options in terms of their objectives, beam management functionality, and control and user plane latencies is presented in Table 5.2.

These different types of repeater/relay nodes have different implementation complexities and performance tradeoffs. A number of studies on relay nodes can be found in the literature. See, for example, [233, 234, 235, 236, 237, 238] where theoretical performance and bounds on amplify-and-forward and decode-and-forward relay operations have been studied. However, in almost all of these studies, practical constraints for amplify-and-forward operation such as the constraint on the maximum amplification gain for stability and noise figure differences between amplify-and-forward and decode-and-forward operation have not been captured. In addition, the performance differences between conventional and smart amplify-and-forward repeaters as well as the impact of different levels of side control information on the smart repeater's performance have not been studied. For decode-and-forward relay nodes which can function with half- or full-duplex capabilities, the impact of spatial reuse in the case of multi-user scheduling has also not been studied. In this chapter, we present an in-depth analysis of performance tradeoffs for different types of repeater/relay nodes with several practical constraints modeled.

5.2.1 IAB VIA DECODE-AND-FORWARD RELAYING

Backhaul and access can operate on different frequency bands, which is called as *out-band* operation. Example scenarios of out-band include a sub-7 GHz access and

Table 5.2
Comparisons across different relay options

	Simple repeaters	Smart repeaters	IAB
Objective	Coverage extension	Coverage extension Interference reduction	Coverage extension Interference reduction
Control plane protocol latency	Not applicable	L1 Downlink Control Information (DCI)-based (1 slot) L2 Medium Access Control channel-Control Element (MAC-CE)-based (few ms)	RRC-based (> 10 ms) L1 DCI-based (1 slot) L2 MAC-CE-based (few ms)
Beam management	Fixed beams for TX and RX	Beams designed for interference reduction and coverage extension	
User plane operational mode	Amplify-and-forward		Decode-and-forward
User plane latency	Smaller than a cyclic prefix. < 10 ns for one hop since only RF/IF forwarding is possible.	RF/IF forwarding 10s of ns for one hop. Additional delay of 1 slot for beam synthesis and control	A few slots for decoding, re-encoding and forwarding. Scheduling delays may be incurred in forwarding

mmWave frequency-based backhaul. Static partitioning of the resources guarantees minimal cross-link interference between access and backhaul. On the other hand, backhaul and access can share the same frequency bands (called as *in-band* operation) and the resource can be dynamically managed according to local traffic and channel conditions. Such flexibility can improve resource utilization. In addition, because of elevated heights of base stations, directional beams of mmWave access and backhaul are usually directed at different elevation angles, making it possible to spatially multiplex access and backhaul simultaneously over the same frequency. In either out-band or in-band operations, it is desired to share a set of common design principles between access and backhaul to leverage technology, standardization and product development. This approach is called *Integrated Access and Backhaul* (IAB).

There are many aspects of IAB defined in 3GPP covering physical, medium-access, and upper layer protocols, resource management between backhaul and access links, and network architecture. Covering all these aspects of IAB in detail is beyond the scope of this book and the reader is referred to 3GPP standards and other papers for this purpose [229]. The goal here is to outline a simple method for end-to-end performance characterization of an IAB link, to enable contrasting it against other alternatives such as amplify-and-forward repeaters.

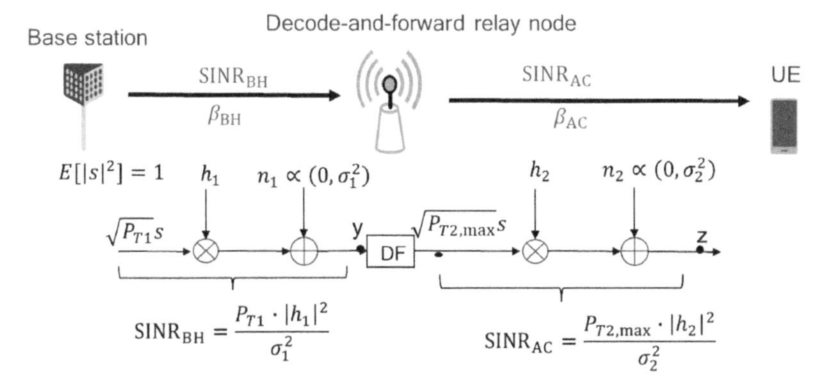

Figure 5.3 System model for a two hop IAB connection.

The system model for a decode-and-forward relay node with downlink operation is shown in Figure 5.3 and is described below:

$$y = h_1 \sqrt{P_{T1}} s + n_1 \tag{5.1}$$
$$z = h_2 \sqrt{P_{T2,\text{max}}} s + n_2. \tag{5.2}$$

We use the following notations in this study:

- s: Signal with unit power
- P_{T1}: Transmit power of the base station
- $P_{T2,\text{max}}$: Maximum transmit power of relay node
- h_1 and h_2: Post-beamformed channel states for backhaul and access links, respectively
- n_1 and n_2: Interference and noise with Gaussian distribution (zero mean and variances of σ_1^2 and σ_2^2 for backhaul and access links, respectively)
- SINR_{BH} and SINR_{AC}: SINRs for backhaul and access links, respectively
- β_{BH} and β_{AC}: Fraction of time-domain resources allocated for backhaul and access links, respectively.

The backhaul and access SINRs can be calculated as follows:

$$\text{SINR}_{\text{BH}} = \frac{P_{T1}|h_1|^2}{\sigma_1^2}, \quad \text{SINR}_{\text{AC}} = \frac{P_{T2,\text{max}}|h_2|^2}{\sigma_2^2}. \tag{5.3}$$

The decode-and-forward relay node can operate based on one of the following two modes:

- *Full-duplex decode-and-forward mode* (*FDDF*): In this mode, backhaul and access links can operate at the same time with $0 < \{\beta_{\text{BH}}, \beta_{\text{AC}}\} \leq 1$. Let $C(\text{SINR})$ denote the rate achieved at an operating point of SINR. It can be shown that the maximum achievable rate is [234]:

$$C(\text{SINR}_{\text{FDDF}}) = \min\Big(C(\text{SINR}_{\text{BH}}),\, C(\text{SINR}_{\text{AC}})\Big). \tag{5.4}$$

The corresponding end-to-end effective SINR, denoted as $\mathsf{SINR}_{\mathsf{FDDF}}$, can be obtained by using the inverse rate function.

- *Half-duplex decode-and-forward mode* (*HDDF*): In this mode, backhaul and access links are time-multiplexed with $\beta_{\mathsf{BH}} + \beta_{\mathsf{AC}} \leq 1$. It can be shown that the optimum resource allocation is achieved when $\frac{\beta_{\mathsf{BH}}}{\beta_{\mathsf{AC}}} = \frac{C(\mathsf{SINR}_{\mathsf{AC}})}{C(\mathsf{SINR}_{\mathsf{BH}})}$ and the resulting achievable rate is half of the harmonic mean of $C(\mathsf{SINR}_{\mathsf{BH}})$ and $C(\mathsf{SINR}_{\mathsf{AC}})$. That is,

$$C(\mathsf{SINR}_{\mathsf{HDDF}}) = \frac{1}{\frac{1}{C(\mathsf{SINR}_{\mathsf{BH}})} + \frac{1}{C(\mathsf{SINR}_{\mathsf{AC}})}} \tag{5.5}$$

$$= \frac{1}{2} \cdot \text{Harmonic mean}\Big(C(\mathsf{SINR}_{\mathsf{BH}}),\ C(\mathsf{SINR}_{\mathsf{AC}})\Big). \tag{5.6}$$

The harmonic mean can be written as:

$$C(\mathsf{SINR}_{\mathsf{HDDF}}) \approx \alpha \cdot C(\mathsf{SINR}_{\mathsf{FDDF}}) \tag{5.7}$$

for some α satisfying $1/2 \leq \alpha \leq 1$. The approximation in (5.7) is based on the property of harmonic means:

$$\alpha \approx \begin{cases} \frac{1}{2} & \text{if } \frac{C(\mathsf{SINR}_{\mathsf{BH}})}{C(\mathsf{SINR}_{\mathsf{AC}})} \approx 1 \\ 1 & \text{if } \frac{C(\mathsf{SINR}_{\mathsf{BH}})}{C(\mathsf{SINR}_{\mathsf{AC}})} \gg 1 \text{ or } \frac{C(\mathsf{SINR}_{\mathsf{BH}})}{C(\mathsf{SINR}_{\mathsf{AC}})} \ll 1. \end{cases} \tag{5.8}$$

5.2.2 SIMPLE AND SMART REPEATERS VIA AMPLIFY-AND-FORWARD RELAYING

Simple RF repeaters are well-known and have been used extensively for filling coverage holes in prior generations of wireless systems [239, 240]. There are two main aspects worth exploring when it comes to applying RF repeaters at mmWave frequencies. First, mmWave systems are expected to mostly be TDD-based systems and RF repeaters for TDD systems have not been studied extensively in the past. Knowledge of the TDD structure (or a lack thereof) can impact the extent of stable amplification gain used at the repeater. Second, the presence or absence of UE-specific beamforming can make a significant difference to the quality of links when an RF repeater is used. These aspects of amplify-and-forward repeaters as applied to mmWave systems are outlined next.

For the amplify-and-forward repeater, a unified system model is established for both conventional and smart repeaters with downlink operation, as shown in Figure 5.4. Here, the amplify-and-forward repeater is characterized by an amplification gain G and the following parameters:

- G_{max} is the maximum allowed amplification gain taking stability constraints into account. The amplify-and-forward repeater adjusts the amplification gain G to achieve a target output transmission power $P_{T2,\mathsf{max}}$ unless it is limited by G_{max}.

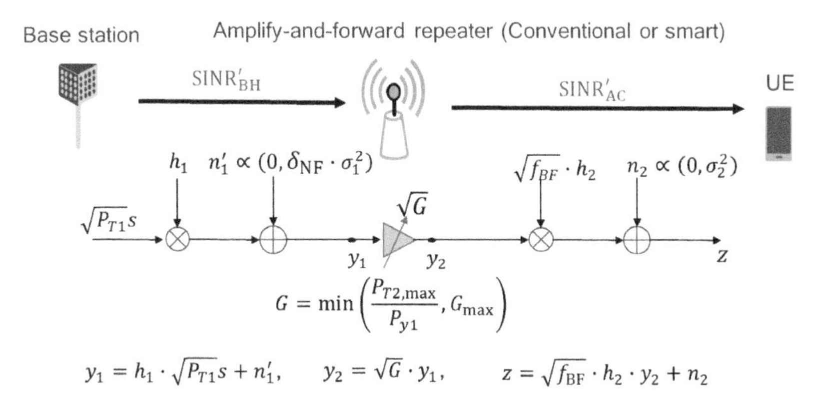

Figure 5.4 System model for the downlink of amplify-and-forward repeaters.

- δ_{NF} is the noise figure of an amplify-and-forward node relative to a decode-and-forward relay node and captures the noise figure differences between the RF chains of amplify-and-forward and decode-and-forward nodes, if any. Further, this parameter would depend on the actual implementation of both nodes.
- $f_{BF} \leq 1$ is the beamforming gain with respect to the maximum array gain over an access link. Note that $f_{BF} = 1$ means no beamforming loss and a larger f_{BF} is associated with smaller loss. It is expected that when an access link is used with a small cell or an IAB node, the maximum array gain can be achieved by UE-specific beamforming. For amplify-and-forward repeaters, especially for the simple repeater, UE-specific beamforming may not be feasible since the scheduling decisions are not known at the repeater. Therefore, some kind of compromise is needed on beamforming so as to accommodate an area of users. This factor is intended to capture the loss from such a compromise. This allows us to work with the ideal SINR and scale it as needed.

For a conventional repeater, without knowledge of beam steering direction toward the UE, a fixed broad beamwidth beam is typically used for the access link to cover all possible directions and thus the beamforming loss is $f_{BF} < 1$. On the other hand, for a smart repeater with knowledge of the UE's direction, a narrow beamwidth beam can be formed toward the UE's direction without any beamforming loss (that is, $f_{BF} = 1$). Note that this beamforming loss factor is only for the access link. For a backhaul link, it is assumed that the relay node is stationary and a narrow beamwidth beam can always be formed over the backhaul link during deployment. Thus, no beamforming loss over a backhaul link is assumed.

It can be shown that the final received signal z at the UE side can be represented as:

$$z = \sqrt{f_{BF}} h_2 \sqrt{G} \cdot \left(h_1 \sqrt{P_{T1}} s + n_1'\right) + n_2. \tag{5.9}$$

In order to readily compare the performance of amplify-and-forward repeater with a decode-and-forward relay node, the above system model is normalized by the gain and re-stated as:

$$\widetilde{z} = s + \frac{1}{\sqrt{\mathsf{SINR}_{\mathsf{BH}}/\delta_{\mathsf{NF}}}}\widetilde{n}_1 + \frac{1}{\sqrt{f_{\mathsf{BF}} f_{\mathsf{P}} f_n \mathsf{SINR}_{\mathsf{AC}}}}\widetilde{n}_2 \tag{5.10}$$

where $\widetilde{n}_1$ and $\widetilde{n}_2$ are normalized interference and noise quantities with unit variance. Also, $\mathsf{SINR}_{\mathsf{BH}}$ and $\mathsf{SINR}_{\mathsf{AC}}$ are the SINRs of the backhaul and access links for the decode-and-forward relay node, respectively. Here, parameters f_{P} and f_n are explained below:

- $f_{\mathsf{P}} \leq 1$ is the power loss factor resulting from having a finite gain in the repeater that sometimes prevents realizing the target transmission power $P_{T2,\mathsf{max}}$. The power loss factor depends on the received power P_{y1} at the repeater node

$$P_{y1} = \sigma_1^2 \cdot (\mathsf{SINR}_{\mathsf{BH}} + \delta_{\mathsf{NF}}). \tag{5.11}$$

If $P_{y1} \geq \frac{P_{T2,\mathsf{max}}}{G_{\mathsf{max}}}$, there is no loss and $f_{\mathsf{P}} = 1$. Otherwise, $f_{\mathsf{P}} < 1$. Therefore, we have

$$f_{\mathsf{P}} = \min\left(1, \frac{P_{y1} G_{\mathsf{max}}}{P_{T2,\mathsf{max}}}\right) \tag{5.12}$$

$$= \min\left(1, \frac{\sigma_1^2 \cdot (\mathsf{SINR}_{\mathsf{BH}} + \delta_{\mathsf{NF}}) G_{\mathsf{max}}}{P_{T2,\mathsf{max}}}\right). \tag{5.13}$$

- $f_n \leq 1$ is the noise forwarding loss factor. Note that the transmission power of the amplify-and-forward repeater includes both the signal and noise parts. This loss factor captures the ratio of signal power to the total transmission power of the relay node and represents the proportion of the repeater's output power that is carrying the signal. That is,

$$f_n = \frac{|h_1|^2 P_{T1}}{P_{y1}} = \frac{\mathsf{SINR}_{\mathsf{BH}}}{\mathsf{SINR}_{\mathsf{BH}} + \delta_{\mathsf{NF}}} \leq 1. \tag{5.14}$$

Define $\mathsf{SINR}_{\mathsf{BH}}^{'} = \frac{\mathsf{SINR}_{\mathsf{BH}}}{\delta_{\mathsf{NF}}}$ and $\mathsf{SINR}_{\mathsf{AC}}^{'} = f_{\mathsf{BF}} f_{\mathsf{P}} f_n \mathsf{SINR}_{\mathsf{AC}}$. It can be shown that the end-to-end effective downlink SINR for the amplify-and-forward node is half of the harmonic mean of $\mathsf{SINR}_{\mathsf{BH}}^{'}$ and $\mathsf{SINR}_{\mathsf{AC}}^{'}$:

$$\mathsf{SINR}_{\mathsf{AF,DL}} = \frac{1}{\frac{1}{\mathsf{SINR}_{\mathsf{BH}}^{'}} + \frac{1}{\mathsf{SINR}_{\mathsf{AC}}^{'}}} \tag{5.15}$$

$$\approx \alpha^{'} \cdot \min\left(\frac{\mathsf{SINR}_{\mathsf{BH}}}{\delta_{\mathsf{NF}}}, f_{\mathsf{BF}} f_{\mathsf{P}} f_n \mathsf{SINR}_{\mathsf{AC}}\right) \tag{5.16}$$

where $1/2 \leq \alpha^{'} \leq 1$ with the approximation justified as in (5.7). The resulting achievable rate is given as $C(\mathsf{SINR}_{\mathsf{AF,DL}})$.

Compared with a conventional amplify-and-forward repeater, a smart amplify-and-forward repeater has side information provided by the network to improve the

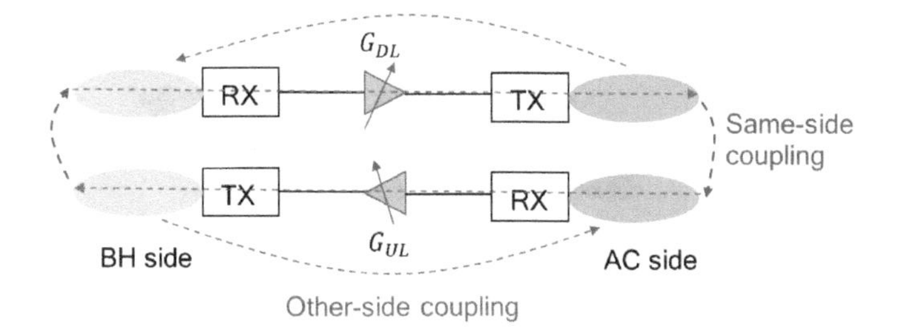

Figure 5.5 Signal coupling in amplify-and-forward repeaters can limit the stable gain of operation.

performance over the conventional repeater. In this section, we focus on the impact of two types of side control information for the smart repeater: TDD downlink or uplink configuration information and the scheduled beam's properties or information associated with the scheduled UE for the access link (discussed previously). For a conventional repeater, due to lack of knowledge of the TDD downlink/uplink configuration of the system, it has to turn on two RF chains for both the downlink and uplink directions in every slot, as illustrated in Figure 5.5, regardless of whether a slot is a downlink slot or an uplink slot. With both the RF chains on, the transmitted signal will loop back into (or interfere with) to the receiver side via a same-side coupling between the two RF chains (the red loop) and the other-side coupling within each RF chain (the blue loop), as illustrated in Figure 5.5. This can lead to unstable oscillations if the amplification gain exceeds a maximum gain limit $G_{\max}$. The gain limit $G_{\max}$ depends on the coupling matrix between the transmitter and the receiver. Note that the same-side coupling with aligned beam directions is much stronger than the other-side coupling without aligned beam directions.

If the TDD downlink or uplink configuration is known to a smart repeater, the smart repeater only needs to turn on one RF chain depending on whether the slot is a downlink slot or an uplink slot. In this case, there is only other-side antenna coupling, which is generally weaker than the same-side antenna coupling. Therefore, a smart repeater with knowledge of TDD downlink or uplink configuration can operate at a much higher maximum stable gain than a conventional repeater. If the network were to operate in a static TDD configuration, then the repeater can be provided that knowledge at configuration. This would still require the repeater to maintain synchronization to the donor (that is, the source) base station.

5.2.3 COMPARISONS VIA NUMERICAL STUDIES

In the numerical studies, we consider a deployment of base stations and relay nodes over a Manhattan grid (of streets and avenues) as pictorially illustrated in Figure 5.6. Here, the base stations are placed at the intersections along every even street and the relay nodes are placed at the intersections along every odd street. Each base station has four sectors covering the east, west, north and south directions. Each relay node

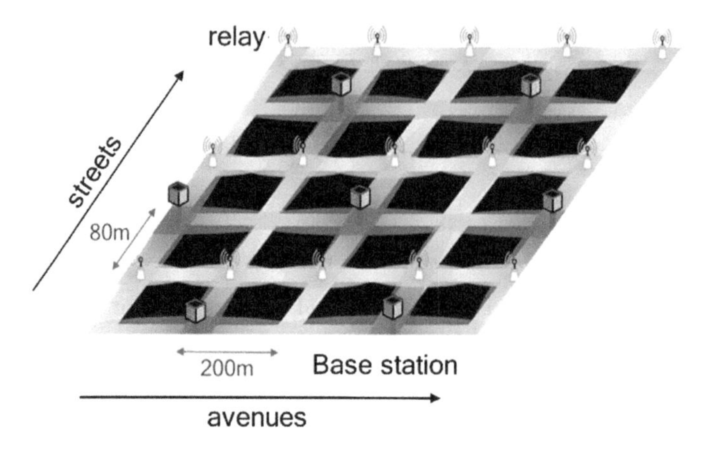

Figure 5.6 Simulation topology for comparison of different relay node-types.

connects to one base station from which it receives the largest power via one of its two backhaul sectors, pointing to the north and south directions. It then provides service to the UEs along the adjacent street via its two access sectors pointing to the east and west directions. It can be observed that if the relay nodes are not deployed, there will be no coverage at odd streets except in the areas close to the intersections.

We consider a simulation area of 2000×2000 square meters with a total of 84 base stations and 156 relay nodes. In this setup, 840 UEs (or an average of 10 UEs per base station) are randomly dropped outdoor and along the streets and avenues. For a UE, the serving node (a base station or a relay node) is determined as a node from which the received power at the UE is the largest. If the serving node is a base station, the UE is called a *direct UE* of that base station. On the other hand, if the serving node is a relay node which is connected to a base station via a backhaul link, the UE is called an *indirect UE* of that base station. Detailed simulation parameters are shown in Table 5.3 for this study. It is assumed that there are buildings along the sides of streets and avenues and the signals are diffracted by the building modeled using the single knife-edge diffraction model [27].

In the results shown here, it is assumed that $G_{\text{max}} = 50$ dB for the conventional repeater and $G_{\text{max}} = 70$ dB for the smart repeater. Another important side control information is the scheduled beam (or equivalently, cluster direction) for the access link. For a conventional repeater, due to lack of scheduled beam information, a fixed broad beamwidth beam is used on the access link to cover all the potential beam directions with a smaller array gain (that is, $f_{\text{BF}} < 1$). But for a smart repeater, if the scheduled beam information can be provided by the network, the smart repeater can form a narrow beamwidth beam toward the scheduled UE with the maximum antenna array gain (that is, $f_{\text{BF}} = 1$). Note that the TDD downlink or uplink configuration can be semi-static which does not change over the time-scale of scheduling slots. On the other hand, the scheduled beam information can be dynamic and can change over the

Table 5.3
Simulation assumptions

Topology	Manhattan grid with 84 base stations and 156 relay nodes Inter-(avenue, street) distance is $(200, 80)$ m (Avenue, street) width is $(14, 8)$ m
Antenna array	Base station is 16×4 per sector Relay is 4×1 per backhaul sector Relay is 16×4 per access sector UE is 2×1
Frequency planning	$f_c = 28$ GHz and bandwidth $= 800$ MHz
Large-scale channel parameters	Backhaul: PLE of 2.0 if $d < 200$ m and 3.2 otherwise Access: PLE of 2.0 if $d < 30$ m and 3.2 otherwise Shadow fading of 8 dB (access) and 4 dB (backhaul) Knife-edge diffraction model from [27] PA power = 7 dBm
Relay parameters	$G_{max} = 50$ dB (repeater), $G_{max} = 70$ dB (smart repeater), $\delta_{NF} = 1$ dB
Beamforming	Fixed broad beamwidth beam in access for conventional repeaters DFT beam for azimuth and elevation in other cases
Link association	One hop based on received power
Scheduler	Round-robin Spatial reuse between direct and indirect UEs for HDDF relay scheme
Inter-cell interference	Based on azimuth beam pattern and fixed elevation gain No interference between access and backhaul links

time-scale of scheduling slots and has a larger control signaling overhead. In order to understand how much performance improvement can be achieved with different levels of side information, the simulation results presented consider two types of smart repeaters: semi-smart repeater with only TDD downlink or uplink configuration information and smart repeater with both TDD downlink or uplink configuration information and scheduled beam information. Compared with an amplify-and-forward repeater that has only RF front end components, a decode-and-forward relay node has a relatively larger implementation complexity and latency because it needs additional digital components to decode the received packet and then transmit it on the next hop. It could also entail additional resource management complexity across multiple hops especially in a half-duplex setting.

It is assumed in our study that the elevation pattern is common to the two types of repeaters and the maximum array gain over azimuth can be achieved with the use of 16 antennas (that is, $10\log_{10}(16) = 12$ dB). Along the descriptions of Chapter 4.2.2, a static broad beamwidth beam pattern that balances azimuthal coverage with peak gain is designed for the conventional (simple) repeater and the semi-smart repeater. The beam pattern has a peak gain of 4 dB and therefore will have a beamforming loss factor of $f_{BF} \leq -8$ dB relative to a smart repeater and a decode-and-forward relay node with the maximum array gain.

Another concept to experiment within the multi-hop decode-and-forward framework is "spatial reuse." When spatial reuse is allowed, the donor base station (or the parent cell in IAB terminology) can schedule a transmission to another UE or relay node when one of its child nodes is forwarding a packet to a different UE. Such spatial reuse can improve the resource utilization of the network at the expense of additional interference being created. However, with the narrow beamwidth beamforming of mmWave systems that generally limits interference, spatial reuse can be advantageous. This is seen in some of the results presented in this section.

In Figures 5.7(a, b), the CDFs of end-to-end effective SINR of indirect UEs and direct UEs over different simulation scenarios are presented. In the simulation, inter-cell interference has been explicitly modeled based on the angle of arrival/departure and beam patterns. With all types of relay nodes, the SINR performance is improved compared to the case without relay nodes. In Figure 5.8(a), CDFs of achieved rates over scheduled slots (spectral efficiency) for all indirect UEs are shown for different simulation scenarios. It can be seen that:

- *Achievable rates of indirect UEs follow the ordering*: Conventional repeater < Semi-smart repeater < Smart repeater < Full-duplex relay.
- There is a cross-over point between the performance of smart repeater and half-duplex relay (not so prominent) and semi-smart repeater and half-duplex relay (very prominent). In the amplify-and-forward case, the harmonic mean of $\mathsf{SINR}_{\mathsf{BH}}$ and $\mathsf{SINR}_{\mathsf{AC}}$ is taken before the capacity function $\log_2(1+x)$. In the decode-and-forward case with half-duplex operation, it is the other way around. It can be seen that if δ_{NF}, f_{BF}, f_{P} and f_n are all equal to 1, then the amplify-and-forward scheme always yields a strictly better spectral efficiency than the decode-and-forward scheme with half-duplex operation. In practice, the backhaul is noisy (that is, $f_n < 1$). Thus, there can be a cross-over between the two cases. However, since the backhaul transmit power is much higher, the SINR is dominated by $\mathsf{SINR}_{\mathsf{AC}}$. In most typical cases, such a cross-over might happen early as a function of the effective SINR or rate and can be difficult to notice.
- The achievable rates of indirect UEs for cases of full-duplex relay operation with and without spatial reuse are almost the same, with the latter being slightly better (due to less interference). Similar observations hold for half-duplex relay operation with and without spatial reuse.

In Figure 5.8(b), CDFs of sector throughputs including both direct and indirect UEs are shown for all simulation scenarios. It can be seen that:

- *Sector throughputs follow the ordering*: Conventional repeater < Semi-smart repeater < Smart repeater < Full-duplex relay with no spatial reuse < Full-duplex relay with spatial reuse.
- We can compare full-duplex relay operation with and without spatial reuse. In contrast with Figure 5.8(a) which shows that the latter has a slightly higher achievable rate for indirect UEs, Figure 5.8(b) shows that the former has a significantly higher sector throughput, due to improvement in direct UE throughputs. Similar observations hold for half-duplex relay operation with and without spatial reuse.

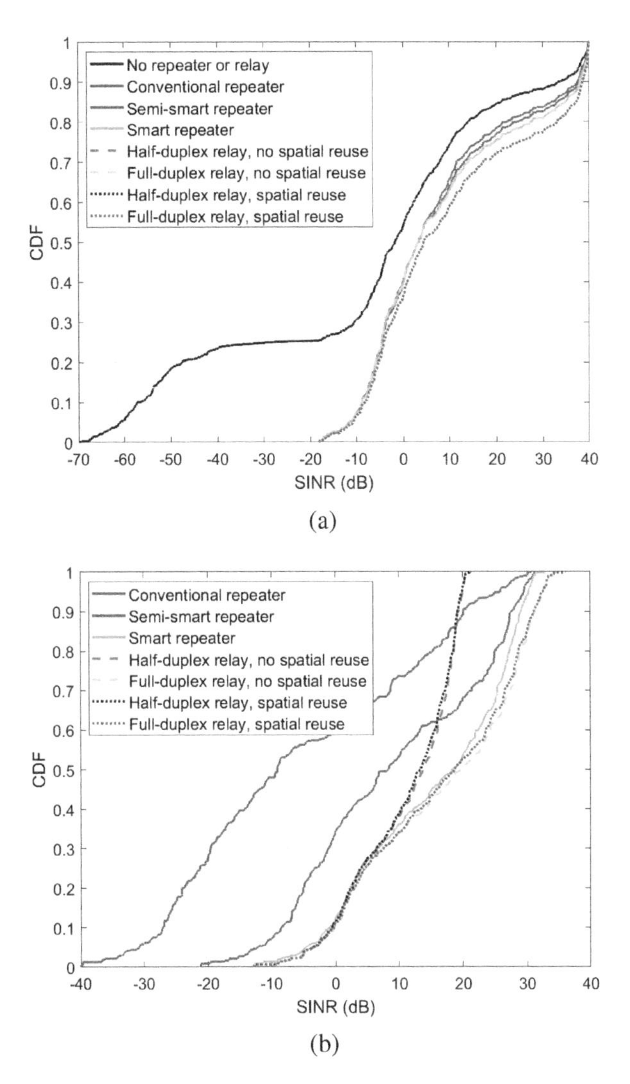

Figure 5.7 SINR CDFs for all UEs under different relaying methods: (a) Indirect UEs and (b) direct UEs.

- The half-duplex relay operation with no spatial reuse performs worse than a smart repeater, but there is a cross-over point between the performance with spatial reuse and a smart repeater. In the case of spatial reuse, the half-duplex penalty for decode-and-forward relay nodes can be compensated partially.

To summarize, relays of various types can all provide good performance and coverage gains in mmWave systems. The simple or conventional repeater with its limited

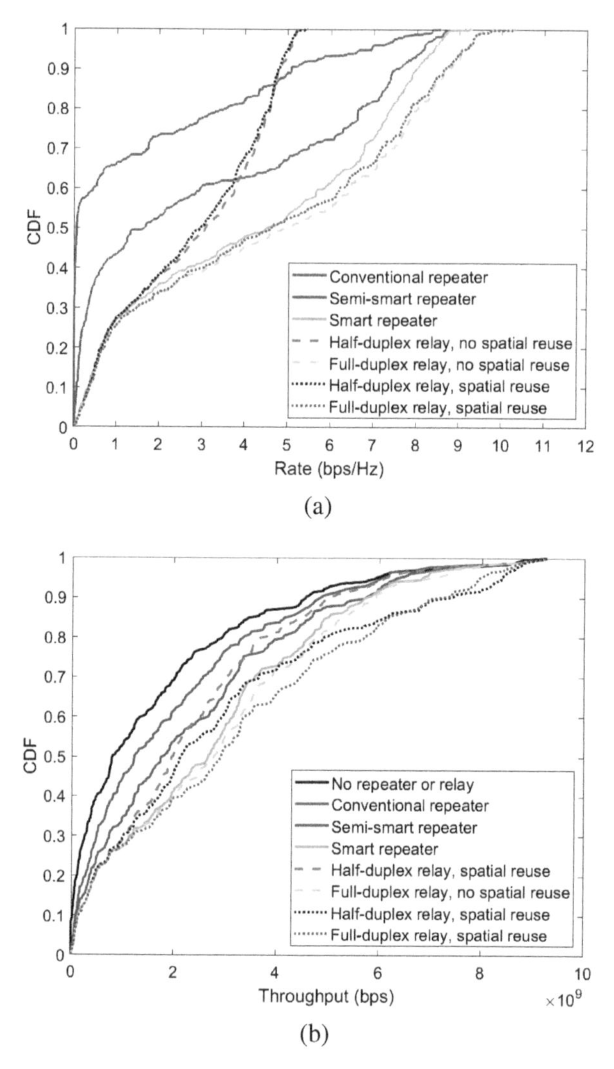

Figure 5.8 (a) Rate CDFs for relay-connected UEs under different relaying methods. (b) Sector throughput CDFs under different relaying methods.

amplification and absence of per-UE beamforming provides a marginal improvement in outage. However, such repeaters can still provide value in extending coverage into relatively uncovered areas. The simple repeaters do not require any dynamic signaling from the donor base stations. On the other hand, the semi-smart repeater learns or is provided the TDD frame structure and is expected to track the system timing. This capability allows the semi-smart repeater to only turn amplification on for either the downlink or the uplink slot (whichever is active). This feature allows the repeater to operate at a higher stable amplification and further extend coverage compared with the simple repeater. Note that just as in the case of a simple repeater, the semi-smart repeater will not implement per-UE beamforming.

In contrast to the semi-smart repeater, the smart repeater is given TDD knowledge and dynamic information about the scheduled UE and the associated beam to use. Therefore, the smart repeater can (in many instances) achieve the same RSRP as a decode-and-forward repeater. The need for dynamic signaling between the donor base station and the smart repeater implies additional complexity of operations compared with either the simple repeater or the semi-smart repeater. Finally, the decode-and-forward repeater has the best performance and the highest complexity among all the options considered. Unlike the other relay types considered which only amplify the incoming signal in the data path, the decode-and-forward relay performs a full decoding and re-encoding prior to transmissions. When the reception and transmission are done in-band, care needs to be taken to avoid self-interference that can severely degrade the incoming signal. A half-duplex version of decode-and-forward accomplishes this objective by simply avoiding transmission and reception at the same time. Of course, this comes at a steep penalty in resource utilization efficiency. Full-duplex decode-and-forward is the best performing relay, but in addition to the aforementioned signaling and resource coordination support, care needs to be taken to provide sufficient isolation or cancellation of the self-interfering signals.

These relay nodes play a powerful role in expanding mmWave coverage and performance. It is important to stitch them into the network topology in a cost-optimal manner toward realizing a certain performance objective, and this aspect is covered in detail when deployment considerations are studied.

5.3 INTERFERENCE MANAGEMENT

The coverage analysis in Chapter 5.2.3 shows that SINR is high for most devices leading to high data rates given the large bandwidths available at mmWave frequencies. Further, most of the devices are neither in outage nor where the SINR is dominated by SNR. This fraction depends on the network density. Network densification is required for robust coverage via macro-diversity. On the other hand, a large number of base stations deployed in a small area may lead to excessive interference where SINR is dominated by interference (that is, signal-to-interference ratio or SIR).

We study the distribution of SNR and SINR with the Manchester, UK deployment considered in Chapter 5.1.1 assuming 54 sub-7 GHz macros and 77 mmWave small cells. Figures 5.9(a, b) plot these distributions in the mmWave and sub-7 GHz networks, respectively. As can be seen from Figure 5.9(a), the SNR and SINR distributions are close for mmWave, unlike the case of sub-7 GHz as illustrated in Figure 5.9(b). In fact, with all the interfering signals considered, the median gap between SNR and SINR is less than 2 dB. Furthermore, by removing the dominant interfering signal, the SINR can be made to approach SNR. In contrast, in the sub-7 GHz network, there is an ≈ 30 dB median gap between SNR and SINR. Removing even the top *six* dominant interferers in this case still leaves a median gap of ≈ 10 dB between

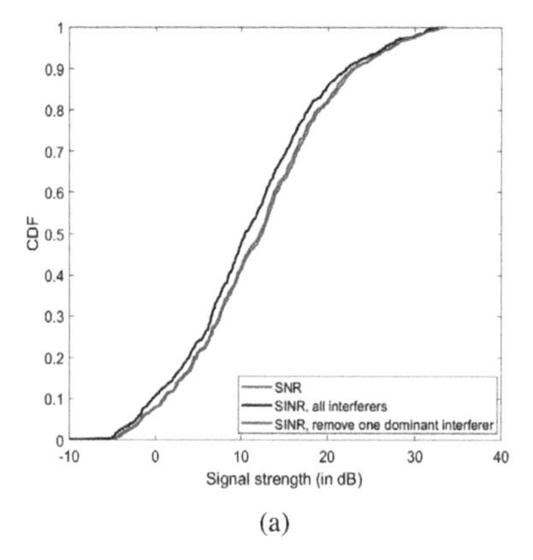

(a)

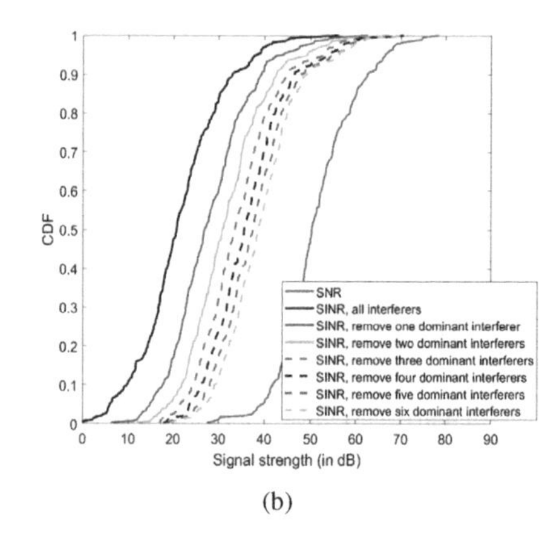

(b)

Figure 5.9 SNR vs. SINR differences between a (a) mmWave and (b) sub-7 GHz network.

SNR and SINR. Thus, the sub-7 GHz network is heavily interference-dominated and this interference is averaged across a large number of paths. While the mmWave network appears to have negligible or minimal interference in most channel realizations, this is also not always the case. Strong interference can be observed in some instantiations in the mmWave setting when a UE is in the cell edge scenario with directional signals interfering within the (narrow) beamwidth of an intended signal. Without incorporating this bimodal behavior of interference in system design, the promise of mmWave networks cannot be fully leveraged in practice. This section studies interference management for densely deployed mmWave cells with the main focus being improvement of the tail distribution of SINR.

Consider a greedy interference management scheme where all the base stations synchronize their transmissions in a slot-by-slot manner. In a slot, the base stations are given distinct priorities. The base station with the top priority first selects one of its served devices. The other base stations then select one of their served devices in a descending order of priorities. When a base station is selecting from amongst the devices it serves, it ensures that:

- At its device, the ratio of the signal power and the interference power from any previous base station is above X (a pre-configured signal strength threshold) and
- At the device selected by any previous base station, the ratio of the signal power and the interference power from the base station is above X.

If the base station cannot find any device to meet the above two conditions, then the base station does not transmit in the present slot. After the base station successfully selects the device, it steers the beam to the selected device. Figure 5.10 illustrates the greedy interference management scheme in a scenario with two base stations. In the present slot, base station 1 is of higher priority than base station 2 and selects device 1 for transmission. The transmit (TX) beam 1 and receive (RX)

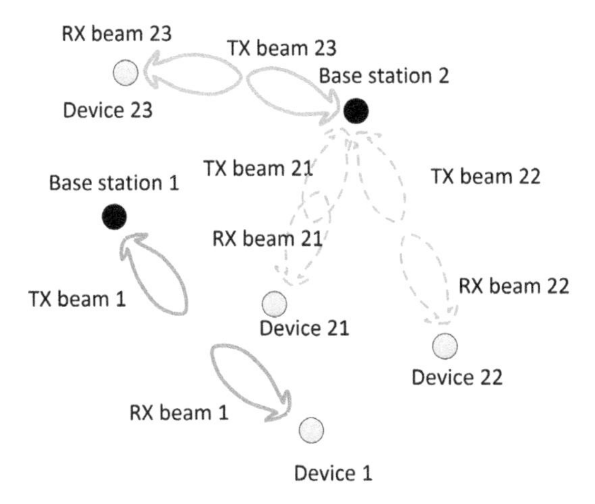

Figure 5.10 Illustration of the greedy interference management scheme.

beam 1 to be used by base station 1 and device 1 are shown in the figure. If base station 2 selects device 21, then it will employ TX beam 21 and the SIR at device 1 will drop below X because of the interference from base station 2. Thus, device 21 cannot be selected. If base station 2 selects device 22, then device 22 will emply RX beam 22 and the SIR at device 22 will be below X because of the interference from base station 1. Thus, device 22 cannot be selected. Base station 2 selects device 23, because the SIR is greater than X at device 1 and at device 23 when TX beam 23 and RX beam 23 are used.

Note that X is a configurable design parameter. A larger value of X results in more protection of SINR of the selected links at the cost that more base stations are not allowed to transmit and this reduces spatial reuse. In addition, the more devices a base station serves, the more likely the base station can find a device that meets the two conditions. The beamforming directivity or beamwidth also affects spatial reuse. In general, eliminating dominant interference improves the SINR tail distribution. The greedy interference management scheme can also be implemented in a distributed manner with the help of over-the-air measurements of beams used at the base stations and devices.

5.4 DEPLOYMENT CONSIDERATIONS

We have built up a comprehensive understanding of propagation, RF, and system design aspects, and with a variety of node-types proposed to address coverage challenges in mmWave systems. What remains to be done is to define a methodology to stitch the pieces together into a high-performing network deployment. However, mmWave deployments for mobile communications had never been done prior to

5G-NR and it is important to understand what differences arise in planning these deployments. At the outset, we make the following points:

- As described in Chapter 2.2.4, mmWave systems are highly sensitive to blockage and depending on the type of blockage, the signal undergoes severe attenuation (e.g., due to foliage, body/hand) or gets completely blocked (e.g., by buildings). Diffraction losses being significantly higher compared with sub-7 GHz systems, blockage plays a substantial role in determining network performance. The consequence of these propagation challenges is that statistical models are poor indicators of performance in a specific region or area. Statistical models provide a coarse characterization of performance but are not that helpful in planning a network. The question then is: how well does the physical environment need to be modeled to capture these blockage effects well?
- Many densification options exist such as small cells, simple or smart repeaters, integrated access and backhaul, out-band fronthaul or backhaul, etc. Sub-7 GHz deployments are typically dominated by macro cells initially, followed by more macro or small cells for densification and repeaters sporadically being used to fill coverage holes or to overcome out-to-in penetration losses. Unlike these deployments, mmWave deployments need to consider these additional node options from the beginning to achieve a satisfactory level of performance. How then should site location and node-type be determined? Can site selection and node-type selection be done jointly?
- Finally, mmWave systems will more often than not be used in conjunction with existing or new sub-7 GHz deployments. What then is the best deployment strategy for mmWave when a sub-7 GHz network is also present (non-standalone deployments)? How does population density variation across a region affect the deployment strategy? Given a certain bandwidth ratio between mmWave and sub-7 GHz, what densification strategy is optimal and how much should mmWave cover? How sensitive is the mmWave benefit to the presence of hotspots and deployment of mmWave nodes with hotspot consideration?

All of the above points are addressed next. We start by recognizing the tremendous advances made in *digital twin* creation via ML approaches and take advantage of that framework for characterizing the physical environment. We then characterize the environment using ray tracing with isotropic transmitters and receivers. This allows us to post-process with a variety of node-types and their associated antenna, beam properties and characteristics. We also show how end-to-end SNR can be calculated for multi-hop links so that they can be seamlessly integrated with single-hop links. Finally, the resultant network graph is used to formulate and solve an optimization problem. The output of the optimization problem is the identification of the subset of transmitter locations to be selected along with their node-types (e.g., small cells, repeaters, etc.). We also show how formulations for coverage or throughput can be made and solved. Finally, we extend the methodology to the case of joint design of mmWave and sub-7 GHz to answer questions on how the two systems complement each other and the level of mmWave coverage at which the gains are substantial.

5.4.1 DIGITAL TWIN CREATION

The term *digital twin* is used to describe a digital representation of the physical environment. In the context of mmWave deployment planning, this term means that aspects of the physical environment that are *relevant* to propagation prediction are represented in some fashion that can be consumed by a ray tracing software. As an example, the digital twin could include buildings and foliage as three-dimensional objects and capturing these representations have a profound impact on mmWave propagation. Another example is the representation of utility or light poles at the street level as three-dimensional objects suitable for mounting small cells, repeaters, or other nodes. All the objects need to be placed in locations that are relatively consistent with each other and their dimensions need to be properly captured. Since mmWave signals are sensitive to blockage, the relative locations need to be fairly precise. For example, a 1 m relative position error between a pole and a building can significantly change conclusions about whether the building blocks the pole or not. Further, as seen in earlier sections, blockage/diffraction losses from buildings can be significant at mmWave frequencies. Similarly, for foliage, the simplest approximation in modeling would be to use a dB/m loss value based on the tree type and then evaluate the length of the signal/ray that passes through foliage to figure out the loss due to foliage. This implies that the dimensions of the trees need to be known for the digital twin.

Geographic Information System (GIS) data for the purpose of digital twin creation is sometimes readily available and published by local city or state governments [241]. There are also crowd-sourced, open-source information on streets, buildings, parks, etc., that can be leveraged in creating the digital twin [5]. Satellite imagery is available for most of the earth (sometimes freely) and street-level imagery may be available for purchase as well. Google Earth (GE) and Google Street View (GSV) are often good sources of information provided the images are relatively recent and (especially for street-level views) taken with sufficient spatial sampling rates for this problem. In some instances, high-quality data obtained from lidar, camera, etc., may be purchased from various sources or custom-collected for network planning [242]. Since deep learning for object detection is a very mature technology, it can be leveraged for the purpose of creating the digital twin for propagation prediction. Well-known architectures such as those based on Convolutional Neural Networks (CNNs) and their variants can be applied to this problem with a high degree of success and reliability. In some instances where GIS databases are available, they can be used as priors to direct the image retrieval defining the camera pan, tilt or zoom required to improve the detection probability. However, when multiple sources of data are used to create the digital twin, errors can creep into the model from at least the following sources:

- Mismatched dates for data collection
- Missing data
- Localization errors and
- Coordinate system differences.

Figure 5.11 Scene of a roadway with trees segmented via the instance segmentation algorithm.

In particular, for aerial imagery-based data, projection issues can lead to errors in localization of objects of interest.

In an example illustration of the above proposals, in Figure 5.11, a typical roadway in NJ is presented with instances of deciduous trees lining the side walks and utility poles. Tree canopies/structures are identified with ML techniques (instance segmentation) with different score levels assigned to the classification reflecting the algorithm's confidence on the identified shape being a tree. Similarly, road markers on the edge of the roadway are identified and these road markers are illustrated in Figure 5.12. These road markers are used as a measuring source to make a height estimate for the trees. For this, we observe that the true width of the road should remain invariant to the camera angle used to obtain the street-level image. Thus, by using the invariance relationship:

$$\frac{\text{Number of pixels that make the road width}}{\text{True width of road}} = \frac{\text{Number of pixels that make the tree height}}{\text{True height of tree}} \tag{5.17}$$

we can identify the tree's true height estimate and draw bounding boxes around the tree structure. The identified bounding boxes are presented in Figure 5.12. A pole finder algorithm is also run and bounding boxes around the utility poles are also drawn in Figure 5.13.

A good source of tree heights is aerial color maps of greenery, which is often available with GIS data. Another way to determine tree heights is to calibrate the height in the image based on detection of objects with known heights (e.g., the two-arm electricity pole in Figure 5.11) or *a priori* known estimates of certain objects in the image (e.g., lamp head dimensions). These can be realized using well-known

Figure 5.12 Typical scene of a roadway with road markers and bounding boxes around instances of trees.

techniques such as pinhole camera geometry or its variants. Objects such as poles can be detected either via triangulation of images or from single street-level images. Other street objects relevant to mmWave communications are street signs, billboards, etc. Window detection via ML techniques is also useful for fixed wireless access applications.

In a further application of the above proposals, lamp or utility poles can be isolated/detected using street-level images with ML techniques and the pole heads'

Figure 5.13 Typical scene of a roadway with bounding boxes around instances of utility poles.

Figure 5.14 An example digital twin appropriate for propagation modeling.

locations can be further refined. Given that this process can lead to a large number of false alarms in identification of utility poles, a combination of street-level and aerial views of a pole (when available) can greatly improve the localization accuracy of poles. While a 1–2 meter accuracy is more than adequate for path loss prediction (for mmWave network analysis deploying relays), localization accuracy is important to ensure that backhaul topologies do not become disconnected as a result of inaccurate position information. By combining or fusing the street-level and aerial views, we can localize and identify the location of the true poles in a map.

Beyond the pole identification process, identification of streetscape such as tree canopies, foliage, buildings, etc., is important from a propagation modeling perspective. ML techniques can be used to localize with bounding boxes around objects of interest. From these localization efforts, a digital twin can be created where the urban streetscape including poles, buildings and foliage are described. As a particular illustration of this effort, the digital twin created for the Manchester, UK deployment is presented in Figure 5.14.

5.4.2 PROPAGATION MODELING

Propagation modeling comprises path loss characterization that includes both distance and shadowing losses as well as small-scale parameter determination such as power-angular-delay profile. For the purpose of mmWave network planning, the small-scale parameters (other than the directional information on AoD/AoA for a few dominant paths) are ignored. The modeling of reflection losses and delay/angular spread requires proper characterization of the electromagnetic properties of the reflecting surface. While ML techniques on images, lidar or radar data can be used to ascertain such information, incorporating such information in propagation modeling via a ray tracing software can make it extremely slow. It is simpler and faster therefore to model reflections as being purely specular with a loss added based on the material properties of the surface. Diffraction losses can be modeled as single or multiple knife edge diffractions [243]. For a given physical geometry, diffraction losses are much higher at higher carrier frequencies and thus, this mode of propagation is weak at mmWave. As such, to speed up the propagation prediction for mmWave, one may choose to ignore the effects of diffraction and as a conservative planning approach only focus on LOS and specular reflections.

Another challenge pertaining to propagation modeling at mmWave frequencies is the modeling of beamforming. The analog/hybrid beamforming codebook design, antenna gain modeling, number of sectors, etc., are all important attributes to properly model and these vary based on the node-type involved. For example, a small cell node may have a rich/better granularity codebook with good elevation and azimuth coverage over three sectors, while a simple repeater may have only one or two sectors with a static beam pattern. A smart repeater may have only one or two sectors, but may have a codebook that is similar to that of a small cell. Both repeater types may have amplification capability which needs to be carefully modeled. Finally, a macro cell may have a larger antenna array and therefore a codebook with more beams and higher peak array gain per beam. The problem of network planning for mmWave frequencies can thus be quite challenging.

To speed up the path loss calculations, several options can be explored. First, we can choose to run the ray tracing software in isotropic mode to capture the raw geometry of the area of interest. This can take the form of finding rays from all the potential sites (e.g., all the street-level poles) to all the points of interest. The individual node-type's RF/beamforming constraints can then be superimposed on the rays while also optimizing the boresight directions in azimuth and elevation. Second, advanced techniques from generative modeling can be used for this purpose to improve speed and transferability from one region to another. For example, see Radio UNet [244] which showcases such an approach using the classical UNet image-to-image neural network. Using Radio UNet, one can go from an image of the map to an image of the coverage. An even simpler alternative is to use formula-based approaches for path loss prediction in network planning, but due to the large variations in mmWave propagation, this is unlikely to yield reliable results.

Foliage modeling would require characterizing the length of ray segments that traverse through foliage along with a loss model for each type of foliage. The digital twin creation can identify not just the size and shape of tree plumage, but can also identify the tree type if desired. Loss models on tree type can then be applied, especially since evergreen trees tend to have much greater loss per meter than deciduous trees [32]. As will be shown subsequently, foliage loss characterization is important to get an accurate estimate of network cost and topology.

5.4.3 OPTIMIZATION FORMULATION

Having created a digital twin and run propagation evaluation for candidate transmit sites and receive locations, the problem then is to determine the optimal topology. That is, to determine which transmit locations should be selected and what transmit node-type should be used at each location. Should it be a small cell, a repeater, an IAB node, or should the site not be included at all in the topology? It is readily seen that this problem can be posed as an optimization problem on a graph with the vertices representing the transmitter or receiver nodes and the link weights representing the SNRs or equivalent signal strength metrics. This representation also extends to the case of multi-hop links including those that use repeaters, with the main difference being that for the repeater end-to-end link SNRs have to be computed for each end-to-end path individually.

Since each node-type is associated with a different cost and performance trade-off, the optimization problem aims to solve for the minimum cost objective while ensuring a minimum percentage of UEs achieve a target SNR or *vice versa*. In some versions of the formulation, the user performance constraint can be expressed as a throughput constraint. The difference between this formulation and the SNR target-based formulation is that the throughput formulation explicitly takes resource constraints into account. This means that the solved topology[2] is also guaranteed to have at least one feasible resource allocation (that is, scheduler) that meets the requirements. Finally, the variables being solved for are binary variables indicating whether a candidate base station location is included in the topology or not and (if included) what node-type it is.

The problem of placing base stations and relays to meet a target coverage of traffic points (TPs) with minimum cost is formulated as an integer linear program (ILP). The number of discrete variables in the ILP is approximately the sum of the number of TPs and the number of feasible backhaul links. The ILP can be proven equivalent to a mixed-ILP (MILP) where the number of discrete variables is reduced by the number of TPs leading to faster convergence to a global optimum by branch-and-bound algorithms. A case study is presented to demonstrate the complexity reduction by the proposed MILP approach. Since solving for the optimal solution to the heterogeneous base station deployment is in general computationally infeasible for large dimensionality, existing studies like the greedy algorithm generate feasible solutions efficiently in a low-complexity manner. However, the quality of the feasible solutions cannot be assessed without establishing reasonably tight converse bounds. Note that linear programming relaxations of ILP and MILP are quite loose in general.

Therefore, we take a different approach from the previous studies in the following way. The network coverage problem involving base stations and relays is modeled as an ILP and instead of focusing on developing heuristic algorithms which yield only achievable bounds, we transform the ILP problem into an equivalent MILP problem with lower complexity. This approach results in faster convergence of the achievable and converse bounds obtained by the standard branch-and-bound algorithm [245–247] used in many commercial and open-source MILP solvers (e.g., MATLAB®, Gurobi, CPLEX, COIN-OR, HiGHS, etc.). A real-world case study will be provided to illustrate the gain in speed of convergence obtained by the proposed

[2]It is not the intent of this chapter to provide an in-depth mathematical analysis of this optimization problem and solution. Therefore, a brief synopsis that illustrates one such problem formulation is provided here.

MILP approach after we present the problem formulation and the main result in the subsequent sections.

This section provides the necessary setup and the formulation of the optimization problem.

5.4.3.1 Constants

Let $\mathscr{M} = \{1,2,\ldots,M\}$, $\mathscr{N} = \{1,2,\ldots,N\}$ and $\mathscr{K} = \{1,2,\ldots,K\}$ be the indexed sets of base stations, relays and traffic points (TPs), respectively. The TPs represent sampled points in an area of interest and can be regarded as potential locations of UEs. A direct link (i,k) between a base station $i \in \mathscr{M}$ and a TP $k \in \mathscr{K}$ is said to be connected if the downlink and uplink SNR are at least γ_{DL} and γ_{UL}, respectively. Similarly, a two-hop link (i,j,k) where a base station i communicates with a TP k through a relay $j \in \mathscr{N}$ is said to be connected if the end-to-end downlink and end-to-end uplink SNR are at least γ_{DL} and γ_{UL}, respectively.

Define the connectivity indicator for link (i,k) as $c_{ik} = \mathbf{1}\{(i,k) \text{ is connected}\}$ where $\mathbf{1}\{\cdot\}$ denotes the indicator function. Similarly, define the two-hop connectivity indicator for link (i,j,k) as $d_{ik}^{(j)} = \mathbf{1}\{(i,j,k) \text{ is connected}\}$. To simplify notations, let

$$\mathbf{C} = [c_{i,k}]_{(i,k)\in\mathscr{M}\times\mathscr{K}} \tag{5.18}$$

denote the $M \times K$ one-hop connectivity matrix for the access links. Also, let

$$\mathbf{D}_j = [d_{i,k}^{(j)}]_{(i,k)\in\mathscr{M}\times\mathscr{K}} \tag{5.19}$$

denote the $M \times K$ two-hop connectivity matrix for the two-hop links that pass through relay j. Intuitively speaking, the entry in row i and column k of $\mathbf{C}$ indicates whether small cell i and TP k can communicate directly with sufficient SNR, and the entry in row i and column k of $\mathbf{D}_j$ indicates whether small cell i and TP k can communicate indirectly through relay j with sufficient end-to-end SNR. Let

$$\boldsymbol{f}^{\mathrm{base}} = [f_1^{\mathrm{base}}\ f_2^{\mathrm{base}}\ \ldots\ f_M^{\mathrm{base}}] \tag{5.20}$$

and

$$\boldsymbol{f}^{\mathrm{rel}} = [f_1^{\mathrm{rel}}\ f_2^{\mathrm{rel}}\ \ldots\ f_N^{\mathrm{rel}}] \tag{5.21}$$

be the vectors corresponding to the deployment costs of the M base stations and the N relay nodes, respectively.

5.4.3.2 Variables

Define the $M \times N$ binary selection matrix for backhaul links as

$$\mathbf{L} = [\ell_{ij}]_{i\in\mathscr{M},\ j\in\mathscr{N}} \tag{5.22}$$

where ℓ_{ij} indicates whether the backhaul link (i,j) is selected. In addition, we define the $1 \times M$ binary base station selection vector

$$\mathbf{x}^{\mathrm{base}} = [x_1^{\mathrm{base}}\ x_2^{\mathrm{base}}\ \ldots\ x_M^{\mathrm{base}}] \tag{5.23}$$

and the $1 \times N$ binary relay selection vector

$$\mathbf{x}^{\text{rel}} = [x_1^{\text{rel}}\ x_2^{\text{rel}}\ \ldots\ x_N^{\text{rel}}] \tag{5.24}$$

where x_i^{base} indicates whether base station i is selected and x_j^{rel} indicates whether relay j is selected. From (5.24), we are restricted to the constraint that each column of $\mathbf{L}$ has at most one 1 (that is, each base station connects to at most one relay). In order to characterize the set of covered TPs, we define the $1 \times K$ binary TP coverage vector

$$\mathbf{y} = [y_1\ y_2\ \ldots\ y_K] \tag{5.25}$$

where y_k indicates whether TP k is covered by either a base station or a relay.

5.4.3.3 Constraints

We impose the following constraints on the optimization problem.

- *Relay selection*: Since each selected relay is fully characterized by $\mathbf{L}$, we have

$$\mathbf{x}^{\text{rel}} = \underbrace{[1\ 1\ \ldots\ 1]}_{1\times M} \cdot \mathbf{L}. \tag{5.26}$$

- *Base station selection induced by backhaul links*: Let $\mathbf{L}_j = [\ell_{1j}\ \ \ldots\ \ \ell_{Mj}]^T$ denote the j-th column of $\mathbf{L}$. Since each selected backhaul link has to be supported by the corresponding base station, it follows that the base station must also be selected which implies that

$$\mathbf{x}^{\text{base}} \geq \mathbf{L}_j^T \tag{5.27}$$

  for each $j \in \mathcal{N}$ where the inequality applies to each corresponding element of both vectors.
- *Coverage constraint*: Since each covered TP must be connected to either a selected base station or a selected relay, we have

$$\mathbf{y} \leq \mathbf{x}^{\text{base}} \cdot \mathbf{C} + \sum_{j=1}^{M} \mathbf{L}_j^T \cdot \mathbf{D}_j. \tag{5.28}$$

  Intuitively speaking, the term $\mathbf{x}^{\text{base}} \cdot \mathbf{C}$ indicates whether a TP is covered by a selected base station and the term $\mathbf{L}_j^T \cdot \mathbf{D}_j$ indicates whether a TP is covered by relay j. Therefore, each selected TP indicated by one in the binary vector $\mathbf{y}$ must be covered by either a base station or a relay according to (5.28). In particular, if relay j is selected, then $\mathbf{L}_j^T \cdot \mathbf{D}_j$ is a binary row vector indicating the set of TPs covered by relay j. On the other hand, if relay j is not selected, both $\mathbf{L}_j^T$ and $\mathbf{L}_j^T \cdot \mathbf{D}_j$ equal the zero vector which indicates that none of the TPs are covered by relay j.

5.4.3.4 Objective Function

We are interested in placing base stations and relays to cover a target fraction (denoted as α) of TPs with minimum cost. Let

$$\beta(\mathbf{x}^{\text{base}}, \mathbf{x}^{\text{rel}}) = \boldsymbol{f}^{\text{base}} \cdot (\mathbf{x}^{\text{base}})^T + \boldsymbol{f}^{\text{rel}} \cdot (\mathbf{x}^{\text{rel}})^T \tag{5.29}$$

be the network cost objective function. Fix a target network coverage value $\alpha \in [0,1]$. Then, the objective function can be written as

$$\min \beta(\mathbf{x}^{\text{base}}, \mathbf{x}^{\text{rel}}) \tag{5.30}$$

subject to the network coverage constraint

$$\frac{1}{N}\sum_{k=1}^{N} y_k \geq \alpha. \tag{5.31}$$

If $\alpha < 1$, not all TPs have to be covered and the optimal solution yields an optimal collection of selected base stations and selected relays covering a subset of TPs with minimum cost. Note that the ILP is nondeterministic polynomial-time (NP)-complete due to the following argument: If we consider the special case where the TP coverage vector $\mathbf{y}$ is set to a fixed constant and no relays are present (that is, $N = 0$), the ILP becomes equivalent to the set cover problem which is well known to be NP-complete [248].

5.4.3.5 Main Result

The standard branch-and-bound algorithm [245–247] is typically used for solving ILP and MILP problems and the computational complexity increases with the number of discrete variables. For example, an MILP problem with one binary and multiple continuous variables can be solved by first creating two branches corresponding to the two values of the binary variable followed by running polynomial-time algorithms to solve the resultant two LP problems involving no discrete variables. Therefore, in order to reduce the computational complexity of the ILP in Chapter 5.4.3, we are motivated to relax as many binary variables as possible without affecting the optimal objective value. The following result shows that relaxing the N variables of the TP coverage vector $\mathbf{y}$ does not affect the objective value. That is, given the ILP in Chapter 5.4.3 with objective function in (5.30), the ILP and the MILP resulting from relaxing the K TP coverage variables $\mathbf{y}$ to continuous variables between 0 and 1 have the same optimal objective value.

To establish this result, note that it is well-known that relaxing variables of an ILP cannot yield a worse objective value. Therefore, it suffices to demonstrate that any optimal solution of the relaxed problem with objective function (5.29) can be mapped to a feasible solution of the original problem with the same objective function. This implies that the objective value of the relaxed problem is no better than that of the original problem.

Suppose we are given an optimal solution $(\mathbf{L}, \mathbf{x}^{\text{base}}, \mathbf{x}^{\text{rel}}, \bar{\mathbf{y}})$ of the relaxed problem with objective (5.30) where $\bar{\mathbf{y}} = [\bar{y}_1\ \bar{y}_2\ \dots\ \bar{y}_N]$ is a vector of continuous variables between 0 and 1 and $(\mathbf{L}, \mathbf{x}^{\text{base}}, \mathbf{x}^{\text{rel}})$ are binary variables. Based on the given solution, we construct a modified solution with variables $(\mathbf{L}, \mathbf{x}^{\text{base}}, \mathbf{x}^{\text{rel}}, \mathbf{y}^*)$ where $\mathbf{y}^* = [y_1^\star \dots y_K^\star]$ is defined through the following relationship:

$$y_k^* \triangleq \begin{cases} 0 & \text{if } \bar{y}_k = 0, \\ 1 & \text{if } 0 < \bar{y}_k \leq 1 \end{cases} \tag{5.32}$$

for all $k \in \mathcal{K}$ and the other variables remain unchanged. Since the optimal solution $(\mathbf{L}, \mathbf{x}^{\text{base}}, \mathbf{x}^{\text{rel}}, \bar{\mathbf{y}})$ satisfies all the constraints of the original problem, it follows that the modified solution satisfies (5.26) and (5.27) of the original problem. Moreover, the TP coverage constraint (5.28) is also satisfied because every positive value in $\mathbf{x}^{\text{base}} \cdot \mathbf{C}$ and every positive value in $\sum_{j=1}^{M} \mathbf{L}_j^T \cdot \mathbf{D}_j$ are at least one and hence the inequality is guaranteed by the relaxed problem. That is,

$$\bar{\mathbf{y}} \leq \mathbf{x}^{\text{base}} \cdot \mathbf{C} + \sum_{j=1}^{M} \mathbf{L}_j^T \cdot \mathbf{D}_j \tag{5.33}$$

together with (5.32) implies that

$$\mathbf{y}^* \leq \mathbf{x}^{\text{base}} \cdot \mathbf{C} + \sum_{j=1}^{M} \mathbf{L}_j^T \cdot \mathbf{D}_j, \tag{5.34}$$

which is the TP coverage constraint of the original problem. In addition, the network coverage constraint (5.31) is also satisfied since $\sum_{k=1}^{N} y_k^* \geq \sum_{k=1}^{N} \bar{y}_k$ by (5.32).

Consequently, the modified solution derived from the relaxed problem is feasible for the original problem where the values of the objective function (5.29) evaluated at the modified solution for both the relaxed problem and the original problem are equal to the objective value of the relaxed problem. As a result, the objective value of the relaxed problem is no better than that of the original problem. Since variable relaxation can only improve the objective value of the original problem, it follows that relaxing the TP coverage vector $\mathbf{y}$ does not improve the objective value.

The above result allows us to transform the original ILP problem into an MILP problem with K fewer binary variables which leads to a significant reduction in computational complexity. A few important observations regarding the generalization of the above approach are mentioned below. Some other interesting properties are also briefly mentioned.

- *Greedy Algorithms*: We now make a note on the use of ILP to solve the aforementioned network design problem. Given the one-hop connectivity matrix $\mathbf{C}$ and the N two-hop connectivity matrices $\{\mathbf{D}_j,\ j \in \mathcal{N}\}$, our objective is to find the minimum deployment cost to meet a target network coverage $\alpha \in (0, 1]$ (that is, to cover αK TPs). A straightforward heuristic approach is to implement the following greedy algorithm:

1. At the first stage, deploy a small cell on a pole such that the small cell can cover most TPs according to $\mathbf{C}$.
2. At each subsequent stage, deploy a small cell or a relay on a pole such that the small cells and the relays that have been deployed maximize the number of covered TPs per unit cost according to $\mathbf{C}$ and $\{\mathbf{D}_j,\ j \in \mathscr{N}\}$.
3. The greedy algorithm stops at a stage where the number of covered TPs exceeds αN.

Although the greedy algorithm can find a feasible solution quickly, it yields only a sub-optimal solution without the knowledge of its quality. In contrast, the MILP approach described in Chapter 5.4.3 always yields an achievable bound and a converse bound at any given time and the two bounds will eventually converge given a sufficiently long run-time. The greedy approach may be employed for very large network design problems and/or topologies with a single node-type while the MILP approach may provide substantial reduction in deployment cost for targeted areas (like downtown areas in a city) and/or when many node-types need to be considered.

- *Formulation for joint design of sub-7 GHz and mmWave networks*: The main challenge with extending the ILP formulation to the case of mixed technology (such as sub-7 GHz and mmWave) is in balancing between the vastly different capabilities of the two networks. Sub-7 GHz networks provide excellent coverage with limited bandwidth while mmWave networks provide significantly more bandwidth with some coverage limitation. One way to tackle this problem in the aforementioned formulation is to redefine the notion of coverage. Rather than coverage being defined as the minimum SINR being met, we rather look at achievable throughput as the primary metric and map that to a minimum SINR needed on a per-cell basis. Note that this minimum SINR will now be the maximum of that required to maintain the lowest MCS (or for control channel decoding) and the calculated value for meeting data channel throughput. The deployment planning would therefore accumulate as many TPs as needed such that the throughput requirements are met. The calculation of the throughput itself then becomes a function of the bandwidth (and MIMO capability) of the individual technology as well as the loading (that is, TPs assigned to that cell). In this manner, by going from required throughput to maximum allowed TPs per cell to the target SINR per cell, one can bring the disparate technology capabilities to a common metric and reuse most of the optimization formulation described above.
- *Formulation with guaranteed throughput constraints*: In some instances such as when mmWave networks are used in conjunction with Service Level Agreements (SLAs) for throughputs, the problem that needs to be formulated and solved is one of designing a network so that a given set of locations are guaranteed a target throughput. This is in contrast to the original problem and the one undertaken for joint sub-7 GHz and mmWave design. In those problems, the entire outdoor area is sampled with some regularity and either an SINR target is used to define coverage (original formulation) or a throughput with activity factor assumptions for the region (joint sub-7 GHz and mmWave problem). In contrast, in this formulation,

we assume that the locations are given and the required throughputs (not necessarily all equal) are also given. The ILP then finds the least cost solution that meets the individual throughput requirements while ensuring that a feasible time scheduler exists. The time scheduling constraints are expressed by constraining the total time fraction across users belonging to a cell to be less than 1.

- *Formulation for Fixed Wireless Access (FWA) use-case*: The use of mmWave technology for fixed wireless access as a cable replacement solution has been gaining traction. In addition to the base station placement problem, the formulation here can be generalized to include a number of candidate locations at each building/residence at which fixed wireless services need to be provided. With a modification where we ensure that a building/residence is deemed covered if at least one of the candidate locations for the building is covered, the optimization formulation can be directly used. This constraint ensures that each building is counted only once when total coverage is being calculated and that coverage implies at least one of the candidate locations associated with a building/residence is covered.
- *Formulation for diversity to blockage*: In an environment with a high likelihood of dynamic blockers to the signal, it may be beneficial to provide some spatial diversity to mitigate blockage. A convenient (but somewhat sub-optimal) method of accomplishing this would be to first run the MILP to find a network design for a target coverage. Then, from the connectivity matrix, we eliminate the serving links (by setting those entries to zero) as well as any links that would not be sufficiently separated in AoA from the primary link. The latter constraint is to ensure that a dynamic blocker, typically close to the UE, will not block the diversity link as well. Furthermore, we set the binary variables for the selected positions for the primary coverage to be unity. One can then re-run the optimization problem with the objective being set to maximize the number of links covered with the incremental cost for obtaining diversity being a constraint. An example target would be to set the primary link coverage target to 80% and maximizing the percentage of links with diverse coverage subject to the incremental cost for diversity not exceeding 15% of the original cost. It is possible to formulate and run a single optimization problem that includes the diversity constraint but the complexity of this problem is quite high.
- *Extension to "brown-field" deployment use-case*: The formulation can be readily extended to the "brown-field" case where an existing network footprint exists and the design needs to account for coverage provided by the existing footprint. This is easily done in the formulation by forcing the binary variables for the existing locations to unity and either making the cost of those locations zero or a very small value relative to the cost of adding a new location. The case-study of joint sub-7 GHz and mmWave design shown below assumes that an existing sub-7 GHz network footprint exists and the question being addressed is the best way to densify the network to increase throughput.

5.5 ILLUSTRATIVE RESULTS

A number of important observations regarding network planning will be made in this section. Two different geographical regions are used to illustrate a variety of results in this section.

- *Philadelphia, USA*: An approximately 2.4 sq km region in the central business district of Philadelphia is used for illustrating various results pertaining to mmWave-only network designs.
- *Manchester, UK*: An approximately 6.7 sq km region of Manchester is used for illustrating various results pertaining to joint mmWave and sub-7 GHz network designs. The region includes the main areas of the town, some important stadium areas and the connecting area in between those two features. A number of regions (as shown subsequently) are designated as "hotspot" zones with a higher population density than in the "non-hotspot" zones. Network designs that prioritize mmWave base station deployments in the hotspots are undertaken.

5.5.1 IMPACT OF FOLIAGE ON MMWAVE NETWORK PLANNING

A significant number of measurement results have indicated the additional sensitivity of propagation at mmWave frequencies to foliage compared with propagation at sub-7 GHz bands. Some illustrative examples of the additional penetration losses experienced by mmWave signals traversing through foliage can be found in [244]. To demonstrate the importance of capturing foliage presence in the digital twin, network designs are compared for the case where foliage is ignored relative to the case where the foliage is captured (even if only approximately) in the digital twin. It is seen in the SNR coverage map illustrated in Figures 5.15(a, b) that the network design accounting for foliage requires significantly more small cells (57) to attain 85% coverage in the Philadelphia deployment case as compared with when foliage is not accounted for. In the latter case, only 37 small cells are required for 85% coverage. This study illustrates the importance of area-specific propagation modeling at mmWave frequencies and the importance of correctly capturing foliage in the digital twin model. Fortunately, such foliage data can be obtained and curated for this purpose. Since foliage tends to grow over time, it is important in the propagation modeling to allow for some margin in the foliage contours and heights.

5.5.2 USE OF MULTIPLE NODE-TYPES IN MMWAVE NETWORK DESIGNS

As was outlined earlier, several different node-types have been proposed for improving mmWave coverage. These include IAB, simple repeaters, and smart repeaters. While the characteristics of these relay nodes have been well studied, it is important to understand the usage of these nodes in a practical deployment context. In turn, this comes down to a formulation that takes their relative performance vs. cost into account so that overall a cost-optimal deployment can be achieved while meeting the

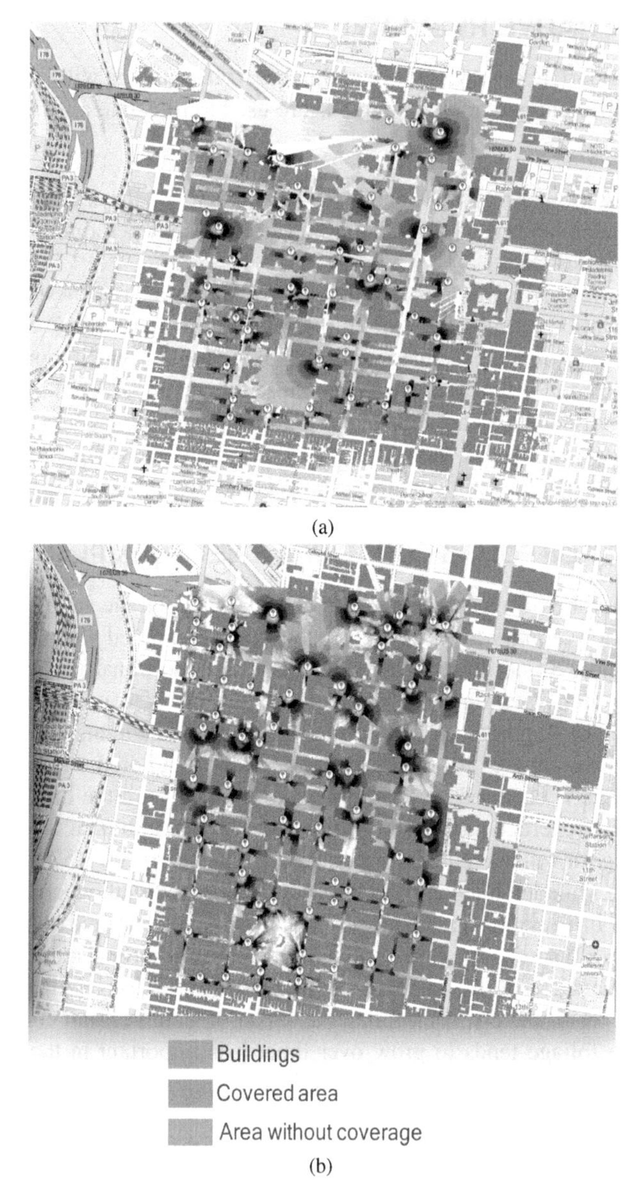

Figure 5.15 SNR coverage map when foliage (a) is not accounted for and (b) is accounted for. Map data from OpenStreetMap with license agreement covering this usage available at `https://www.openstreetmap.org/copyright`.

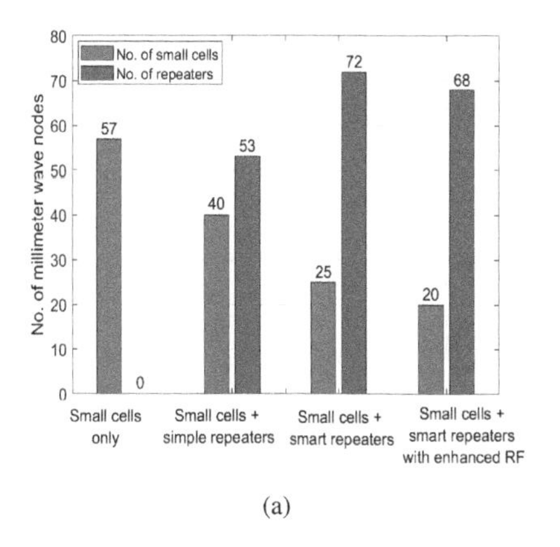

(a)

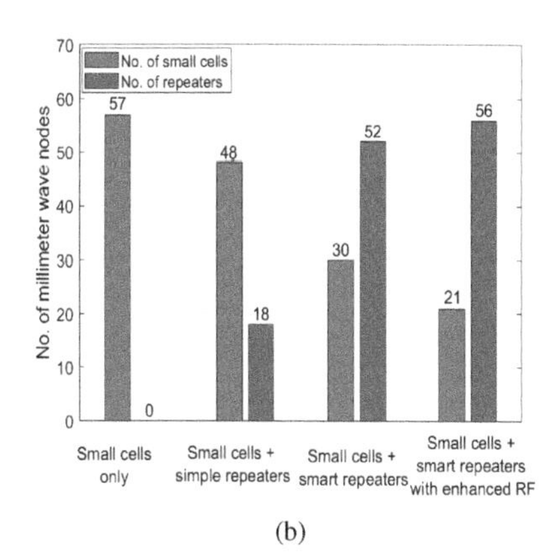

(b)

Figure 5.16 Illustration of network designs with small cell to repeater cost ratio of (a) 5 to 1 and (b) 3 to 1.

performance requirements. To quickly recapitulate the capabilities assumed for these node-types, we have the following:

- IAB provides a decode-and-forward relay functionality. An IAB node has flexibility in beamforming to/from a UE just as a small cell node would. For the purpose of modeling, it is important to properly account for the resource usage (when shared) between the hops in an IAB setting. Half- and full-duplex assumptions may be considered.
- Simple repeater provides an amplify-and-forward relay functionality. A simple repeater is the "RF repeater" well-known for many generations of wireless technologies. It takes in an analog RF signal, amplifies it, and transmits an RF signal forward. Crucially, it is assumed that the simple repeater has no information to allow per-UE beamforming and therefore may have a reduced link budget compared with a small cell or an IAB node.
- Smart repeater also provides an amplify-and-forward functionality, but includes a side control channel to the donor. The side control channel allows the donor node to provide information regarding the TDD structure, beamforming to use on the forwarding link, etc. Compared with a simple repeater, the smart repeater is therefore expected to have higher amplification per UE, but unlike an IAB, it can still forward the amplified input noise through to the destination.

Different network designs can be conceived with the characteristics of these node-types and associated cost ratios relative to using a small cell. Figure 5.16 illustrates some example designs for the Philadelphia scenario with different cost ratio assumptions between small cells and repeater nodes. Figure 5.16(a) shows different designs achieved when the repeater costs a fifth of a small cell while Figure 5.16(b) corresponds to the case when repeaters cost a third of a small cell. From left to right in

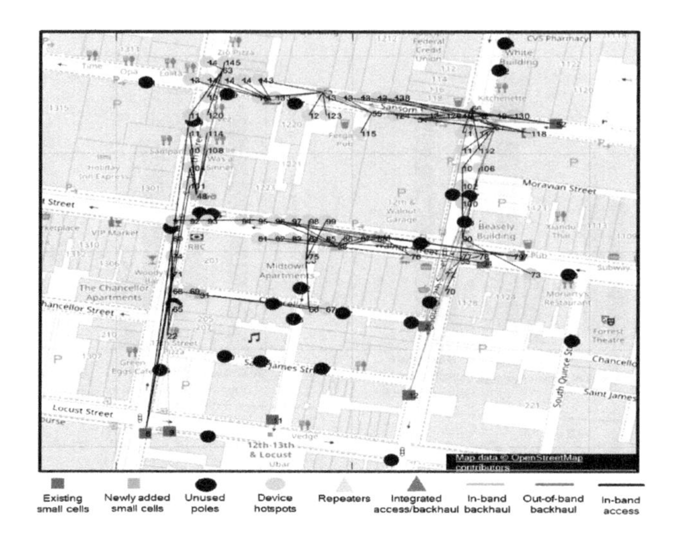

Figure 5.17 Illustration of a throughput-based cost-optimal network deployment that utilizes the richness of the mmWave solution suite ranging from repeaters to IAB to out-band fronthaul. Map data from OpenStreetMap with license agreement covering this usage available at `https://www.openstreetmap.org/copyright`

each set of bar graphs is increasing repeater capability. A simple repeater is assumed to have a static beam facing the service (UE) side and limited stable amplification when TDD information is unknown. In contrast, a smart repeater assumes higher stable amplification as well as per-UE beamforming on the service side. An enhanced smart repeater is for the case when the repeater has the same sized antenna array as the small cell and inherits all other properties of a smart repeater. From these bar graphs, we observe that the number of repeaters used increases as the repeater capability increases as well as its cost reduces (relative to the small cell).

Network designs for mmWave frequencies can also be focused on delivering a high target throughput to hotspot areas which could benefit the most from deployment of mmWave technology rather than area coverage as considered in the examples so far. With throughput being the target, resource allocation constraints need to be properly accounted for in the optimization formulation. In doing so, relay nodes like IAB, out-band fronthaul, etc., can also play an important role in ascertaining the network topology as they tradeoff resource usage and performance in different ways than analog repeaters. A detailed discussion of the problem formulation and solution for this case is beyond the scope of this book, but an illustrative example that shows how these additional node options can provide great value in deployments is provided in Figure 5.17. For this study, one can notice that the cost-optimal solution emerging from the optimization problem includes new small cells, IAB, smart repeaters, and out-band fronthaul. The solutions make intuitive sense in that repeaters are used to "bend" the signal into spaces with fewer hotspots while out-band IAB or fronthaul solutions are preferred in places where a high performance is needed. This example provides a simple illustration that the richness of the solution set available for mmWave can be fully brought to bear for cost-optimal real-world deployments.

5.5.3 PERFORMANCE OF SUB-7 GHZ AND MMWAVE JOINT DEPLOYMENTS

While comparing and contrasting the performance of sub-7 GHz and mmWave bands is informative in understanding these two technologies, they can powerfully complement each other in real-world practical deployments. Sub-7 GHz networks provide extensive and robust coverage, but typically have smaller bandwidths to offer and are more limited by interference. In contrast, mmWave coverage is more sensitive both to blockage and to the ability to track users' beams in a dynamic/mobile environment, but offer significantly larger bandwidths and interference rejection from the use of narrow beamwidth beams. Therefore, it seems reasonable to expect that the two technologies when deployed optimally could provide the extremely reliable coverage of sub-7 GHz *and* the high throughputs of mmWave frequencies (especially in regions of high demand).

In this section, the following results are presented:

- User throughput scaling with the density of base stations per unit area for mmWave and sub-7 GHz
- Sensitivity of throughput scaling to non-uniformity of user density (presence of "hotspots")
- Scaling of signal and interference terms with base station density per unit area for mmWave and sub-7 GHz
- Complementing the results of Chapter 4.5.2, impact of higher-order MU-MIMO in sub-7 GHz and an assessment of its ability to counter the bandwidth advantage of mmWave frequencies.

Results are provided for the Manchester case-study outlined earlier and cover downlink as well as uplink. A table of parameter values and assumptions used in these studies is also provided in Table 5.1 with the results for each study in a separate section.

The baseline deployment is that of 54 sub-7 GHz macro base stations deployed on building rooftops to cover the entire region disregarding the user population and focusing just on reliable area coverage. The question to resolve is whether a densification strategy using mmWave small cells or one that continues to deploy sub-7 GHz macros is a superior strategy. The conclusion depends on how the signal and interference scale, the relative bandwidth between sub-7 GHz and mmWave and the support/effectiveness of MU-MIMO. All these aspects are explored next.

5.5.3.1 Impact of Base Station Density on Throughput Scaling

Figure 5.18(a) shows the normalized throughput as a function of the base station density for the option of sub-7 GHz macro-based densification and mmWave-based densification. The throughput normalization is with respect to the baseline deployment of 54 sub-7 GHz macro base stations. In addition to the performance with sub-7 GHz and mmWave-based densification, another scenario denoted as "ideal offload" is also illustrated. This scenario represents the throughput scaling that would be achieved

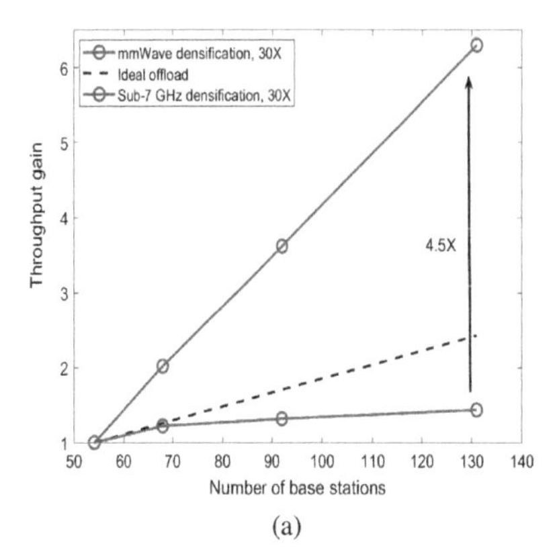

(a)

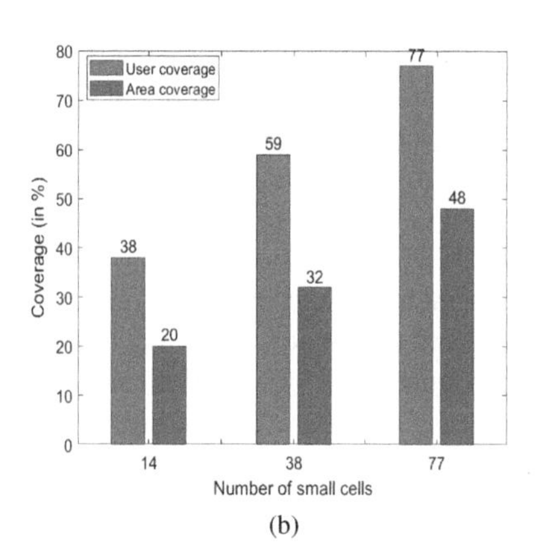

(b)

Figure 5.18 (a) Median throughput scaling as a function of the number of base stations. (b) User concentration in hotspots makes mmWave deployment valuable.

if the added base stations result in an ideal rebalancing of the load/demand. Thus, if the base stations were doubled, the "ideal offload" would indicate a normalized throughput gain of 2. The reason to benchmark against this metric is to expose any differences in signal vs. interference scaling that may be happening with different deployments.

Figure 5.18(a) shows that the sub-7 GHz-based densification underperforms the ideal offload whereas the mmWave-based densification significantly outperforms the ideal offload. At an added density of 77 base stations, the normalized throughput gain of mmWave is ~ 4.5 times more than what an equivalent number of added sub-7 GHz macro base stations would provide. The reasons for this will be made clear in the next few sections.

The throughput scaling results have been shown for the case where the hotspot areas have 30 times more population than the non-hotspot areas. This is a particularly favorable scenario for mmWave as relatively few mmWave base stations can provide the enhanced user coverage. As a reference, Figure 5.18(b) shows the contrast between the area coverage vs. the user coverage of mmWave for the three added density values. Thus, with a concentration of users in hotspots typical of urban population centers, the use of mmWave can provide significant benefit in user experience even without covering a very large area. Furthermore, the effect of offloading traffic from sub-7 GHz to mmWave also reduces the load on the sub-7 GHz network and thereby enhances the performance of sub-7 GHz users who may not be able to get a strong mmWave signal.

5.5.3.2 Impact of User Concentration in Hotspots

The advantages of densifying with mmWave are particularly strong when there are hotspots of demand, as is typical of urban centers. The greater the concentration of

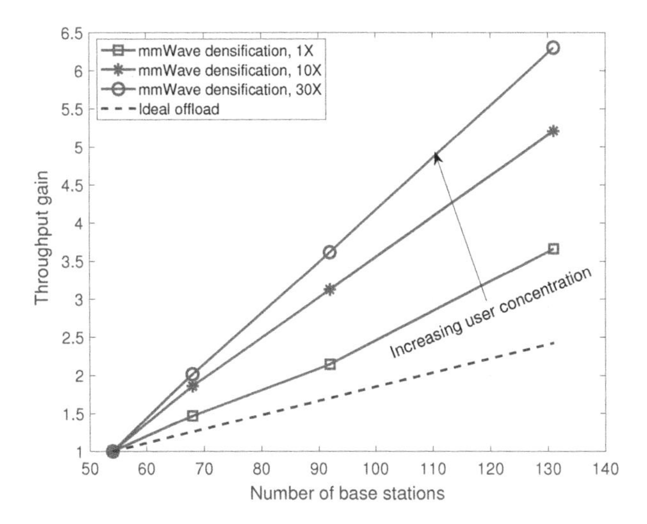

Figure 5.19 Scaling law for mmWave with different user concentrations.

users in an area, the greater the need for mmWave's defining features like bandwidth, interference resilience, etc. Figure 5.19 shows the normalized throughput ratio relative to the baseline for different levels of hotspot concentration ranging from uniform (1X) to 10 or 30 times the population density in hotspot areas relative to non-hotspot areas. This study shows that the maximum advantage with mmWave occurs when it is deployed optimally in hotspot-oriented areas.

5.5.3.3 Signal and Interference Scaling

Fundamentally, there are three effects at play when it comes to understanding the reason for a much better scaling performance with mmWave relative to sub-7 GHz-based densification. The first is bandwidth. In these results, it is assumed that mmWave systems have 800 MHz bandwidth compared with sub-7 GHz which has 100 MHz of bandwidth. As noted earlier, sub-7 GHz can go to much higher spectral efficiencies than mmWave. As outlined in the table earlier in this section, the peak spectral efficiency per layer for sub-7 GHz is 7.4 bps/Hz/layer with a maximum of 4 layers, while mmWave supports a peak spectral efficiency of 5.5 bps/Hz/layer with a maximum of 2 layers. At the peak, this reduces mmWave's rate gain over sub-7 GHz to a factor of around 3 with lower ratios expected at more typical spectral efficiencies of operations. The remaining gain has to come from some other sources. The second source of benefit is the gain in the signal strength with densification. With similar path loss models, this gain should be similar between sub-7 GHz and mmWave.

The third and most important benefit comes from the scaling of interference plus noise as base station densification increases. Due to the narrower beamwidth beams in mmWave systems, this provides a significant gain compared with sub-7 GHz where mmWave behaves noise-dominated in contrast to interference-limited

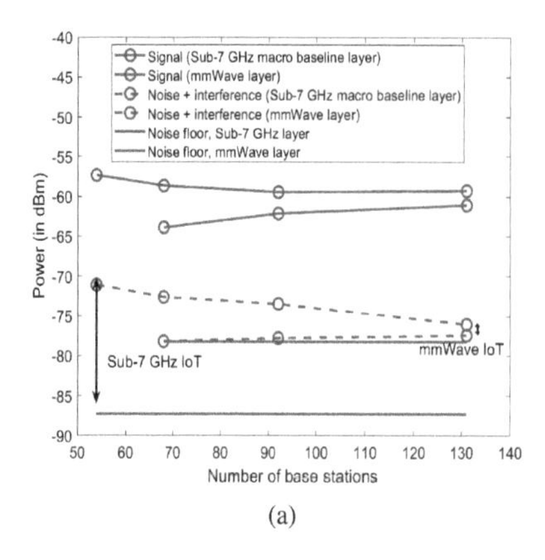

(a)

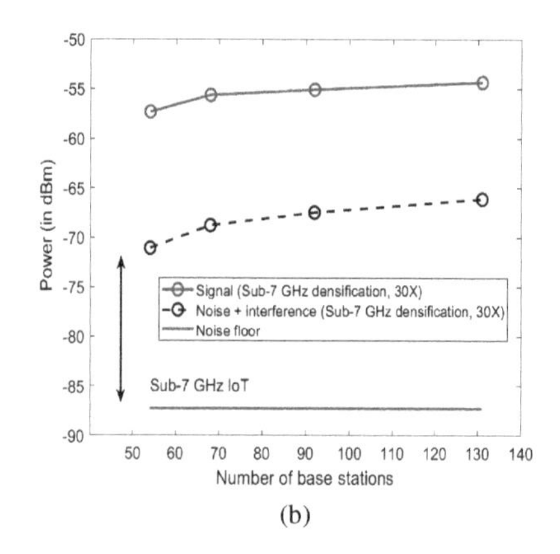

(b)

Figure 5.20 Signal and interference scaling with densification on a baseline sub-7 GHz deployment under (a) mmWave-based and (b) sub-7 GHz-based deployments.

behavior over much larger range of densities. With the signal growing potentially at about the same rate as interference, the SINR for sub-7 GHz does not improve much (if at all) with density. On the other hand, mmWave users see a substantial SINR gain as the density increases. While both technologies benefit from the spatial reuse of spectrum with densification, mmWave alone gets the added boost of an improved SINR as well. For the results shown in this section, mmWave accomplishes this even with a far smaller base station antenna aperture compared with sub-7 GHz. In particular, the sub-7 GHz deployment consists of 8 vertical and 12 horizontal cross-polarized antenna elements geared for small cell deployment over poles. On the other hand, the mmWave deployment consists of 8 vertical and 24 horizontal cross-polarized antenna elements geared for macro deployment over rooftops. This leads to an aperture/panel size of 36 cm $\times$ 55 cm at sub-7 GHz and 15 cm $\times$ 5 cm at mmWave frequencies.

Figure 5.20(a) shows the scaling of the signal and interference terms for sub-7 GHz and mmWave users with mmWave-based densification. We notice that the total interference plus noise term is barely higher than just the noise term for mmWave. On the other hand, the sub-7 GHz users experience a 12–15 dB total interference plus noise term relative to just the noise term. Further, the figure shows that there is little growth in median interference as the density of mmWave base stations increase and this is due to the narrow beamwidth beamforming. On the other hand, Figure 5.20(b) shows the scaling of the signal and interference terms for users with sub-7 GHz-based densification. We notice that the total interference plus noise term relative to just noise increases from $\sim$ 15 dB for the baseline case to $\sim$ 21 dB at the highest density. That is, the interference grows as strongly as the signal term and therefore (statistically) there are no SINR gains to be had with sub-7 GHz-based densification. This is in contrast to mmWave-based densification.

A final note on the scaling of SINR itself. Path loss models often used in practice correspond to a piece-wise linear characterization wherein the PLE is 2 till a certain distance threshold and > 2 beyond this distance. Furthermore, path loss models often distinguish between LOS and NLOS paths. When considering the scaling of signal and interference, different transition points of the signal and (dominant) interference terms with densification can cause the SINR to vary and this is true for both sub-7 GHz and mmWave path loss models. An intuitive way to think about this is in terms of the following regions (with increasing base station density).

- *Region I*: Here, the density of base stations is very low leading to the performance to be noise-limited and interference is negligible compared with noise. As the density increases, the user finds a closer base station and may also transition from an NLOS path to an LOS path. As a result, the signal term improves a lot and the SINR improves commensurately.
- *Region II*: As the density keeps increasing beyond Region I, the interference starts becoming larger than noise. Furthermore, the interference may also start transitioning from NLOS paths to LOS paths for many users. If the signal has already transitioned from NLOS to LOS earlier in density (as would be reasonable to expect), then the interference in this region grows faster than signal and the SINR can actually reduce. Another factor that may contribute to interference growing faster than the intended signal is the growing number of interfering terms. Note that there is almost always still a net user throughput benefit of densification due to higher spectral reuse in the region.
- *Region III*: When the density increases further, the system becomes even more interference-limited and both signal and interference operate in the LOS regime experiencing similar growth in their values. In this region, the SINRs tend to be flat vs. base station density and the user experiences a pure spectral reuse (or lower loading per cell) gain. Also, the flattening in this region could also be due to the idle base station effect starting to kick in offsetting the more aggressive interference growth, if Region II does happen earlier. Otherwise, it could be that both the signal and interference grow at roughly the same rate.
- *Region IV*: At very high density values, there is a greater occurrence of idle base stations and therefore the signal term grows without necessarily a commensurate increase in the dominant interference term. In this region, one can expect a modest growth in SINR again and this will be an additional source of gain in addition to offloading. In the limit where the average number of users per cell is less than 1, further densification will not yield any offloading gain, but rather only a potential SINR gain. The SINR gain will improve the rate and therefore the throughput gain will again transition similar to a logarithmic relationship. One does not expect too many systems to operate in Region IV as it can be wasteful of system resources.

One way to think about sub-7 GHz vs. mmWave using the above lens is that mmWave spends a far greater time in Region I compared to sub-7 GHz and hence is more favorable for densification. Another point to note is that not all transitions can occur in every deployment scenario, but this general trend will hold.

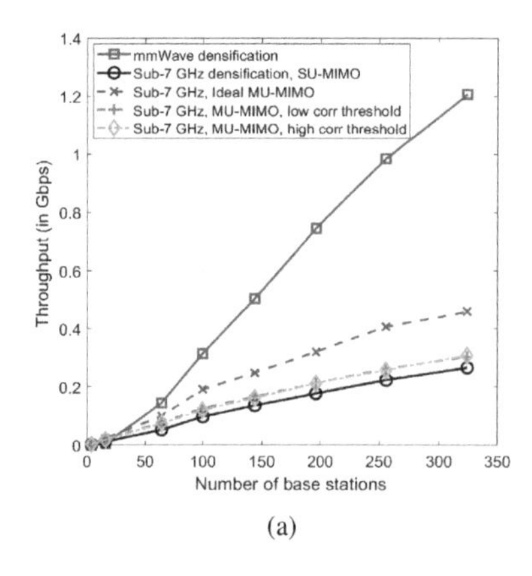

(a)

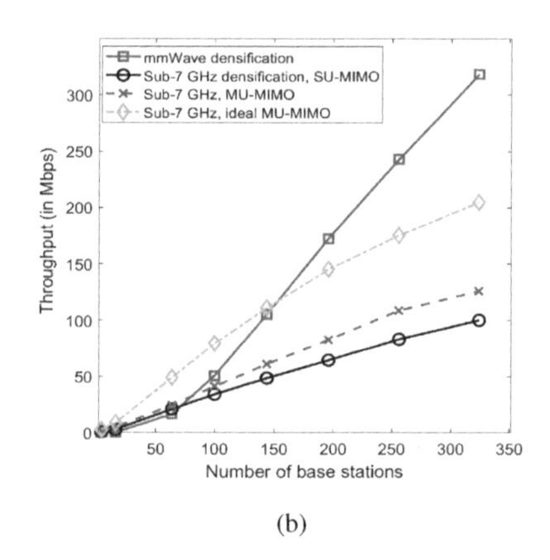

(b)

Figure 5.21 Comparison of throughput growth versus base station density for sub-7 GHz and mmWave densifications for (a) downlink throughput and (b) uplink throughput.

5.5.3.4 Impact of MU-MIMO in Sub-7 GHz Networks

Millimeter wave bands have a substantial advantage over sub-7 GHz in terms of available bandwidth for offering service. As sub-7 GHz systems move toward massive MIMO, the possibility of higher-orders of SU- and MU-MIMO become likely. The question then is whether higher-order (rank ≥ 4) MU-MIMO can offset the bandwidth advantage of mmWave systems. In this section, it is shown that even in highly idealized assumptions of MU-MIMO operation at sub-7 GHz, the mmWave bandwidth advantage still remains strong.

In contrast to other results shown earlier, we now rely on a statistical simulation to conduct this study. We use a square grid of approximately the same area as the results in the previous section. Base stations are placed in rows and columns inside this region with a wrap-around for the central region to avoid edge effects. The 3GPP UMi model [18] is used for modeling both sub-7 GHz and mmWave systems.

Different assumptions are made on MU-MIMO for sub-7 GHz: an "ideal" curve where there is no modeling of inter-stream interference and two realistic curves where inter-stream interference is counted and UEs are co-scheduled if pairwise channel correlation is below a certain threshold. The low and high correlation thresholds for user selection correspond to SNRs of -9 and -14 dB, respectively. While up to 4 UEs can be paired, we are typically limited by the number of available UEs in a sector as the base station density increases. The SU-MIMO-only case is also shown for reference. These performance plots are characterized in Figure 5.21(a). For the downlink, as long as the base station density is greater than 30, mmWave systems start outperforming even ideal sub-7 GHz MU-MIMO. As a reference, base station density of 150 corresponds to an approximate inter-site distance of ~ 250 m, which is well within the typical deployments of urban areas. In contrast, the uplink crossover point happens around 150 base stations, which is much later than the crossover observed on the downlink.

5.5.3.5 Uplink Performance

Finally, we consider uplink performance between mmWave and sub-7 GHz systems and continue to use the simple system simulation framework used for the MU-MIMO comparison of the previous section (UMi scenario). Compared with the downlink, one can reasonably expect the gains from mmWave over sub-7 GHz systems to be lower. First, the total radiated power on the uplink of a mmWave system compared with downlink is a lot smaller because the UE would have far fewer RF chains with a lower PA rating per antenna. As an example, if the base station has 256 antenna elements and the UE has 4, even assuming that the PA capabilities are the same, the total radiated power of the uplink would be lower by $\sim$ 18 dB. Since mmWave systems are noise-limited, this can severely impact the uplink link budget and performance compared with the downlink. Second, for MU-MIMO on the uplink at sub-7 GHz, there is no power distribution across links as each UE will radiate at a certain power of its own accord. The absence of power splitting among the links as would happen with downlink MU-MIMO could provide a higher MU-MIMO gain on uplink compared with the downlink. However, with the additional uplink transmissions due to MU-MIMO, interference could start limiting this gain.

In Figure 5.21(b), median uplink throughput is considered for mmWave and sub-7 GHz densifications where three different sub-7 GHz schemes are further analyzed. These three schemes include the ideal case, SU-MIMO and MU-MIMO-based. Clearly, we see that the use of MU-MIMO enhances the performance limited by SU-MIMO. Nevertheless, the mmWave densification can perform better than even an ideal sub-7 GHz densification assuming that the base station density is sufficiently large.

5.6 APPENDIX

We now consider how standardization support can enable the deployment of mmWave systems in practice. While there are a number of references such as [1, 3] for standardization aspects, we provide a brief summary of this support for beam management aspects which is the focus of deployment considerations in this chapter.

5.6.1 LEGACY BEAM MANAGEMENT ASPECTS FOR MOBILITY IN STANDARDS

Beam management is the core component of mmWave transmissions and it is also increasingly important at sub-7 GHz frequencies as beamformed transmissions with a larger number of antenna elements are considered at these frequencies.

In the downlink of 5G systems, hierarchical beam management starting from initial acquisition to beam refinement is characterized as the P-1, P-2, P-3 process in 3GPP specifications [203]. In the P-1 process, the base station and the UE perform a

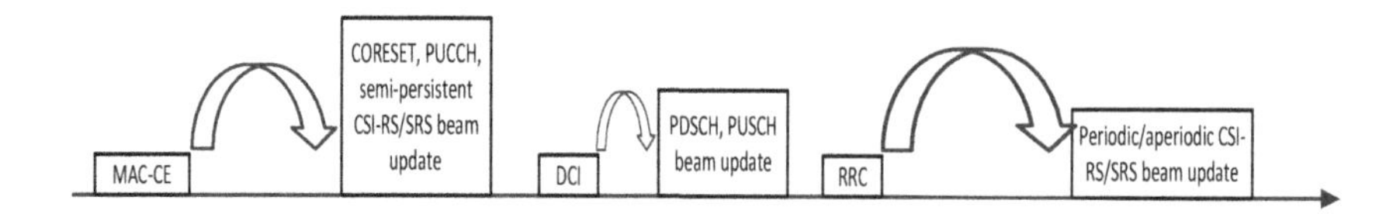

Figure 5.22 Beam management as seen in Rel. 15 with the MAC-CE, DCI and RRC-based options.

beam sweep over a set of broader beamwidth beams at both ends for establishing a link. In the P-2 process, the best beam determined at the UE in the P-1 process is kept fixed, and a set of narrower beamwidth beams around the best beam is determined at the base station and is swept to refine the base station's transmit beam. With such a refined beam determined in P-2 kept fixed at the base station end, a beam refinement process at the UE end is then perfomed in the P-3 process where a set of narrower beamwidth beams around the best beam at the UE side is beam swept. In practice, hierarchical beam management can be flexibly performed with different beamwidth-gain tradeoffs. While an equivalent version of hierarchical beam management (U-1, U-2, U-3 process) can be performed on the uplink, this process suffers from the limited link budget/total radiated power available at the UE in 5G systems. Further, while the downlink beam management can be a broadcast setup, uplink beam management has to be unicast making it further unattractive in practice.

From a signaling perspective, the synchronization signal block (SSB) is used by the UE to identify the beams that are suitable for connecting with its serving cell(s). That is, SSBs serve as reference signals for the P-1 process. The SSB is transmitted periodically, with a typical periodicity of 20 ms used in practical deployments. The base station can transmit up to 4, 8 or 64 different SSBs in one SSB burst set for the sub-3 GHz, sub-7 GHz and mmWave bands, respectively. By measuring the signal strengths of the SSBs as captured by the rank-1 power across polarizations (that is, the RSRP), the UE can identify the suitable beams for its communication with the base station for both the serving cell and for inter-cell mobility considerations (handover). Additional signals such as CSI-RSs with wider bandwidths can be used for the P-2 and P-3 processes and also for inter-cell mobility.

In Rel. 15, beam switching signaling is done on a per-physical channel basis as illustrated in Figure 5.22. On the downlink side, there are PDSCH, PDCCH and CSI-RS for channel state feedback. On the uplink side, there are PUSCH, PUCCH and SRS for beam switching applications. To switch a beam in a channel, the beam switch command will be sent from the base station to the UE. This command may be carried in the DCI of PDCCH, MAC-CE of PDSCH, or RRC signaling corresponding to PDSCH. PDCCH-based beam switching has the lowest latency (e.g., 250 us corresponding to two slots at 120 kHz subcarrier spacing). On the other extreme, RRC-based beam switching has the highest latency (e.g., 10 ms). MAC-CE-based beam switching is of intermediate beam switching latency. Another difference between these three approaches is in terms of synchronization between the base station and the UE. For DCI and MAC-CE-based beam switching, the timing of beam switching is synchronous between the base station and the UE due to the use of a Layer 1 acknowledgment channel. For RRC-based beam switching, there is an uncertainty period between the base station and the UE on when to switch to a new beam.

On the other hand, beam switching for carrier aggregation is on a per-cell basis. For example, for a UE using 8 component carriers (CCs), the base station needs to send the beam switching command on a per-CC basis. For intra-band carrier aggregation where the UE typically uses a single beam for the different CCs in the same band, the overhead of beam switching can be unnecessarily large if done on a per-cell basis. Note that the intended objective of per-cell beam switching is for flexibility reasons.

5.6.2 ENHANCEMENTS TO ADDRESS MOBILITY

As 3GPP standard specifications evolve, beam management has been enhanced in multiple ways:

- Lower latency beam switching time across channels/signals
- Lower overhead for carrier aggregation
- Lower interruption time for inter-cell mobility
- Higher reliability for both control and data channels and
- Higher-rank support for data channels (PDSCH and PUSCH).

More enhancements are in consideration such as higher flexibility for full-duplex operation at the base station and overhead/power consumption reduction with machine learning-based beam management solutions at both the base station and UE.

In addition, reliability and consistency of beam switching across CCs is also being studied. For example, if MAC-CE is used to switch the beam per CC, unless the UE decodes those PDSCHs carrying the beam switching command via MAC-CE at the same time, when to switch its beam is a practical issue given that there is only one beam. For this scenario, Rel. 16 introduces a CC group-based beam switching where a single beam switching command can be used to switch the beams for all the CCs in a group. In practice, the UE may use a single beam for all relevant channels such as PDCCH, PDSCH, PUCCH and PUSCH. Another simplification in Rel. 16 is to introduce a transmit beam at the UE (for uplink transmissions) following the receive beam (for downlink). In Rel. 17 and 18, a further enhancement is to allow a single beam switching command to switch the beams for all channels via a unified TCI state and such a single command can be based on DCI, as illustrated in Figure 5.23.

In some of the commercial deployments, an integrated distributed unit (DU) and radio unit (RU) are used, which is the Split-Option 2 in an Open-RAN architecture. In this case, each sector of a site corresponds to a DU and each sector may host just a single cell at a given carrier frequency (multiple carriers are also used across different frequencies to support carrier aggregation for UEs). In a typical illustrative example, there are three cells used in each area. When the UE crosses the cell boundary, Layer 3-based mobility is triggered. Layer 3-based mobility involves measurement reports with filtering over 200 ms (for example). RRC signaling is used to

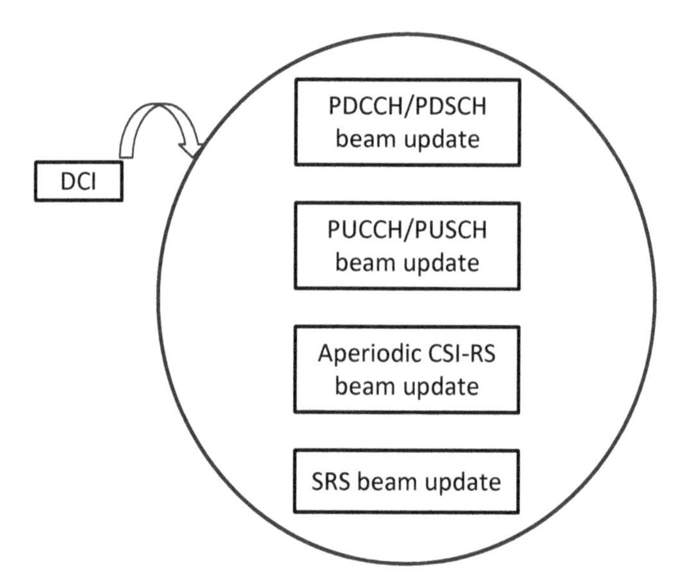

Figure 5.23 Beam management as seen in Rel. 17 where the DCI updates the beams for multiple downlink/uplink channels/RSs.

trigger handover from one cell to another cell. After triggering handover, the UE may go through a random access channel (RACH) procedure, which can be contention-based or contention-free. The first message known as physical random access channel (PRACH), common to both contention-based and contention-free approaches, is transmitted by the UE to the target cell. Such a signal is used by the target cell to know the incoming UE. The signal is used to estimate the timing advance (TA) needed for the UE to transmit in the uplink toward this target cell.

The RACH procedure may incur large latencies due to a number of reasons.

- PRACH opportunities may have longer periodicities (e.g., 40, 80 or 160 ms) due to the need to reduce overheads. At mmWave frequencies, for each downlink SSB, there is an associated uplink PRACH transmission that is paired as the PRACH transmission is used to indicate the downlink beams that the UE may have chosen for receiving subsequent downlink and data. For coverage considerations, PRACH may use more symbols compared with one SSB. For example, one commonly used PRACH format in the existing deployment may use six symbols: it is 50% more compared with one SSB (four symbols in time domain). Many commercial deployments use downlink-heavy slot formats (e.g., DDDSU). As a result, uplink opportunities are limited. To support a larger number of SSBs with a one-to-one mapping between SSB and PRACH occasions, PRACH transmission opportunities are stretched in the time domain leading to longer latencies. In particular, for a 20 ms periodicity of SSB transmissions, PRACH periodicity may be 160 ms. Such a configuration with longer PRACH periodicities may introduce large data interruption time during handover, as there is no data communication after the UE sends PRACH until it is acknowledged.

- Additional latency will be added for contention-based RACH where RACH messages referred to as Messages 2, 3, and 4 are further exchanged between the base station and UE before actual data communications can resume. Such latency introduces undesired effects[3] for applications such as eXtended Reality (XR) which are delay-intolerant. If Layer 3-based handover is introduced (e.g., every 80 ms), data interruption time can be reduced and user experience will be greatly improved by the data pipe between the base station and the UE.

To overcome this large data interruption time with Layer 3-based mobility, Layer 1/Layer 2-based inter-cell mobility has been standardized in Rel. 18. Compared with Layer 3-based mobility, there are a few key enhancements. The first enhancement is that there is Layer 1-based measurement and reporting for candidate cells. With Layer 3-based measurements, the link quality reflects the average channel condition by using filtering with a typical averaging time of 200 ms. In addition, the measurement reporting is via RRC signaling which incurs additional latency compared with an Layer 1-based report. The second enhancement is that a MAC-CE is used to signal the handover command instead of using RRC for Layer 3-based mobility. This will reduce the signaling latency from several tens of ms to single digit ms level. The third enhancement is to move RACH procedure ahead of time: before cell switching/beam switching. One main usage of the RACH procedure for Layer 3-based mobility is to allow target cell to estimate TA for the uplink transmissions of the UE. By moving this procedure ahead of the beam switching time, the latency due to longer periodicity of PRACH is significantly reduced. Finally, one area to further enhance performance is earlier channel state feedback acquisition. By measuring CSI for the new beam to be used in the target cell before beam switching, the base station can use a better MCS/rank for scheduling right after the beam switching command has been issued.

With the incorporation of the above enhancements to beam management both within and across cells, performance of mmWave systems under mobility has been greatly improved. In addition to the larger available bandwidth and low latency from shorter symbol durations, applications such as XR can be fruitfully addressed with improvements to both a specific user as well as in terms of network level considerations such as how many users can be supported.

5.6.3 STANDARDS SUPPORT FOR VARIOUS RELAY NODE-TYPES

A brief synopsis of 3GPP standardization support for the different relay node-types described in Section 5.2 is now provided.

- *Simple and semi-smart repeaters*: 3GPP standardization support for simple repeaters is no different from that of their 4G counterparts. The main aspects

[3]Splitting the computing between the device/edge and the cloud is an efficient way to enable XR applications. The server in the cloud renders the video based on pose information provided by the device via a 5G link between the base station and the device (which in this case is a XR headset). In some of the XR applications, downlink data packets arrive in an almost periodic manner (e.g., every 10 ms), while uplink data packets which carry pose information may arrive in a periodic manner (e.g., every few ms).

specified are emission requirements [233] and electromagnetic compatibility requirements [234]. Nothing else needs to be specified as there is no signaling or provisioning that requires inter-operability between vendors. Semi-smart repeaters are handled in a manner agnostic to the standardization support and would need to conform to the requirements of simple repeaters.

- *Smart repeaters*: In contrast to simple repeaters, for a smart repeater the specifics of a "side-channel" for passing control information needs to be specified. The side channel for dynamic information is supported via a number of alternatives: RRC-signaling (slowest), MAC-CE (moderate) and DCI-based (fastest). Bootstrapping takes place using RRC and subsequent control information is carried on MAC-CE or PDCCH DCI with a special format.
- *IAB*: For all practical purposes, the IAB relay node may be viewed as combining the functionality of a UE (to send to or receive from the parent node on the uplink/downlink) and base station (to send to or receive from the child node on the downlink/uplink). RRC and resource management are done only in the root node and protocols[4] F1-U and F1-C are established to facilitate communication between the parent and child nodes to accomplish these tasks.

[4]F1 is the interface that connects a base station's CU to the base station's DU. This interface is applicable to the CU-DU split base station architecture. The control plane (F1-C) allows signaling between the CU and DU while the user plane (F1-U) allows the transfer of application data.

6 6G Evolution

The evolution of mmWave research from 5G-NR into 6G designs is now the focus of standardization efforts. In this chapter, we provide a brief overview of where 6G is headed.

We start with the spectrum landscape as we evolve from 5G into 6G. As more spectrum is opened up, the main challenges lie in the design of device components that can be used over this spectrum and in managing interference to coexisting services. We explain how the regulatory view on coexistence can evolve in 6G. Since lower cost devices entail the use of wider bandwidth RF components. Hence, we then focus on the challenges in wideband operations that are common to mmWave, sub-THz and cmWave bands and the necessary optimizations at the RF and antenna level these operations lead to.

For UEs that are in the same capability class as today's systems, we focus on device-level optimizations and consider challenges from the perspective of antenna module design and placement. We show how antenna module placement optimized for blockage performance can significantly improve the user experience at the UE via improved rates and better coverage. We then consider the challenges posed by newer capabilities such as full-duplex operation.

As network densification will remain a continuing theme as we evolve into 6G, reducing the energy consumption is necessary for a more sustainable and greener future. Network energy savings and wireless fronthaul are two important features to enable further topology enhancements in 6G. As the use cases served by 6G expand, "sensing as a service" that utilizes both mmWave and upper mmWave band frequencies toward the goal of joint communications and sensing (JCAS) is becoming relevant.

Most (if not all) of these emerging use cases can benefit from the artificial intelligence/machine learning (AI/ML) revolution that has dominated image and video processing advances today. In one particular example of the application of AI/ML techniques to mmWave systems, we showcase how beam prediction and beam management can be simplified. We illustrate the tradeoffs in the application of these techniques at the base station and UE.

6.1 SPECTRUM LANDSCAPE AND EMERGING CHALLENGES

Given the current set of frequency bands that have been defined at 3GPP (see Table 1.1), future spectrum allocation[1] can be in frequencies beyond 71 GHz or in the 7.125–24.25 GHz range.

[1] At 3GPP, FR4 and FR5 are defined as 71–114.25 GHz and beyond 114.25 GHz, respectively. We can loosely call them upper mmWave and sub-THz bands, respectively. FR3 is defined as the 7.125–24.25 GHz band.

DOI: 10.1201/9781032703756-6

Regarding FR4 and FR5, a significant amount of ongoing research to improve mmWave systems, both at the device level as well as in terms of algorithms, will also be beneficial for systems at these bands. Initial product design in this green-field spectrum will follow the approach established for mmWave systems today. For instance, new RFIC chips with on-chip or on-board antenna arrays, finite-precision phase shifters and amplitude/gain control will be implemented. There are many challenges in the design of upper mmWave and sub-THz systems with the key problem being UE power consumption. Simply scaling the power demands of mmWave systems to account for 100s of Gbps data rates (envisioned for the sub-THz bands) results in unrealistically high power consumption. Therefore, 6G will have to provide much better performance in terms of energy efficiency (information bits transferred per Joule expended) compared with 5G [249, 250]. As bandwidth is abundant at sub-THz frequencies, the traditional paradigm of trading off processing power with utilized wireless channel bandwidth can be changed by the following:

- Use of new physical layer waveforms
- Design of new channel codes which tradeoff spectral efficiency for better power/energy efficiency [251, 252]
- New antenna designs utilizing concepts inspired from the optical domain (e.g., lenses [106]) which have the potential to lower RF power requirements
- New approaches for beam management and channel equalization
- New RF front end designs to improve efficiency of elements like PAs and LNAs for the new bands.

From a beamforming perspective, the higher frequencies considered for 6G also present a fundamental challenge starting at the antenna domain. As the area of a well-radiating antenna element is proportional to λ^2, more antenna elements are required at higher frequencies to maintain the same EIRP at the transmitter or the energy gathering area at the receiver as in lower frequencies. Narrower beamwidths due to the use of increased antenna dimensions can lead to reduced interference, but also to increased latencies in beam acquisition, increased power consumption, and higher thermal overheads [147, 73]. Maintaining spherical coverage guarantees at higher bands can also become challenging due to poor radiative capabilities of antenna structures in a form factor UE design. Enhanced hand blockage losses at higher frequencies and the use of beam management to avoid blockages and obstacles also becomes challenging with narrower beamwidths.

Regarding RF front end power efficiency, as the frequency increases, the power efficiency of complementary metal oxide semiconductor (CMOS) transistors decreases, possibly requiring the use of non-CMOS semiconductor solutions beyond 170 GHz. Furthermore, LO phase noise is expected to increase in the higher bands and limit the highest spectral efficiency that can be achieved. Sensitivity to atmospheric conditions, rain, and fog in some bands can become more pronounced demanding innovative system solutions. To overcome coverage holes in deployment, densified networks are envisioned by the use of smart repeaters and relay nodes. These may possibly use passive or active elements (also called as

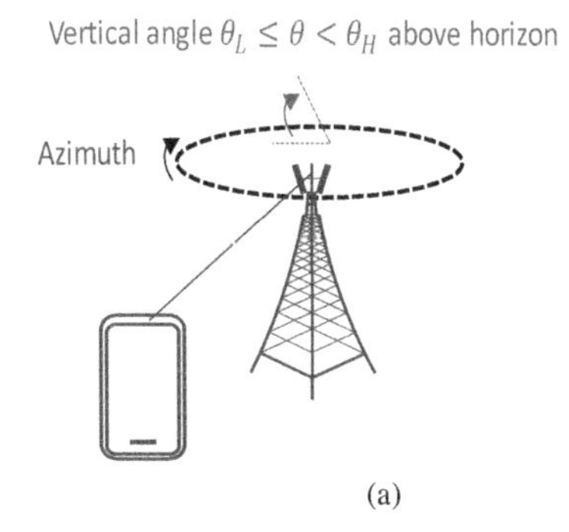

(a)

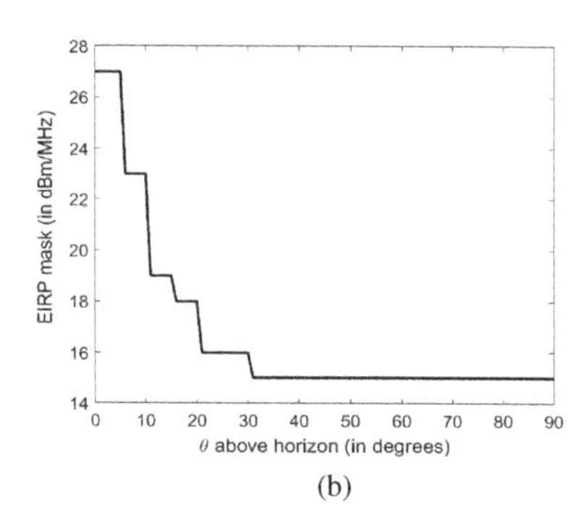

(b)

Figure 6.1 (a) A typical coexistence setup with a base station serving as aggressor node and communicating with a UE. (b) A typical EIRP mask regulating emissions above the horizon.

reconfigurable intelligent surfaces or RISs) to engineer a better effective channel matrix. However, the cost-complexity-power-performance tradeoffs of such intermediate nodes needs to be fully understood. The readers are referred to works such as [253, 28, 254, 255, 256, 257, 258, 259, 260, 261] for a discussion of challenges at the upper mmWave and sub-THz bands.

Spectrum sharing is expected to be the dominant theme of spectral evolution at the lower frequency end. From a device design perspective, given that the intermediate frequencies of FR2 RFIC chips are in FR3 (see discussion in Chapter 3.4.2) and a number of terrestrial and satellite services dominate these frequencies, coexistence issues take prominence. That is, as the RF signal is down-converted to IF, the signal at this IF can interfere with coexisting services at these frequencies. Note that a more recent coexistence issue of interest has been the one between airplane radio altimeters (RAs) operating in the 4.2–4.4 GHz range and C-band cellular services operating at up to 3.98 GHz. Leakage of radiation from cellular services to RAs can lead to safety issues and poor performance of RAs.

From a deployment perspective, Figure 6.1(a) presents a typical coexistence setup wherein a base station serves as an aggressor node and is communicating with a UE. A potential victim node at an elevation angle (e.g., a satellite) experiences interference from the aggressor node. To mitigate this interference, one way forward is to limit the maximum emissions (or EIRP) over the set of elevation angles via regulatory requirements. A typical example of such an "EIRP mask" specified for elevation angles above the horizon is presented in Figure 6.1(b) where the emissions are averaged over the azimuth plane [262, p. 466]. By carefully limiting the emissions from potential aggressor nodes to be within a specified EIRP mask, coexistence can be guaranteed at potential victim nodes. However, a more aggressive pursuit of such regulatory requirements can hurt the intended communications performance of these networks. Thus, a careful compromise between mitigating interference and enhancing intended communications needs to be investigated.

Other challenges include how the RF and antenna design at mmWave bands can be leveraged for good performance at the upper frequencies of FR3. In particular, design of antenna elements/arrays at these frequencies can either be based on leveraging antenna module solutions from FR2 contingent on the available aperture at the UE, or can be based on discrete antenna elements from FR1. The cost vs. performance tradeoffs between these two alternatives need to be well-understood. The readers are referred to a recent work in [263] for understanding the opportunities and challenges at the FR3 band.

6.2 CHALLENGES IN WIDEBAND OPERATIONS

Given the large amounts of spectrum expected to be available in 6G, wideband operations is a common theme across these frequencies. While the digital front end part of today's cellular networks can be optimized for wideband operations, optimization of the RF front end part is far more challenging.

Typical challenges with wideband operations include the following:

- Channel structural changes across a wide bandwidth such as notches and peaks due to constructive/destructive interference of signals across different layers of materials (see, for example, Chapter 2.2.3.1 for similar behavior at mmWave frequencies)
- Non-linear behavior of RF components over wideband operation (e.g., PAs, LNAs, phase shifters, etc.)
- Limitations in terms of spectral efficiencies due to RF impairments such as phase noise
- Precoder optimization and channel state feedback for wideband systems [228].

We now focus on one specific challenge relevant in antenna array design. Operation of an antenna array at a single carrier frequency can be easily optimized by a careful design of the inter-antenna element spacings relative to that carrier frequency. However, operation at a single carrier frequency is an ideal assumption given the wideband nature of signaling in upper mmWave systems. Wideband transmissions lead to mismatches between intended and observed directions of the main lobe, side lobes, nulls, etc. These are typically called as *beam squinting* effects [25, 62, 264]. To reduce the impact of beam squinting, inter-antenna element spacing optimization is thus an important component of antenna array design for both legacy (lower mmWave) as well as upper mmWave systems.

If the inter-antenna element spacing is made large (relative to a baseline spacing of $\lambda/2$ at a certain carrier frequency), the antennas are decoupled from (or uncorrelated with) each other. This can be a distinct advantage for higher-rank transmissions with the antenna array provided the channel is sufficiently rich to afford this capability. On the other hand, larger inter-antenna element spacings create repetitions in the array pattern called as *grating lobes* or *fringes*, which constitute interference in unintended directions. From Chapter 4, we reproduce the array gain at an angle θ for a ULA structure below:

$$\mathsf{SNR}(\theta) = \frac{A^2}{N\sigma^2} \cdot \left| \frac{\sin\left(\frac{\pi N d \cdot (\sin(\theta) - \sin(\theta_0))}{\lambda}\right)}{\sin\left(\frac{\pi d \cdot (\sin(\theta) - \sin(\theta_0))}{\lambda}\right)} \right|^2. \tag{6.1}$$

From (6.1), note that both the numerator and denominator of $\mathsf{SNR}(\theta)$ are zero if

$$\frac{\pi d \cdot (\sin(\theta) - \sin(\theta_0))}{\lambda} = n\pi, \; n \neq 0. \tag{6.2}$$

In other words, for a θ satisfying (6.2), a peak gain of N is seen. Rearranging this equation, we have

$$d = \frac{n\lambda}{\sin(\theta) - \sin(\theta_0)}. \tag{6.3}$$

Thus, we have

$$d \geq \frac{\lambda}{|\sin(\theta) - \sin(\theta_0)|} \geq \frac{\lambda}{|\sin(\theta)| + |\sin(\theta_0)|} \geq \frac{\lambda}{1 + |\sin(\theta_0)|}. \tag{6.4}$$

For the boresight direction ($\theta_0 = 0^\circ$), we require that $d \geq \lambda$ and for the endfire direction ($\theta_0 = 90^\circ$), we require that $d \geq \lambda/2$. In other words, if $d \geq \lambda$, a grating lobe is seen for any scanning angle whereas $\lambda/2 < d \leq \lambda$ leads to a grating lobe at a certain scanning angle. To illustrate this issue, we assume a ULA structure and plot $\mathsf{SNR}(\theta)$ as a function of θ for $N = 16$ with different choices of d/λ in Figure 6.2 for $\theta_0 = 0^\circ$. In general, there are $2d/\lambda$ repetitions of the basic beam pattern over the $(-180^\circ, 180^\circ)$ angular region.

In contrast, if the inter-antenna element spacings are made small (relative to $\lambda/2$), adjacent antennas can mutually couple with each other resulting in an inability to fully reap the degrees of freedom possible with the antenna dimensions (e.g., full array gain). For spacings below $\lambda/2$, plotting the beam pattern as described in (6.1) shows a widened main lobe whereas the side lobes are at the same level as with $d = \lambda/2$. This observation suggests that the total energy radiated by these set of beam weights is enhanced for the smaller antenna spacings. The main reason for this discrepancy is that the formula in (6.1) poorly estimates the SNR with increased mutual coupling. In these settings, instead of using (6.1), one must resort to electromagnetic simulations (e.g., based on HFSS [166]) that incorporate mutual coupling.

For a robust performance avoiding the tradeoffs between grating lobes and mutual coupling, it is typical to uniformly space antennas at a $\lambda/2$ separation. This works for the case for a linear array architecture. On the other hand, a planar array is expected to provide both azimuth and elevation coverage (via 3D/FD MIMO beamforming [23]). At the base station, coverage is typically that of a sector which is 90°-120° in azimuth and 30°-60° in elevation. Given the typically narrow elevation coverage required for a base station, from (6.4), the problem of grating lobes

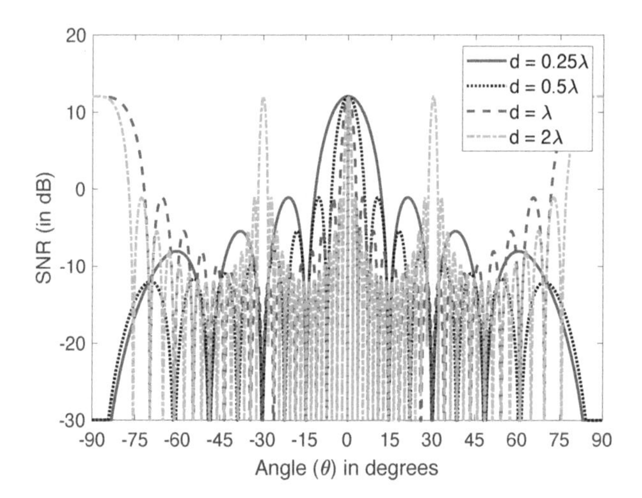

Figure 6.2 Grating lobe beam pattern with different inter-antenna element spacings.

can be avoided[2] within the intended elevation coverage area despite increasing the inter-antenna element spacings in the elevation domain beyond 0.5λ. If $d > \lambda/2$, in addition, increased elevation gain can be accrued due to the larger size of the radiator. Thus, many practical base station designs use a 0.5λ spacing for the antenna elements in the azimuth domain and even up to a 0.8λ spacing ($\pm 20°$ coverage) for the antenna elements in the elevation domain. Planar array designs at the UE (if they are used) continue to use a 0.5λ spacing in both azimuth and elevation domains since real-estate and aperture size become crucial design issues.

Another complication in inter-antenna element spacings is that antenna elements are expected to work seamlessly over the wide bandwidths at mmWave carrier frequencies and beyond (multiples of GHz in some scenarios). For example, from Table 1.1, we see that the n257 band covers 26.5–29.5 GHz whereas n258 band covers 24.25–27.5 GHz. As carrier frequencies increase, unlicensed as well as licensed coverage are expected in the WiFi bands from 57–71 GHz. Further focus in 3GPP specifications is also in the sub-THz regime beyond 114.25 GHz (for example, the 141–148.5 GHz band is of broad interest). In these scenarios, it is typical to choose the inter-antenna element spacing to be $\lambda/2$ for a specific carrier frequency within the wideband with $> \lambda/2$ spacing or $< \lambda/2$ spacing at many frequencies within this wideband. The precise choice of the carrier frequency with $\lambda/2$ spacing is then a subject of careful optimization based on multiple competing issues such as gain seen over the wideband, number of antenna elements desired (which in turn depends on the MAPL), form factor constraints, packaging technology used, etc.

One common design choice is to set the inter-antenna element spacing to $\lambda/2$ for the highest carrier frequency of interest, thus totally avoiding grating lobes. A

[2]However, as we noted in Chapter 6.1 on coexistence issues, interference to coexisting services in the unintended elevation coverage region should also be considered.

better design choice is to select the inter-antenna element spacing for the highest carrier frequency according to (6.4) to avoid grating lobes within only the coverage area intended. With a spacing of $d = \lambda_0/2$ and phase shifter weights set for this frequency as:

$$\phi_i = \frac{2\pi(i-1)d\sin(\theta)}{\lambda_0} = \pi(i-1)\sin(\theta), \tag{6.5}$$

we have the following gain at a general λ:

$$\mathsf{SNR}(\theta,\lambda) = \frac{A^2}{N\sigma^2} \cdot \left| \frac{\sin\left(\pi N \cdot (\sin(\theta) - \sin(\theta_0)\lambda_0/\lambda)\right)}{\sin\left(\pi \cdot (\sin(\theta) - \sin(\theta_0)\lambda_0/\lambda)\right)} \right|^2. \tag{6.6}$$

If $d = \lambda_0/2$ is set for the highest carrier frequency of interest, we have $\lambda > \lambda_0$ and a gain of N is seen for $\mathsf{SNR}(\theta,\lambda)$ at

$$\theta = \sin^{-1}\left(\sin(\theta_0)\lambda_0/\lambda\right). \tag{6.7}$$

Nulls in $\mathsf{SNR}(\theta,\lambda)$ are seen at

$$\pi N \cdot (\sin(\theta) - \sin(\theta_0)\lambda_0/\lambda) = n\pi, \; n \neq kN, \; n \neq 0 \tag{6.8}$$

$$\Longrightarrow \theta = \sin^{-1}\left(\sin(\theta_0) \cdot \frac{\lambda_0}{\lambda} + \frac{n}{N}\right), \; n \neq kN, \; n \neq 0. \tag{6.9}$$

A grating lobe (of gain N) is seen at $n = kN$ in the above equation:

$$\theta = \sin^{-1}\left(\sin(\theta_0) \cdot \frac{\lambda_0}{\lambda} + k\right), \; k \in \mathbb{N}, \; k \neq 0. \tag{6.10}$$

Side lobes of $\mathsf{SNR}(\theta,\lambda)$ correspond to

$$\pi N \cdot (\sin(\theta) - \sin(\theta_0)\lambda_0/\lambda) = n\pi/2, \; n \,\mathsf{is\,odd}, \; n \geq 3 \tag{6.11}$$

$$\Longrightarrow \theta = \sin^{-1}\left(\sin(\theta_0) \cdot \frac{\lambda_0}{\lambda} + \frac{n}{2N}\right). \tag{6.12}$$

Thus, as λ changes (with $\lambda > \lambda_0$), directions corresponding to peak gain, nulls, side lobes and grating lobes all change. These observations correspond to aliasing in the spatial domain. The antenna gain at the lowest carrier frequency of interest is smaller and dependent on d used at the highest carrier frequency of interest. The frequency range of interest, associated gains over this range and deviations from intended beam designs (including grating lobes) determine the optimal inter-antenna element spacing. All the antenna array design as well as RF challenges need to be amicably solved for wideband operations at higher carrier frequencies.

6.3 ANTENNA MODULE PLACEMENT OPTIMIZATION

As described in Chapter 3, multiple antenna modules are necessary at the UE side to guaranteed good performance of mmWave systems. However, the use of a large

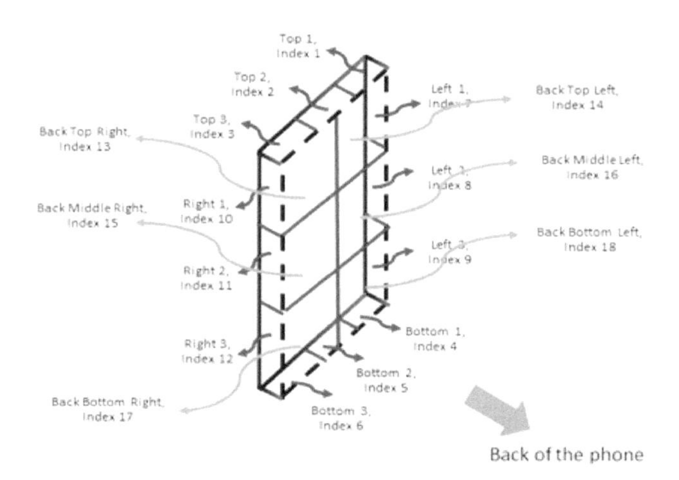

Figure 6.3 UE's edges and back face partitioned into $L = 18$ possible locations for the placement of a single antenna module.

number of antenna modules leads to increased cost and area/real-estate consumed at the UE, as well as increased power and thermal considerations. Further, with competition across antenna elements designed for different applications (e.g., BlueTooth), optimal antenna module placement at mmWave frequencies is a heavily contested issue [76]. Thus, many original equipment manufacturers (OEMs) are considering the use of a small number of antenna modules (e.g., two) to meet mmWave performance requirements.

Given that hand/body blockage is a serious detriment in the performance of mmWave systems, good antenna module placements that are robust to blockage are essential for enhancing the user experience at mmWave carrier frequencies (and beyond). At this moment of mmWave system commercialization, this problem is addressed in a qualitative manner based on tradeoffs in terms of UE layout/form factor considerations, a qualitative assessment of likely hand holding modes, available real estate for antenna module placement optimization, etc. We now propose a quantitative/ML framework consisting of six steps toward the objective of determining good antenna module placements.

- **Step 1 (Candidate Antenna Module Locations):** As illustrated in Figure 6.3, we begin by partitioning the UE's surface into different candidate locations. In each of the candidate locations, a distinct antenna module (controlled by an RFIC chip) can be located. With L such partitions and K potential antenna module locations, we have $\binom{L}{K}$ possible UE designs with K antenna modules in the UE that can be studied from a performance tradeoff perspective.
- **Step 2 (Representative Hand Holding Grips):** We then consider many example/illustrative hand holding grips/use cases adopted by users and which are representative of usage with the current generation of UEs. These hand holding

use cases encompass the portrait mode, the landscape mode, as well as non-portrait/non-landscape (or miscellaneous) modes of operation. Each hand holding use-case blocks different locations/parts of the UE and thus has different impacts on the design of the UE based on the candidate antenna module locations.

- **Step 3 (Distribution Function of Phone Orientation and Likelihood of Each Hand Holding Grip):** We look at user experience (UX) studies on how users typically hold their phones. The UX studies typically involve a survey sampling a large user base/population and its behavior in terms of:
  1. Hand positions and grips (whether the user is left- or right-handed, number of fingers, part(s) of the phone that are best suited for sensor locations, etc.)
  2. Modes of usage (portrait, landscape, etc.)
  3. Application types used (voice call, video downloading/viewing, gaming mode, etc.)
  4. Angle between the plane of the phone and the global horizontal plane, etc.

  From these survey studies, we can estimate a likelihood/probability that a random user will hold the UE in these different hand holding grips.
- **Step 4 (Spherical Coverage in No Blockage/Blockage Scenarios):** For each one of the candidate antenna module locations, we can compute the array gain coverage over a sphere around the UE (spherical coverage) in scenarios with and without blockage. These computations can be based on measurements from an offline beam characterization procedure in an anechoic chamber, or based on simulations with a hand blockage model used to capture the impairments of blockage.
- **Step 5 (Priors on Path Loss and AoA/ZoA):** We generate statistical models for the AoA/ZoA of the dominant clusters between the base station and the UE, along with their path loss from representative deployment studies of practical networks in outdoor and indoor settings. These deployment studies correspond to practical base station placements (e.g., stand-alone or co-located with a sub-7 GHz macro) and ISDs that capture the level of densification of the network. These studies also capture the impact of physical blockages in the network (e.g., buildings, vehicles, humans, etc.). The dominant cluster could correspond to an LOS or an NLOS scenario and is hence deployment-dependent.
- **Step 6 (Computing the Effective Spherical Coverage):** We localize the spatial attention of the array gain to that part of the sphere corresponding to the AoA/ZoA of the dominant reflections in the NLOS channel or the LOS direction (that is, the dominant MPCs in the channel). We do this based on the statistical models for the AoA/ZoA and the models on the likelihood of the different hand holding grips in either portrait-heavy, landscape-heavy or equal weightage for portrait and landscape orientations. For each candidate antenna module location pair, we make a localized estimate of the effective spherical coverage. Good antenna module locations are then identified based on KPIs such as uplink outage or uplink throughput at different percentile points.

We now provide some illustrative examples of how the above approach can be utilized for antenna module placement optimization. Towards this goal, we assume that each edge of the UE is partitioned into three regions and the back face of the UE

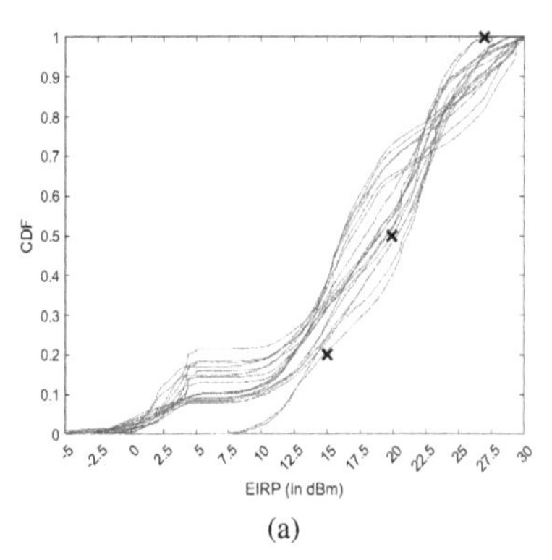

(a)

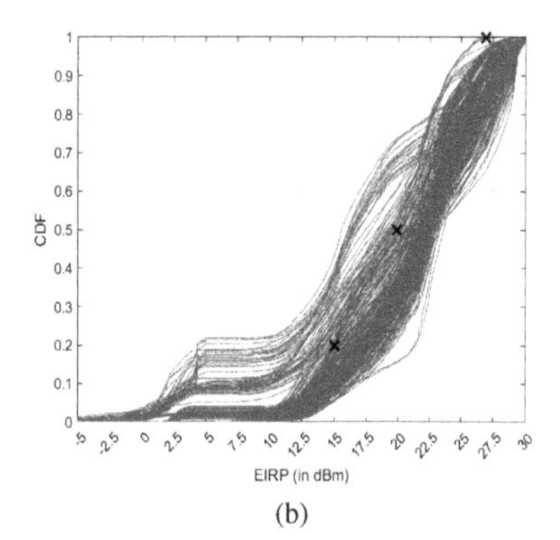

(b)

Figure 6.4 Effective spherical coverage with (a) 18 devices with a single antenna module and (b) 153 devices with two antenna modules.

is partitioned into six regions as illustrated in Figure 6.3. Assuming that the UE has a single antenna module, there are $L = 18$ possible locations for this antenna module and correspondingly, 18 CDF curves that capture the effective spherical coverage. For the hand holding modes, we consider a representative holding in portrait mode and a representative holding in landscape mode with each being given equal weightage. In addition to the realized EIRP over the sphere plotted in Figure 6.4(a), we also display (in black crosses) the EIRP requirements for the current generation of UEs as mandated by a typical network operator [75] at the 100-th, 50-th and 20-th percentile levels. From Figure 6.4(a), we observe that no single antenna module location passes the EIRP requirements at the 20-th percentile level, whereas at the 50-th percentile, only four single antenna module locations pass the EIRP requirements.

In the two module case, Figure 6.4(b) illustrates the EIRP with priors on UE orientation, AoA/ZoA, hand holdings, etc. for all the $\binom{L}{K} = \binom{18}{2} = 153$ antenna module location pairs. From this figure, we observe that the EIRP requirements eliminate some of the poor antenna module placements which are to the left of the EIRP requirements, but a significant number of antenna module placements are useful in meeting the requirements. Thus, the study of field performance (e.g., with blockage and with priors on AoA/ZoA) and comparing them with the existing EIRP requirements implicitly lead to the use of two antenna modules in good placements at the UE, without such a requirement being explicitly mandated by network operators.

Carefully parsing the trends in Figure 6.4, we note that from a UE design perspective, when there is a degree of freedom associated with the use of a pair of antenna modules, placing them proximate to each other is a *bad* idea. This is because both modules can be simultaneously blocked with certain types of hand holdings/use-cases. On the other hand, separating them as far apart as possible, akin to the design of sub-7 GHz UE design, is also a *bad* idea. This is because the main determinant of good performance at mmWave carrier frequencies (and beyond) is not the separation between modules, but their relative orientation in mitigating the effect of different hand holdings. This study showed that ensuring that one of the two modules provides good coverage for portrait-type hand holdings and another module provides good coverage for landscape-type hand holdings is a better design idea. One such example is the use of the top edge for one of the modules and the middle strip of the back face (including perhaps the middle portion of the left/right long edge) for the other module. Such an investigation of antenna module placements for other carrier frequencies is important and relevant in the context of 6G.

6.4 FULL-DUPLEX COMMUNICATIONS

The topic of full-duplex communications has been explored in the context of sub-7 GHz frequencies for many years; see, e.g., [265, 266] and the references therein. In the context of mmWave networks, full-duplex has received recent interest as seen in [267] and [268]. Despite the possible upside of doubling time-frequency resources compared with half-duplex systems, realizing the potential of full-duplex communications in practical systems requires solving many challenges. These include:

- Self-interference from imperfect antenna isolation between the transmitter and the receiver
- Reflections in the environment due to clutter
- Challenges in self-interference cancellation
- Interference coordination across downlink and uplink transmissions on neighboring cells
- Realizing dynamic downlink and uplink transmissions as needed by the traffic conditions.

With rapid advancements in RF technology especially at mmWave frequencies, both the UE and network infrastructure are evolving to operate using multiple panels with good isolation properties. As illustrated in Figure 6.5, in full-duplex operation,

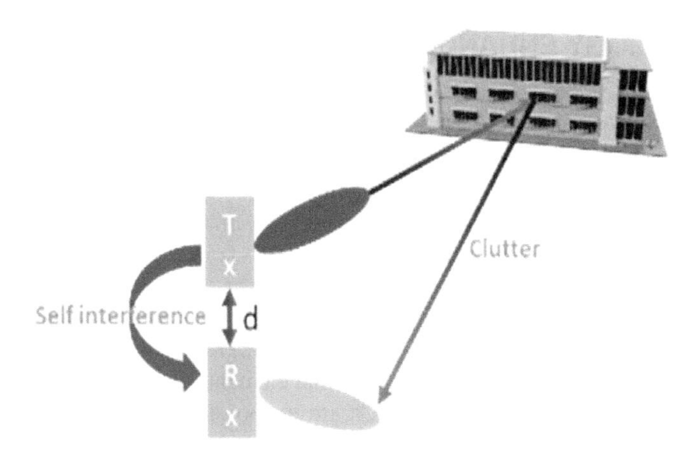

Figure 6.5 Illustration of coupling and clutter impact in a full-duplex system.

there are two types of interference for the receiver in addition to the interference from other transmitters. The first type of interference is the coupling from its co-located transmitter (e.g., the output of PA coupled to the input of LNA of the receiver). The second type of interference is clutter where the transmitted signal may be reflected (e.g., by a building) to the receiver. Due to the shorter distance between the two arrays compared with the clutter, the interference caused by the clutter has longer propagation delay. The coupling (or spatial isolation) is a function of:

- Distance between the transmit and receive arrays
- Placement of the transmit and receive arrays (e.g., front and back placement vs. top and bottom placement of the two subarrays)
- Frequency reuse across the transmit and receive operations (e.g., fully overlapping tranmit/receive spectrum or sub-band operation in an FDM context)
- Material between the transmit and receive arrays (e.g., an absorber placed between them can help increase spatial isolation)
- Transmit and receive beamforming directions.

The clutter reflects the signal from the transmitter to its co-located receiver like in a radar operation. The signal strength observed by the receiver depends on the distance between the transmitter and the clutter as well as the distance between the receiver and the clutter. It also depends on the directions of transmission and reception. With careful selection of beam weights, the clutter can be suppressed. In one particular illustration, the base station operating in full-duplex mode may pair users that are spatially well separated to minimize the impact of clutter. On top of all these mitigation techniques, self-interference cancellation in the digital domain can be used to reduce the interference due to the coupling and/or clutter [269]. Further, the frequency domain can provide additional isolation beyond the spatial isolation.

The readers are referred to 3GPP specifications for full-duplex systems in [270] and [271, 272] for full-duplex system designs at 60 GHz and over-the-air studies on their practical viability.

6.5 TOPOLOGY ENHANCEMENT

Enhancing network robustness as elucidated in Chapter 5 via densification with different types of repeaters and IAB nodes is likely to be an ongoing theme in 6G. However, as described in Chapter 3.4.3, increased power consumption is a significant problem in mmWave systems. In fact, the total power consumption at mmWave frequencies is dominated by the RF front end. The increased power consumption with dense networks (or more number of small cells) is likely to be a bottleneck on sustainability and a greener future.

Thus, a typical use-case of evolution into 6G is for energy or power savings as the base station, IAB node and/or the UE deactivate(s) redundant resources. The redundant resources may be cells, beams, antenna elements, subarrays, antenna modules/panels, bandwidth parts or BWPs, CCs, etc. at appropriate time-scales to accommodate the time-varying traffic load and mobility requirements. UEs can be scheduled with the most energy-efficient resource within a radio access technology (RAT) or distributed to the most energy-efficient RAT in a dynamic and use-case dependent manner. Measurement and search can consume significant power at mmWave carrier frequencies. Hence, these events/opportunities can be optimized by proper scheduling of internal UE resources that are most likely to succeed/result in good RSRPs/SINRs as inferred/predicted by different techniques at the device.

The frontier of network energy savings can also be advanced via the pursuit of energy-efficient fronthaul solutions. 5G systems have introduced significant architectural flexibility by disaggregating the classic base station into three logical nodes: the central unit (CU), the distributed unit (DU) and the radio unit (RU). Fronthaul refers to the connecting links between DUs and RUs, while backhaul pertains to the connections between CUs and the core network. Conventionally, fiber-based connections have been the mainstream technology for fronthaul. However, there are several motivations for considering wireless fronthaul as an alternative. These include:

- *Cost-effectiveness*: Wireless fronthaul offers potential cost savings compared to laying physical fiber especially in areas where it is difficult or expensive to deploy fiber cables. This is particularly important when expanding network coverage to rural or remote areas.
- *Scalability and density*: With the advent of 5G and beyond, networks must support a growing number of devices and higher data rates. Wireless fronthaul is better equipped to scale and support denser small cell deployments accommodating the increasing demands of connected devices and applications.
- *Dynamic network configuration*: Wireless fronthaul enables dynamic reconfiguration and optimization of the network. Operators can easily adjust the placement of DUs and RUs to optimize network performance and resource utilization providing greater adaptability to dynamic operational conditions.
- *Flexibility and rapid deployment*: Compared to laying wired fronthaul infrastructure, wireless fronthaul offers more flexibility and faster deployment options, reducing both time and costs associated with installing fiber cables.

Technologies such as fronthaul compression, leveraging new higher band spectrum, optimizing high rank LOS-MIMO solutions and utilizing RISs are possible ways to meet the enormous capacity needs of fronthaul. The readers are referred to works such as [273, 274] for discussion on this area.

6.6 JOINT COMMUNICATIONS AND SENSING (JCAS)

While the design objectives of 5G are geared toward eMBB, URLLC and mMTC (see Chapter 1 for a discussion), it is expected that 6G air interface design will support a wider variety of use cases beyond communications. Such use cases include XR/Metaverse, "sensing as a service," flexible/full-duplexing, etc.

In particular, the large bandwidths available at mmWave, upper mmWave, and sub-THz bands open up opportunities for JCAS. One key objective of JCAS is to reuse the existing communications infrastructure for sensing. In this context, sensing denotes the process of scanning (illumination) of angular and range domains achieved typically via beamformed transmissions that are strong enough to reach certain ranges and travel back to the transmitter. The signal arriving at the sensing receiver antenna contains information about the environment and is processed to extract information about the existence, range, speed, and direction of the objects in the environment. The required scanning is enabled by the proper use of time, frequency and spatial resources. Particularly, the sensing receiver uses the properties of the radio signal's time, frequency and spatial resources as well as the differences in time, phase and amplitude between the sent and received signals to extract the information on the underlying objects.

The term JCAS, in the 6G context, refers to the introduction of sensing capability as an integral part of the 6G communications network. The evolving ecosystem can enable and enhance cooperation between communication and sensing services in the following ways:

- Communications-assisted sensing which uses RF and/or non-RF sensor inputs to improve the existing Quality-of-Experience (QoE) for services and enabling services for new use cases. These may include gaming, positioning, automotive, unmanned/uncrewed aerial vehicle (UAV) tracking, etc. More specific use-cases can include gesture/motion recognition for health care monitoring, contextual information acquisition for position/speed estimation and cooperative sensing for object detection.
- Sensing-assisted communications which leverages RF and non-RF sensor inputs to improve communications-related operations. Some typical objectives include enhanced radio resource management, beam management, blockage detection and enhanced performance with mobility, etc.

A wireless communications network which is typically designed for communications objectives has to be adapted for sensing use cases. Challenges in this domain include:

- Extremely low latency reception (on the order of microseconds) of the backscattered signal in contrast to the relatively higher latencies in communications (milliseconds)
- Design of JCAS signals to meet the requirements on range resolution, unambiguous range, velocity range and resolution
- Incorporation of AI/ML techniques and solutions in sensor fusion.

The readers are referred to works such as [275, 276, 277, 278, 279, 280] for a brief summary of the opportunities and challenges in this area.

6.7 AI/ML-ASSISTED MMWAVE SOLUTIONS

With the increased interest in AI/ML-based solutions in diverse sets of applications [281], there has been a strong motivation to pursue these approaches for mmWave communications. In general, AI/ML approaches can do better than traditional model-based algorithmic paradigms under certain conditions:

- The model connecting the input(s) and the output(s) of the application-of-interest is unclear or has a large amount of errors or uncertainties
- The number of parameters in the model is large
- The search space of parameters is large so that the number of search parameters with a fine quantization grid can be significant
- Information necessary for learning the appropriate parameter(s) in the model or the appropriate model are not available at a centralized node due to energy/power, thermal, latency, and communications constraints. Further, the information needed for model parameter learning can be significant to average out the randomness associated with the input(s) and the output(s)
- Thus, the cost associated with model learning is significant enough to allow the use of AI/ML approaches instead of traditional paradigms.

With the above guidelines, many mmWave communications applications render themselves amenable to the pursuit of AI/ML-assisted solutions in Rel. 18 (and beyond). We now illustrate how AI/ML techniques can be tailored for beam prediction.

Beam management for mmWave systems spans idle mode operation to initial acquisition and radio link failure [1]. In the idle mode operation, the UE camps on a cell by decoding system information, and monitors paging from the network. When there is a request for data transfer either due to data arrival at the UE or a request (page) from the network, the UE begins the initial acquisition procedure and moves to connected state. In the connected state, the UE maintains and tracks the serving beam by measuring SSBs and providing Layer 1 measurement reports to its serving cell. Supported by 3GPP specifications, beam failure recovery procedure is introduced at the UE to monitor the quality of the current serving beam. In case there is a beam failure, the UE may use the recovery procedure to indicate a new beam to its serving cell such that the new beam pair between the serving cell and the UE can be used for further signaling. In case there is no suitable new beam between the current serving cell and the UE, radio link failure procedure can be triggered.

AI/ML-assisted beam prediction can be implemented at the UE, the base station, or at both ends of a link. Further, beam prediction can be on the spatial, temporal, or joint spatial and temporal domains. In spatial beam prediction, the UE may measure a set of wider beamwidth beams, but it can predict the viability of another set of narrower beamwidth beams. In temporal beam prediction, the UE may measure a set of beams at a certain time, and predict the quality of another set of beams at a future time. By using AI/ML-assisted beam prediction, power consumption at both the base station and UE can be minimized. In particular, the base station can reduce reference signal transmissions as the best beam has already been predicted. Similarly, the UE can reduce the time over which the RF front end is turned on for making beam measurements. Table 6.1 provides a list of base station and UE side constraints in implementing AI/ML-assisted beam prediction in practice. These constraints include different aspects such as computational capabilities, knowledge of beam patterns at either side, measurement reporting, and knowledge of beams used by all the UEs across the cell.

Table 6.1
Comparisons between base station and UE side constraints in AI/ML-assisted beam prediction

Issue	Base station side	UE side
Computational capability	More than the UE	Limited
Base station beam weights	Aware	Limited knowledge
UE beam weights	Proprietary and difficult to know	Aware
UE orientation	Difficult to know	Inferred from UE sensors
Measurement reports	Only strongest beams are fed back	Instant and filtered measurements for all beams
Measurements across UEs	Access to historical and location-aware Layer 1 reports	Limited knowledge of historical Layer 1 reports and no knowledge of other UE's feedback
Feedback quality	Quantized feedback, can be possibly missed	Measurements are unquantized

While there are extensive simulation-based beam prediction analyses where data are generated from system level simulation studies (see works such as [282–285] and references therein), there are also real-world prototyping studies using AI/ML approaches at both the base station and UE. For example, in [286], commercial chipsets supporting AI/ML-assisted beam management are used at the UE. With advancements in both AI/ML algorithms as well as hardware/software which are optimized for AI/ML operation, we expect these to significantly reshape and optimize mmWave performance at both the device and network sides.

Applications in which UEs follow predictive mobility paths (e.g., high speed trains, service vehicles, drones or UAVs, etc.) or are confined within a specific geographic area (e.g., indoor use cases such as offices, shopping malls, stadiums, downtown settings, special events deployments, etc.) can exploit AI/ML approaches to enable power savings. For example, prior mobility-based history can be leveraged to appropriately schedule beams based on UE location without a need to do a lot of beam searching or refinement. History may include information such as UE location(s), beam information, realized RSRPs/SINRs, interference caused in the system, blockage events and beam handovers, etc. AI/ML approaches can also be used to manage the overheads such as SSB periods in a network or even manage the increased capacity needs via appropriate resource allocation mechanisms.

Further, densification of networks and the increase in the number of links (massive access) can serve as a natural realm for the study of AI/ML-driven approaches for scheduling, interference coordination and mitigation, and beam and link adaptation. Inter-cell coordination of UEs to either avoid or mitigate interference entails the scheduling or (conditional) handover of the UE of interest to the appropriate cell, selection and/or switching of the appropriate beams at either the base station or the UE, and scheduling of other UEs with an appropriate flexible frame structure (subcarrier spacings and slot format indicators that determine uplink or downlink symbol orderings) to minimize inter-beam interference. Load balancing and coverage optimization can be realized at base stations with optimal UE scheduling across RATs and cells within a RAT, and AI/ML-assisted solutions can be useful toward these objectives. Use of such approaches can allow the personalization and tailoring of solutions for specific user profiles and behaviors (e.g. traffic shaping, route optimization, network energy savings, etc.) and sharing learned experiences with other nodes in the network via federated learning mechanisms.

Bibliography

1. W. Chen, P. Gaal, J. Montojo, and H. Zisimopoulos, *Fundamentals of 5G Communications: Connectivity for Enhanced Mobile Broadband and Beyond*, 1st ed. McGraw-Hill, 2021.
2. A. Zaidi, F. Athley, J. Medbo, U. Gustavsson, G. Durisi, and X. Chen, *5G Physical Layer: Principles, Models and Technology Components*, 1st ed. Academic Press, 2018.
3. E. Dahlman, S. Parkvall, and J. Skold, *5G NR: The Next Generation Wireless Access Technology*, 2nd ed. Academic Press, 2020.
4. C. Johnson, *5G New Radio in Bullets*, 1st ed. Independently published, 2019.
5. OSM, "Open Street Map software," Available: [Online]. `https://www.openstreetmap.org/`.
6. M. G. L. Frecassetti, "ETSI white paper no. 9, E-band and V-band - Survey on status of worldwide regulation," June 2015, Available: [Online]. `https://www.etsi.org/images/files/ETSIWhitePapers/etsi_wp9_e_band_and_v_band_survey_20150629.pdf`.
7. IEEE, "IEEE Std 802.11-2016. Part 11: Wireless LAN Medium Access Control (MAC) and Physical Layer (PHY) Specifications, Chapter 20," Dec. 2016.
8. N. Bhushan, J. Li, D. Malladi, R. Gilmore, D. Brenner, A. Damnjanovic, R. T. Sukhasvi, C. Patel, and S. Geirhofer, "Network densification: The dominant theme for wireless evolution into 5G," *IEEE Communications Magazine*, vol. 52, no. 2, pp. 82–89, Feb. 2014.
9. R. Baldemair, T. Irnich, K. Balachandran, E. Dahlman, G. Mildh, Y. Selen, S. Parkvall, M. Meyer, and A. Osseiran, "Ultra-dense networks in millimeter-wave frequencies," *IEEE Communications Magazine*, vol. 53, no. 1, pp. 202–208, Jan. 2015.
10. T. L. Marzetta, E. G. Larsson, H. Yang, and H. Q. Ngo, *Fundamentals of Massive MIMO*, 1st ed. Cambridge University Press, 2016.
11. Z. Pi and F. Khan, "An introduction to millimeter-wave mobile broadband systems," *IEEE Communications Magazine*, vol. 49, no. 6, pp. 101–107, June 2011.
12. T. S. Rappaport, S. Sun, R. Mayzus, H. Zhao, Y. Azar, K. Wang, G. N. Wong, J. K. Schulz, M. Samimi, and F. Gutierrez, "Millimeter wave mobile communications for 5G cellular: It will work!" *IEEE Access*, vol. 1, pp. 335–349, 2013.
13. S. Rangan, T. S. Rappaport, and E. Erkip, "Millimeter-wave cellular wireless networks: Potentials and challenges," *Proceedings of the IEEE*, vol. 102, no. 3, pp. 366–385, Mar. 2014.
14. X. Lin, J. Li, R. Baldemair, T. Cheng, S. Parkvall, D. Larsson, H. Koorapaty, M. Frenne, S. Falahati, A. Grövlen, and K. Werner, "5G New Radio: Unveiling the essentials of the next generation wireless access technology," *IEEE Communications Standards Magazine*, vol. 3, no. 3, pp. 30–37, Sept. 2019.

15. J. G. Andrews, T. Bai, M. N. Kulkarni, A. Alkhateeb, A. K. Gupta, and R. W. Heath, Jr., "Modeling and analyzing millimeter wave cellular systems," *IEEE Transactions on Communications*, vol. 65, no. 1, pp. 403–430, Jan. 2017.
16. T. S. Rappaport, R. W. Heath, Jr., R. C. Daniels, and J. N. Murdock, *Millimeter Wave Wireless Communications*, 1st ed. Pearson, 2014.
17. M. Shafi, A. F. Molisch, P. J. Smith, T. Haustein, P. Zhu, P. D. Silva, F. Tufvesson, A. Benjebbour, and G. Wunder, "5G: A tutorial overview of standards, trials, challenges, deployment, and practice," *IEEE Journal on Selected Areas in Communications*, vol. 35, no. 6, pp. 1201–1221, June 2017.
18. 3GPP TR 38.901 V14.3.0 (2017-12), "Study on channel model for frequencies from 0.5 to 100 GHz (Rel. 14)," Dec. 2017.
19. J. G. Proakis, *Digital Communications*, 4th ed. McGraw-Hill, 2000.
20. T. S. Rappaport, *Wireless Communications: Principles and Practice*, 2nd ed. Prentice Hall, Upper Saddle River, NJ, 2002.
21. METIS 2020, "METIS channel model, Deliverable D1.4v3," July 2015.
22. A. A. M. Saleh and R. Valenzuela, "A statistical model for indoor multipath propagation," *IEEE Journal on Selected Areas in Communications*, vol. 5, no. 2, pp. 128–137, Feb. 1987.
23. 3GPP TR 36.873 V12.7.0 (2017-12), "Study on 3D channel model for LTE (Rel. 12)," Dec. 2017.
24. 3GPP TS 36.213 V18.1.0 (2023-12), "Evolved Universal Terrestrial Radio Access (E-UTRA); Physical layer procedures (Rel. 18)," Dec. 2023.
25. C. A. Balanis, *Antenna Theory: Analysis and Design*, 3rd ed. Wiley-Interscience, 2005.
26. L. W. Barclay (Editor), *Propagation of Radiowaves*, 2nd ed. Institution of Electrical Engineers, 2003.
27. Radio and Microwave Wireless Systems, "Double knife edge diffraction model," *Univ. of Toronto ECE422 Course Lecture Notes, Prof. Sean V. Hum.*
28. T. S. Rappaport, K. A. Remley, C. Gentile, A. F. Molisch, and A. Zajić (Eds.), *Radio Propagation Measurements and Channel Modeling: Best Practices for Millimeter-Wave and Sub-Terahertz Frequencies*, 1st ed. Cambridge University Press, 2022.
29. Aalto University, AT&T, BUPT, CMCC, Ericsson, Huawei, Intel, KT Corporation, Nokia, NTT DOCOMO, NYU, Qualcomm, Samsung, U. Bristol, and USC, "White paper on '5G channel model for bands up to 100 GHz'," *v2.3*, Oct. 2016.
30. A. Ghosh, T. A. Thomas, M. C. Cudak, R. Ratasuk, P. Moorut, F. W. Vook, T. S. Rappaport, G. R. MacCartney, Jr., S. Sun, and S. Nie, "Millimeter-wave enhanced local area systems: A high data-rate approach for future wireless networks," *IEEE Journal on Selected Areas in Communications*, vol. 32, no. 6, pp. 1152–1163, June 2014.
31. T. S. Rappaport, G. R. MacCartney, Jr., S. Sun, H. Yan, and S. Deng, "Small-scale, local area, and transitional millimeter wave propagation for 5G communications," *IEEE Transactions on Antennas and Propagation*, vol. 65, no. 12, pp. 6474–6490, Dec. 2017.

32. V. Raghavan, A. Partyka, L. Akhoondzadehasl, M. A. Tassoudji, O. H. Koymen, and J. Sanelli, "Millimeter wave channel measurements and implications for PHY layer design," *IEEE Transactions on Antennas and Propagation*, vol. 65, no. 12, pp. 6521–6533, Dec. 2017.
33. S. Sun, T. S. Rappaport, S. Rangan, T. A. Thomas, A. Ghosh, I. Z. Kovács, I. Rodriguez, O. H. Koymen, A. Partyka, and J. Järveläinen, "Propagation path loss models for 5G urban micro and macro-cellular scenarios," *Proceedings of IEEE Vehicular Technology Conference (Spring), Nanjing, China*, vol. 1, pp. 1–5, May 2016.
34. S. Sun, T. S. Rappaport, T. A. Thomas, A. Ghosh, H. C. Nguyen, I. Z. Kovács, I. Rodriguez, O. H. Koymen, and A. Partyka, "Investigation of prediction accuracy, sensitivity, and parameter stability of large-scale propagation path loss models for 5G wireless communications," *IEEE Transactions on Vehicular Technology*, vol. 65, no. 5, pp. 2843–2860, May 2016.
35. I. F. Akyildiz, J. M. Jornet, and C. Han, "Terahertz band: Next frontier for wireless communications," *Physical Communication*, vol. 12, pp. 16–32, Sept. 2014.
36. A. Alammouri, J. Mo, B. L. Ng, J. C. Zhang, and J. G. Andrews, "Hand grip impact on 5G mmWave mobile devices," *IEEE Access*, vol. 7, pp. 60 532–60 544, 2019.
37. A. K. Gupta, J. G. Andrews, and R. W. Heath, Jr., "Macro-diversity in cellular networks with random blockages," *IEEE Transactions on Wireless Communications*, vol. 17, no. 2, pp. 996–1010, Feb. 2018.
38. G. R. MacCartney, Jr., T. S. Rappaport, and S. Rangan, "Rapid fading due to human blockage in pedestrian crowds at 5G millimeter-wave frequencies," *Proceedings of IEEE Global Telecommunications Conference, Singapore*, vol. 1, pp. 1–7, Dec. 2017.
39. D. N. C. Tse and P. Viswanath, *Fundamentals of Wireless Communications*. Cambridge University Press, 2005.
40. V. Raghavan, A. Partyka, S. Subramanian, A. Sampath, O. H. Koymen, K. Ravid, J. Cezanne, K. K. Mukkavilli, and J. Li, "Millimeter wave MIMO prototype: Measurements and experimental results," *IEEE Communications Magazine*, vol. 56, no. 1, pp. 202–209, Jan. 2018.
41. B. H. Fleury, M. Tschudin, R. Heddergott, D. Dahlhaus, and K. I. Pedersen, "Channel parameter estimation in mobile radio environments using the SAGE algorithm," *IEEE Journal on Selected Areas in Communications*, vol. 17, no. 3, pp. 434–450, Mar. 1999.
42. N. Czink and P. Cera, "A novel framework for clustering parametric MIMO channel data including MPC powers," *COST 273 TD(05)104, Lisbon, Portugal*, Nov. 2005.
43. G. R. MacCartney, Jr. and T. S. Rappaport, "A flexible millimeter-wave channel sounder with absolute timing," *IEEE Journal on Selected Areas in Communication*, vol. 35, no. 6, pp. 1402–1418, June 2017.

44. A. F. Molisch, *Wireless Communications: From Fundamentals to Beyond 5G*, 3rd ed. IEEE Press, 2022.
45. C. Gustafson, K. Haneda, S. Wyne, and F. Tufvesson, "On mm-wave multipath clustering and channel modeling," *IEEE Transactions on Antennnas and Propagation*, vol. 62, no. 3, pp. 1445–1455, Mar. 2014.
46. A. M. Sayeed, P. G. Vouras, C. Gentile, A. Weiss, J. T. Quimby, Z. Cheng, B. Modad, Y. Zhang, C. K. Anjinappa, F. Erden, O. Özdemir, R. Müller, D. A. Dupleich, H. Niu, D. G. Michelson, and A. Hughes, "A framework for developing algorithms for estimating propagation parameters from measurements," 2021, Available: [Online]. `https://arxiv.org/pdf/2109.06131.pdf`.
47. J. Ko, Y.-J. Cho, S. Hur, T. Kim, J. Park, A. F. Molisch, K. Haneda, M. Peter, D.-J. Park, and D.-H. Cho, "Millimeter-wave channel measurements and analysis for statistical spatial channel model in in-building and urban environments at 28 GHz," *IEEE Transactions on Wireless Communications*, vol. 16, no. 9, pp. 5853–5868, Sept. 2017.
48. S. Ju, Y. Xing, O. Kanhere, and T. S. Rappaport, "3-D statistical indoor channel model for millimeter-wave and sub-Terahertz bands," *Proceedings of IEEE Global Telecommunications Conference, Taipei, Taiwan*, pp. 1–7, Dec. 2020.
49. A. Paulraj, R. Nabar, and D. Gore, *Introduction to Space-Time Wireless Communications*, 1st ed. Cambridge University Press, 2003.
50. A. J. Goldsmith, *Wireless Communications*, 1st ed. Cambridge University Press, 2005.
51. R. W. Heath, Jr. and A. Lozano, *Foundations of MIMO Communications*, 1st ed. Cambridge University Press, 2019.
52. W. C. Jakes, *Microwave Mobile Communications*, 1st ed. John Wiley, NY, 1975.
53. A. M. Sayeed and B. Aazhang, "Joint multipath-doppler diversity in mobile wireless communications," *IEEE Transactions on Communications*, vol. 47, no. 1, pp. 123–132, Jan. 1999.
54. S. Bhashyam, A. M. Sayeed, and B. Aazhang, "Time-selective signaling and reception for communication over multipath fading channels," *IEEE Transactions on Communications*, vol. 48, no. 1, pp. 83–94, Jan. 2000.
55. A. M. Sayeed, "Deconstructing multi-antenna fading channels," *IEEE Transactions on Signal Processing*, vol. 50, no. 10, pp. 2563–2579, Oct. 2002.
56. A. S. Y. Poon, R. W. Brodersen, and D. N. C. Tse, "Degrees of freedom in multiple-antenna channels: A signal space approach," *IEEE Transactions on Information Theory*, vol. 51, no. 2, pp. 523–536, Feb. 2005.
57. A. M. Sayeed and V. V. Veeravalli, "Essential degrees of freedom in space-time fading channels," *Proceedings of IEEE International Symposium on Personal Indoor and Mobile Radio Communications, Lisboa, Portugal*, vol. 4, pp. 1512–1516, Sept. 2002.
58. P. Schniter and A. M. Sayeed, "Channel estimation and precoder design for millimeter-wave communications: The sparse way," *Proceedings of IEEE Asilomar Conference on Signals, Systems and Computers*, vol. 1, pp. 273–277, Nov. 2014.

59. K. Liu, T. Kadous, and A. M. Sayeed, "Orthogonal time-frequency signaling over doubly dispersive channels," *IEEE Transactions on Information Theory*, vol. 50, no. 11, pp. 2583–2603, Nov. 2004.
60. A. Fish, S. Gurevich, R. Hadani, A. M. Sayeed, and O. Schwartz, "Delay-Doppler channel estimation in almost linear complexity," *IEEE Transactions on Information Theory*, vol. 59, no. 11, pp. 7632–7644, Nov. 2013.
61. D. M. Pozar, *Microwave Engineering*, 4th ed. Wiley, 2011.
62. S. J. Orfanidis, *Electromagnetic Waves and Antennas*. Self-published, 2016, Available: [Online]. `https://www.ece.rutgers.edu/~orfanid/ewa`.
63. W. L. Stutzman and G. A. Thiele, *Antenna Theory and Design*, 3rd ed. Wiley, NY, 2012.
64. European Conference of Postal and Telecommunications Administrations, "Radiodetermination applications in the frequency range 116-260 GHz, CEPT SE24 WI71 doc." *IEEE P802.15 Working Group for Wireless Specialty Networks (WSN)*, Nov. 2020.
65. Sony and Ericsson, "R4-1704866, Frequency response of UE mmWave antennas for the 28 GHz band," *3GPP TSG-RAN WG4 Meeting #83, Hangzhou, China*, June 2017.
66. J. S. Schalch, Z. Zhang, A. Nafe, Y. Yin, and G. M. Rebeiz, "Massively thinned phased arrays with randomized antenna placement and with discrete random thinning," *To be published*, 2021.
67. M. Kohtani, S. Cha, P. Schmalenberg, J. Lee, L. Li, T. Takahata, S. Yamaura, T. Matsuoka, and G. M. Rebeiz, "Thinned array distribution with grating lobe canceller at any scan angle for automotive radar applications," *International Journal of Microwave and Wireless Technologies*, pp. 1–13, Mar. 2024.
68. S. Sanayei and A. Nosratinia, "Antenna selection in MIMO systems," *IEEE Communications Magazine*, vol. 42, no. 10, pp. 68–73, Oct. 2004.
69. R. W. Heath, Jr., S. Sandhu, and A. J. Paulraj, "Antenna selection for spatial multiplexing systems with linear receivers," *IEEE Communications Letters*, vol. 5, no. 4, pp. 142–144, Apr. 2001.
70. A. F. Molisch and M. Z. Win, "MIMO systems with antenna selection," *IEEE Microwave Magazine*, vol. 5, no. 1, pp. 46–56, Mar. 2004.
71. A. Gorokhov, D. A. Gore, and A. J. Paulraj, "Receive antenna selection for MIMO spatial multiplexing: Theory and algorithms," *IEEE Transactions on Signal Processing*, vol. 51, no. 11, pp. 2796–2807, Nov. 2003.
72. V. Raghavan, T. Luo, O. H. Koymen, and J. Li, "Antenna selection for upper millimeter wave and THz bands," *Proceedings of IEEE Asilomar Conference on Signals, Systems and Computers, Pacific Grove, CA*, vol. 1, pp. 1–6, Nov. 2020.
73. V. Raghavan, M. A. Tassoudji, Y.-C. Ou, K. Ravid, O. H. Koymen, and J. Li, "Fundamental limitations of large antenna arrays for millimeter wave systems," *Proceedings of IEEE Asilomar Conference on Signals, Systems and Computers, Pacific Grove, CA*, vol. 1, pp. 1–6, Nov. 2019.

74. 3GPP TS 38.101-2 V18.4.0 (2023-12), "User Equipment (UE) radio transmission and reception; Part 2: Range 2 Standalone (Rel. 18)," Dec. 2023.
75. Verizon Wireless, "Requirements for 5GNR NSA Mobility (V15.0)," Oct. 2023.
76. V. Raghavan, M.-L. Chi, M. A. Tassoudji, O. H. Koymen, and J. Li, "Antenna placement and performance tradeoffs with hand blockage in millimeter wave systems," *IEEE Transactions on Communications*, vol. 67, no. 4, pp. 3082–3096, Apr. 2019.
77. Sony and Ericsson, "R4-1800889, Impact of UE back cover material on peak EIRP at mmWave," *3GPP TSG-RAN WG4 Meeting AH-1801, San Diego, CA*, Jan. 2018.
78. Qualcomm, "R4-1800369, On UE packaging loss, Simulation results," *3GPP TSG-RAN WG4 Meeting AH-1801, San Diego, CA*, Jan. 2018.
79. J. Park and W. Hong, "Antenna-on-Display (AoD) for millimeter-wave 5G mobile devices," *Proceedings of IEEE International Symposium on Antennas and Propagation and USNC-URSI Radio Science Meeting*, vol. 1, pp. 1–2, July 2019.
80. Sony and Ericsson, "R4-1706562, UE EIRP for mmWave 28 GHz," *3GPP TSG-RAN WG4 Meeting #83, Hangzhou, China*, May 2017.
81. Sony, "R4-1707330, On UE mmWave antennas," *3GPP TSG-RAN WG4 Meeting #84, Berlin, Germany*, Aug. 2017.
82. Ericsson and Sony, "R4-1609590, UE reference architecture for NR," *3GPP TSG-RAN WG4 Meeting #81, Reno, NV*, Nov. 2016.
83. Qualcomm, "R4-1712381, Spherical coverage of realistic design," *3GPP TSG-RAN WG4 Meeting #85, Reno, NV*, Nov.-Dec. 2017.
84. Apple and Intel, "R4-1713850, Consideration of EIRP spherical coverage requirement," *3GPP TSG-RAN WG4 Meeting #85, Reno, NV*, Nov.-Dec. 2017.
85. Sony, "R4-1711424, UE power class and spherical coverage for mmWave 28 GHz," *3GPP TSG-RAN WG4 Meeting #84-bis, Dubrovnik, Croatia*, Oct. 2017.
86. Sony and Ericsson, "R4-1800888, UE spherical coverage at mmWave 28 GHz," *3GPP TSG-RAN WG4 Meeting AH-1801, San Diego, CA*, Jan. 2018.
87. J. Helander, K. Zhao, Z. Ying, and D. Sjöberg, "Performance analysis of millimeter-wave phased array antennas in cellular handsets," *IEEE Antennas and Propagation Letters*, vol. 15, pp. 504–507, 2016.
88. R. Trandafir, "An examination of the 5G radio in Google Pixel 6 Pro," Oct. 2021, Available: [Online]. `https://www.techinsights.com/blog/teardown/examination-5g-radio-google-pixel-6-pro`.
89. K. Sahota, "5G mmWave radio design for mobile," *IEEE 5G Summit, Honolulu, HI*, June 2017, Available: [Online]. `http://5gsummit.org/hawaii/docs/slides/D2_%239_Sahota_5g_060617.pdf`.
90. W. Hong, K.-H. Baek, and S. Ko, "Millimeter-wave 5G antennas for smartphones: Overview and experimental demonstration," *IEEE Transactions on Antennas and Propagation*, vol. 65, no. 12, pp. 6250–6261, Dec. 2017.

91. W. Hong, Z. H. Jiang, C. Yu, J. Zhou, P. Chen, Z. Yu, H. Zhang, B. Yang, X. Pang, M. Jiang, Y. Cheng, M. K. T. Al-Nuaimi, Y. Zhang, J. Chen, and S. He, "Multibeam antenna technologies for 5G wireless communications," *IEEE Transactions on Antennas and Propagation*, vol. 65, no. 12, pp. 6231–6249, Dec. 2017.
92. 3GPP TR 38.803 V14.2.0 (2017–09), "Study on new radio access technology: Radio frequency (RF) and co-existence aspects (Rel. 14)," Sept. 2017.
93. Samsung, Apple, and Intel, "R4-1711036, Consideration of EIRP spherical coverage requirement," *3GPP TSG-RAN WG4 Meeting #84-bis, Dubrovnik, Croatia*, Oct. 2017.
94. Qualcomm, "R4-1804587, On UE spherical coverage with glass packaging," *3GPP TSG-RAN WG4 Meeting #86-bis, Melbourne, Australia*, Apr. 2018.
95. NTT DOCOMO, "R4-1709394, EIRP CDF for mmWave UE," *3GPP TSG-RAN WG4 Meeting NR ad-hoc #3, Nagoya, Japan*, Sept. 2017.
96. Sony, "R4-1802868, UE spherical coverage at mmWave 28 GHz," *3GPP TSG-RAN WG4 #86, Athens, Greece*, Feb.-Mar. 2018.
97. ——, "R4-1807490, UE spherical coverage measurements at mmWave 28 GHz," *3GPP TSG-RAN WG4 Meeting #87, Busan, South Korea*, May 2018.
98. LG Electronics, "R4-1806676, Measurement EIRP levels for spherical coverage of NR UE at FR2," *3GPP TSG-RAN WG4 Meeting #87, Busan, South Korea*, May 2018.
99. K. Zhao, S. Zhang, Z. Ho, O. Zander, T. Bolin, Z. Ying, and G. F. Pedersen, "Spherical coverage characterization of 5G millimeter wave user equipment with 3GPP specifications," *IEEE Access*, vol. 7, p. 4442–4452, 2019.
100. H. Krishnaswamy and H. Hashemi, "Integrated beamforming arrays," *In mm-Wave Silicon Technology, (A. M. Niknejad and H. Hashemi, Eds.), Springer, NY*, pp. 243–295, 2008.
101. C. Kim, T. Kim, and J.-Y. Seol, "Multi-beam transmission diversity with hybrid beamforming for MIMO-OFDM systems," *Proceedings of IEEE Global Telecommunications Workshops, Atlanta, GA*, vol. 1, pp. 1–6, Dec. 2013.
102. T. Kim, J. Park, J.-Y. Seol, S. Jeong, J. Cho, and W. Roh, "Tens of Gbps support with mmWave beamforming systems for next generation communications," *Proceedings of IEEE Global Telecommunications Confernce, Atlanta, GA*, vol. 1, pp. 1–6, Dec. 2013.
103. G. M. Rebeiz and K.-J. Koh, "Silicon RFICs for phased arrays," *IEEE Microwave Magazine*, vol. 10, no. 3, pp. 96–103, May 2009.
104. M. A. N. M. Saif, "Design of tunable beamforming networks using metallic ridge gap waveguide technology," Ph.D. dissertaion, Concordia University, Montreal, Canada, Tech. Rep., 2019.
105. X.-Z. Wang, F.-C. Chen, and Q.-X. Chu, "A compact broadband 4×4 Butler matrix with 360° continuous progressive phase shift," *IEEE Transactions on Microwave Theory and Techniques*, vol. 71, no. 9, pp. 3906–3914, Sept. 2023.

106. J. Brady, N. Behdad, and A. M. Sayeed, “Beamspace MIMO for millimeter-wave communications: System architecture, modeling, analysis and measurements,” *IEEE Transactions on Antennas and Propagation*, vol. 61, no. 7, pp. 3814–3827, July 2013.
107. Y. Zeng and R. Zhang, “Millimeter wave MIMO with lens antenna array: A new path division multiplexing paradigm,” *IEEE Transactions on Communications*, vol. 64, no. 4, pp. 1557–1571, Apr. 2016.
108. S. Vashist, M. K. Soni, and P. K. Singhal, “A review on the development of Rotman lens antenna,” *Chinese Journal of Engineering*, vol. 2014, no. 385385, pp. 1–9, 2014.
109. J. A. Laurinaho, J. Aurinsalo, A. Karttunen, M. Kaunisto, A. Lamminen, J. Nurmiharju, A. V. Raisanen, J. Saily, and P. Wainio, “2-D beam-steerable integrated lens antenna system for 5G E-band access and backhaul,” *IEEE Transactions on Microwave Theory and Techniques*, vol. 64, no. 7, pp. 2244–2255, July 2016.
110. T. Kwon, Y.-G. Lim, B.-W. Min, and C.-B. Chae, “RF lens-embedded massive MIMO systems: Fabrication issues and codebook design,” *IEEE Transactions on Microwave Theory and Techniques*, vol. 64, no. 7, pp. 2256–2271, July 2016.
111. C. A. Fernandes, E. B. Lima, and J. R. Costa, “Dielectric lens antennas,” *In Handbook of Antenna Technologies, (Z. N. Chen, D. Liu, H. Nakano, X. Qing and T. Zwick, Eds.), Springer, Singapore*, pp. 1001–1064, 2016.
112. D. Y. C. Lie, J. C. Mayeda, Y. Li, and J. Lopez, “A review of 5G power amplifier design at cm-wave and mm-wave frequencies,” *Wireless Communications and Mobile Computing*, vol. 2018, no. 6793814, pp. 1–16, 2018.
113. S. Shakib, M. Elkholy, J. D. Dunworth, V. Aparin, and K. Entesari, “A wideband 28-GHz transmit–receive front-end for 5G handset phased arrays in 40-nm CMOS,” *IEEE Transactions on Microwave Theory and Techniques*, vol. 67, no. 7, pp. 2946–2963, July 2019.
114. H. Wang, K. Choi, B. Abdelaziz, M. Eleraku, B. Lin, E. Liu, Y. Liu, H. Jalili, M. Ghorbanpoor, C. Chu, T.-Y. Huang, N. S. Mannem, J. Park, J. Lee, D. Munzer, S. Li, F. Wang, A. S. Ahmed, C. Snyder, H. T. Nguyen, and M. E. D. Smith, “Power amplifiers performance survey 2000-present,” Available: [Online]. `https://ideas.ethz.ch/Surveys/pa-survey.html`.
115. E. A. Karahan, Z. Liu, and K. Sengupta, “Deep-learning-based inverse-designed millimeter-wave passives and power amplifiers,” *IEEE Journal of Solid-State Circuits*, vol, 58, no. 11, pp. 3074–3088, Nov. 2023.
116. J. F. Buckwalter, M. J. W. Rodwell, K. Ning, A. S. H. Ahmed, A. A-Purdue, J. Chien, E. O’Malley, and E. Lam, “Prospects for high-efficiency Silicon and III-V power amplifiers and transmitters in 100-300 GHz bands,” *Proceedings of IEEE Custom Integrated Circuits Conference, Austin, TX*, vol. 1, pp. 1–7, Apr. 2021.
117. E. A. Firouzjaei, “mm-Wave phase shifters and switches,” University of California at Berkeley, UCB/EECS-2010-163, Tech. Rep., 2010.

118. D. Pepe and D. Zito, "Two mm-Wave vector modulator active phase shifters with novel IQ generator in 28 nm FDSOI CMOS," *IEEE Journal of Solid-State Circuits*, vol. 52, no. 2, pp. 344–356, Feb. 2017.
119. S. Y. Kim, D. W. Kang, K. J. Koh, and G. M. Rebeiz, "An improved wideband all-pass I/Q network for millimeter-wave phase shifters," *IEEE Transactions on Microwave Theory and Techniques*, vol. 60, no. 11, pp. 3431–3439, Nov. 2012.
120. N. Hosseinzadeh and J. F. Buckwalter, "A compact, 37% fractional bandwidth millimeter-wave phase dhifter using a wideband Lange coupler for 60-GHz and E-band systems," *Proceedings of IEEE Compound Semiconductor Integrated Circuit Symposium, Miami, FL*, vol. 1, pp. 1–4, Oct. 2017.
121. IEEE Std 1139 – 2022 (Revision of IEEE Std 1139-2008), "IEEE Standard Definitions of Physical Quantities for Fundamental Frequency and Time Metrology—Random Instabilities," June 2022.
122. R. Navid, C. Jungemann, T. H. Lee, and R. W. Dutton, "Close-in phase noise in electrical oscillators," *Proceedings of the SPIE*, vol. 5473, pp. 27–37, May 2004.
123. M. Moeneclaey, "The effect of synchronization errors on the performance of orthogonal frequency-division multiplexed (OFDM) systems," *Proc. COST 254 (Emergent Techniques for Communication Terminals), Toulouse, France*, July 1997.
124. A. G. Armada, "Understanding the effects of phase noise in orthogonal frequency division multiplexing (OFDM)," *IEEE Transactions on Broadcasting*, vol. 47, no. 2, pp. 153–159, June 2001.
125. D. Petrovic, W. Rave, and G. Fettweis, "Effects of phase noise on OFDM systems with and without PLL: Characterization and compensation," *IEEE Transactions on Communications*, vol. 55, no. 8, pp. 1607–1616, Aug. 2007.
126. M. E. Rasekh, M. Abdelghany, U. Madhow, and M. J. W. Rodwell, "Phase noise in modular millimeter wave massive MIMO," *IEEE Transactions on Wireless Communications*, vol. 20, no. 10, pp. 6522–6535, Oct. 2021.
127. Ericsson, "R1-1701161, On DL and UL phase noise tracking RS (PTRS)," *3GPP TSG RAN WG1 #87ah-NR, Spokane, WA*, Jan. 2017.
128. 3GPP TR 38.808 V17.0.0 (2021-03), "Study on supporting NR from 52.6 GHz to 71 GHz (Rel. 17)," Mar. 2021.
129. S. Dutta, C. N. Barati, D. Ramirez, A. Dhananjay, J. F. Buckwalter, and S. Rangan, "A case for digital beamforming at mmWave," *IEEE Transactions on Wireless Communications*, vol. 19, no. 2, pp. 756–770, Feb. 2020.
130. S. Verdú, "On channel capacity per unit cost," *IEEE Transactions on Information Theory*, vol. 36, no. 5, pp. 1019–1030, Sept. 1990.
131. ——, "Spectral efficiency in the wideband regime," *IEEE Transactions on Information Theory*, vol. 48, no. 6, pp. 1319–1343, June 2002.
132. 3GPP TS 38.213 V17.3.0 (2022-09), "Physical layer procedures for control version 17.3.0 (Rel. 17)," Sept. 2022.

133. 3GPP TR 38.840 V16.0.0 (2019-06), "Technical Specification Group Radio Access Network; Study on User Equipment (UE) power saving in NR (Rel. 16)," June 2019.
134. F. D. Neeser and J. L. Massey, "Proper complex random processes with applications to information theory," *IEEE Transactions on Information Theory*, vol. 39, no. 4, pp. 1293–1302, July 1993.
135. I. E. Telatar, "Capacity of multi-antenna Gaussian channels," *European Transactions on Telecommunications*, vol. 10, pp. 2172–2178, 2000.
136. G. J. Foschini and M. J. Gans, "On limits of wireless communications in a fading environment when using multiple antennas," *Wireless Personal Communications*, vol. 6, no. 3, pp. 311–335, Mar. 1998.
137. A. J. Goldsmith, S. A. Jafar, N. Jindal, and S. Vishwanath, "Capacity limits of MIMO channels," *IEEE Journal on Selected Areas in Communications*, vol. 21, no. 5, pp. 684–702, June 2003.
138. A. M. Tulino and S. Verdù, "Random Matrices and Wireless Communications," *Foundations and Trends in Communications and Information Theory*, vol. 1, no. 1, June 2004.
139. A. M. Tulino, A. Lozano, and S. Verdù, "Impact of antenna correlation on the capacity of multiantenna channels," *IEEE Transactions on Information Theory*, vol. 51, no. 7, pp. 2491–2509, July 2005.
140. A. Lozano, A. M. Tulino, and S. Verdù, "Multiple-antenna capacity in the low-power regime," *IEEE Transactions on Information Theory*, vol. 49, no. 10, pp. 2527–2544, Oct. 2003.
141. Nokia and Nokia Siemens Networks, "R1-111031, On advanced UE MMSE receiver modelling in system simulations," *3GPP TSG RAN WG1 #64, Taipei, Taiwan*, Feb. 2011.
142. 3GPP TR 38.843 V16.0.0 (2019-06), "Evolved Universal Terrestrial Radio Access (E-UTRA) and Universal Terrestrial Radio Access (UTRA); Radio Frequency (RF) requirement background for Active Antenna System (AAS) Base Station (BS) radiated requirements (Rel. 15)," June 2020.
143. 3GPP TR 36.897 V13.0.0 (2015-06), "Study on elevation beamforming/Full-Dimension (FD) Multiple Input Multiple Output (MIMO) for LTE (Rel. 13)," June 2015.
144. Samsung, "R1-143884, TXRU modeling and antenna virtualization for FD-MIMO," *3GPP TSG RAN WG1 #78-bis, Ljubljana, Slovenia*, Oct. 2014.
145. S. Hur, T. Kim, D. J. Love, J. V. Krogmeier, T. A. Thomas, and A. Ghosh, "Millimeter wave beamforming for wireless backhaul and access in small cell networks," *IEEE Transactions on Communications*, vol. 61, no. 10, pp. 4391–4403, Oct. 2014.
146. S. Rajagopal, "Beam broadening for phased antenna arrays using multi-beam subarrays," *Proceedings of IEEE International Conference on Communications, Ottawa, Canada*, vol. 1, pp. 3637–3642, June 2012.
147. V. Raghavan, J. Cezanne, S. Subramanian, A. Sampath, and O. H. Koymen, "Beamforming tradeoffs for initial UE discovery in millimeter-wave MIMO

systems," *IEEE Journal of Selected Topics in Signal Processing*, vol. 10, no. 3, pp. 543–559, Apr. 2016.
148. G. H. Golub and C. F. V. Loan, *Matrix Computations*, 3rd ed. Johns Hopkins University Press, 1996.
149. T. Dahl, N. Christophersen, and D. Gesbert, "Blind MIMO eigenmode transmission based on the algebraic power method," *IEEE Transactions on Signal Processing*, vol. 52, no. 9, pp. 2424–2431, Sept. 2004.
150. H. Krim and M. Viberg, "Two decades of array signal processing research: The parametric approach," *IEEE Signal Processing Magazine*, vol. 13, no. 4, pp. 67–94, July 1996.
151. R. O. Schmidt, "Multiple emitter location and signal parameter estimation," *IEEE Transactions on Antennas and Propagation*, vol. AP-34, no. 3, pp. 276–280, Mar. 1986.
152. M. Wax and T. Kailath, "Detection of signals by information theoretic criteria," *IEEE Transactions on Acoustics, Speech and Signal Processing*, vol. ASSP-33, no. 4, pp. 387–392, Apr. 1985.
153. M. H. Hansen and B. Yu, "Model selection and the principle of minimum description length," *Journal of the American Statistical Association*, vol. 96, no. 454, pp. 746–774, 2001.
154. R. Roy and T. Kailath, "ESPRIT: Estimation of signal parameters via rotational invariance techniques," *IEEE Transactions on Acoustics, Speech and Signal Processing*, vol. 37, no. 7, pp. 984–995, July 1989.
155. J. A. Fessler and A. O. Hero, "Space-alternating generalized expectation maximization algorithm," *IEEE Transactions on Signal Processing*, vol. 42, no. 10, pp. 2664–2677, Oct. 1994.
156. D. Ramasamy, S. Venkateswaran, and U. Madhow, "Compressive tracking with 1000-element arrays: A framework for multi-Gbps mm-wave cellular downlinks," *Proceedings of Annual Allerton Conference on Communications, Control and Computing, Allerton, IL*, pp. 690–697, Oct. 2012.
157. B. Recht, M. Fazel, and P. A. Parrilo, "Guaranteed minimum-rank solutions of linear matrix equations via nuclear norm minimization," *SIAM Review*, vol. 52, no. 3, pp. 471–501, Aug. 2010.
158. H. Pezeshki, F. V. Massoli, A. Behboodi, T. Yoo, A. Kannan, M. T. Boroujeni, Q. Li, T. Luo, and J. B. Soriaga, "Beyond codebook-based analog beamforming at mmWave: Compressed sensing and machine learning methods," *Proceedings of IEEE Global Telecommunications Conference, Rio de Janeiro, Brasil*, vol. 1, pp. 1–6, Dec. 2022.
159. M. R. Castellanos, V. Raghavan, J. H. Ryu, O. H. Koymen, J. Li, D. J. Love, and B. Peleato, "Channel reconstruction-based hybrid precoding for millimeter wave multi-user MIMO systems," *IEEE Journal of Selected Topics in Signal Processing*, vol. 12, no. 2, pp. 383–398, May 2018.
160. X. Jiang, M. Cirkic, F. Kaltenberger, E. G. Larsson, L. Deneire, and R. Knopp, "MIMO-TDD reciprocity under hardware imbalances: Experimental results," *Proceedings of IEEE International Conference on Communications, London, UK*, vol. 1, pp. 4949–4953, June 2015.

161. J. Flordelis, F. Rusek, F. Tufvesson, E. G. Larsson, and O. Edfors, "Massive MIMO performance—TDD versus FDD: What do measurements say?" *IEEE Transactions on Wireless Communications*, vol. 17, no. 4, pp. 2247–2261, Apr. 2018.
162. F. Kaltenberger, H. Jiang, M. Guillaud, and R. Knopp, "Relative channel reciprocity calibration in MIMO/TDD systems," *Proceedings of IEEE Future Networks Mobile Summit, Florence, Italy*, pp. 1–10, June 2010.
163. R. Rogalin, O. Y. Bursalioglu, H. C. Papadopoulos, G. Caire, and A. F. Molisch, "Hardware-impairment compensation for enabling distributed large-scale MIMO," *Proceedings of IEEE Information Theory Applications Workshop (ITA), San Diego, CA*, pp. 1–10, Feb. 2013.
164. O. Raeesi, A. Gokceoglu, Y. Zou, E. Björnson, and M. Valkama, "Performance analysis of multi-user massive MIMO downlink under channel non-reciprocity and imperfect CSI," *IEEE Transactions on Communications*, vol. 66, no. 6, pp. 2456–2471, Jan. 2018.
165. V. Raghavan, K. Ravid, O. H. Koymen, and J. Li, "Over-the-air calibration via two-way signaling for millimeter wave systems," *Proceedings of IEEE Asilomar Conference on Signals, Systems and Computers, Pacific Grove, CA*, vol. 1, pp. 1–6, Oct. 2023.
166. Ansys, "High-Frequency antenna response/Structure Simulator (HFSS)," Available: [Online]. `https://www.ansys.com/products/electronics/ansys-hfss`.
167. T. K. Y. Lo, "Maximum ratio transmission," *IEEE Transactions on Communications*, vol. 47, no. 10, pp. 1458–1461, Oct. 1999.
168. V. Raghavan, L. Akhoondzadehasl, V. Podshivalov, J. Hulten, M. A. Tassoudji, O. H. Koymen, A. Sampath, and J. Li, "Statistical blockage modeling and robustness of beamforming in millimeter wave systems," *IEEE Transactions on Microwave Theory and Techniques*, vol. 67, no. 7, pp. 3010–3024, July 2019.
169. Qualcomm, "Press release, July 23, 2018," Available: [Online]. `https://www.qualcomm.com/news/releases/2018/07/23/qualcomm-delivers-breakthrough-5g-nr-mmwave-and-sub-6-ghz-rf-modules-mobile`.
170. S. Kozono, T. Suruhara, and M. Sakamoto, "Base station polarization diversity reception for mobile radio," *IEEE Transactions on Vehicular Technology*, vol. 33, no. 4, pp. 301–306, Nov. 1984.
171. R. G. Vaughan, "Polarization diversity in mobile communications," *IEEE Transactions on Vehicular Technology*, vol. 39, no. 3, pp. 177–185, June 1990.
172. M. Shafi, M. Zhang, A. L. Moustakas, P. J. Smith, A. F. Molisch, F. Tufvesson, and S. H. Simon, "Polarized MIMO channels in 3-D: Models, measurements and mutual information," *IEEE Journal on Selected Areas in Communications*, vol. 24, no. 3, pp. 514–527, Mar. 2006.
173. A. S. Y. Poon and D. N. C. Tse, "Degree-of-freedom gain from using polarimetric antenna elements," *IEEE Transactions on Information Theory*, vol. 57, no. 9, pp. 5695–5709, Sept. 2011.

174. G. Caire and S. Shamai (Shitz), "On the achievable throughput of a multi-antenna Gaussian broadcast channel," *IEEE Transactions on Information Theory*, vol. 49, no. 7, pp. 1691–1706, July 2003.
175. S. Vishwanath, N. Jindal, and A. J. Goldsmith, "Duality, achievable rates and sum rate capacity of Gaussian MIMO broadcast channel," *IEEE Transactions on Information Theory*, vol. 49, no. 10, pp. 2658–2668, Oct. 2003.
176. W. Yu and J. M. Cioffi, "Sum capacity of Gaussian vector broadcast channels," *IEEE Transactions on Information Theory*, vol. 50, no. 9, pp. 1875–1892, Sept. 2004.
177. Q. H. Spencer, C. B. Peel, A. L. Swindlehurst, and M. Haardt, "An introduction to the multi-user MIMO downlink," *IEEE Communications Magazine*, vol. 42, no. 10, pp. 60–67, Oct. 2004.
178. H. Weingarten, Y. Steinberg, and S. Shamai (Shitz), "The capacity region of the Gaussian multiple-input multiple-output broadcast channel," *IEEE Transactions on Information Theory*, vol. 52, no. 9, pp. 3936–3964, Sept. 2006.
179. B. Hassibi and M. Sharif, "Fundamental limits in MIMO broadcast channels," *IEEE Journal on Selected Areas in Communications*, vol. 25, no. 7, pp. 1333–1344, Sept. 2007.
180. D. Gesbert, M. Kountouris, R. W. Heath, Jr., C.-B. Chae, and T. Salzer, "Shifting the MIMO paradigm: From single user to multiuser communications," *IEEE Signal Processing Magazine*, vol. 24, no. 5, pp. 36–46, Oct. 2007.
181. C. Lim, T. Yoo, B. Clerckx, B. Lee, and B. Shim, "Recent trend of multiuser MIMO in LTE-Advanced," *IEEE Communications Magazine*, vol. 51, no. 3, pp. 127–135, Mar. 2013.
182. G. Caire, N. Jindal, M. Kobayashi, and N. Ravindran, "Multiuser MIMO achievable rates with downlink training and channel state feedback," *IEEE Transactions on Information Theory*, vol. 56, no. 6, pp. 2845–2866, June 2010.
183. F. Rusek, D. Persson, B. K. Lau, E. G. Larsson, T. L. Marzetta, O. Edfors, and F. Tufvesson, "Scaling up MIMO: Opportunities and challenges with very large arrays," *IEEE Signal Processing Magazine*, vol. 30, no. 1, pp. 40–60, Jan. 2013.
184. 3GPP TS 38.211 V18.1.0 (2023-12), "Physical channels and modulation (Rel. 18)," Dec. 2023.
185. N. Merhav, G. Kaplan, A. Lapidoth, and S. Shamai (Shitz), "On information rates for mismatched decoders," *IEEE Transactions on Information Theory*, vol. 40, no. 6, pp. 1953–1967, June 1994.
186. V. Raghavan, S. Subramanian, J. Cezanne, A. Sampath, O. H. Koymen, and J. Li, "Single-user vs. multi-user precoding for millimeter wave MIMO systems," *IEEE Journal on Selected Areas in Communications*, vol. 35, no. 6, pp. 1387–1401, June 2017.
187. R. Irmer, H. Droste, P. Marsch, M. Grieger, G. Fettweis, S. Brueck, H.-P. Mayer, L. Thiele, and V. Jungnickel, "Coordinated multipoint: Concepts, performance, and field trial results," *IEEE Communications Magazine*, vol. 49, no. 2, pp. 102–111, Feb. 2011.

188. D. Gesbert, S. V. Hanly, H. Huang, S. Shamai (Shitz), O. Simeone, and W. Yu, "Multi-cell MIMO cooperative networks: A new look at interference," *IEEE Journal on Selected Areas in Communications*, vol. 28, no. 9, pp. 1380–1408, Dec. 2010.
189. H. Dahrouj and W. Yu, "Coordinated beamforming for the multicell multi-antenna wireless system," *IEEE Transactions on Wireless Communications*, vol. 9, no. 5, pp. 1748–1759, May 2010.
190. D. Lee, H. Seo, B. Clerckx, E. Hardouin, D. Mazzarese, S. Nagata, and K. Sayana, "Coordinated multipoint transmission and reception in LTE-Advanced: Deployment scenarios and operational challenges," *IEEE Communications Magazine*, vol. 50, no. 2, pp. 148–155, Feb. 2012.
191. A. Lozano, R. W. Heath, Jr., and J. G. Andrews, "Fundamental limits of cooperation," *IEEE Transactions on Information Theory*, vol. 59, no. 9, pp. 5213–5226, Sept. 2013.
192. M. Sawahashi, Y. Kishiyama, A. Morimoto, D. Nishikawa, and M. Tanno, "Coordinated multipoint transmission/reception techniques for LTE-Advanced," *IEEE Wireless Communications*, vol. 17, no. 3, pp. 26–34, June 2010.
193. G. C. Alexandropoulos, P. Ferrand, J.-M. Gorce, and C. B. Papadias, "Advanced coordinated beamforming for the downlink of future LTE cellular networks," *IEEE Communications Magazine*, vol. 54, no. 7, pp. 54–60, July 2016.
194. A. Barbieri, P. Gaal, S. Geirhofer, T. Ji, D. Malladi, Y. Wei, and F. Xue, "Coordinated downlink multi-point communications in heterogeneous cellular networks," *Proceedings of IEEE Information Theory and Applications Workshop, San Diego, CA*, vol. 1, pp. 7–16, Feb. 2012.
195. D. Kumar, J. Kaleva, and A. Tölli, "Reliable mmWave communication via coordinated multi-point connectivity," *IEEE Transactions on Wireless Communications*, vol. 20, no. 7, pp. 4238–4252, July 2021.
196. C. Fang, B. Makki, J. Li, and T. Svensson, "Hybrid precoding in cooperative millimeter wave networks," *IEEE Transactions on Wireless Communications*, vol. 20, no. 8, pp. 5373–5388, Aug. 2021.
197. D. Maamari, N. Devroye, and D. Tuninetti, "Coverage in mmWave cellular networks with base station co-operation," *IEEE Transactions on Wireless Communications*, vol. 15, no. 4, pp. 2981–2994, Apr. 2016.
198. C. Skouroumounis, C. Psomas, and I. Krikidis, "Low-complexity base station selection scheme in mmWave cellular networks," *IEEE Trans. Commun.*, vol. 65, no. 9, pp. 4049–4064, Sept. 2017.
199. M. Khoshnevisan, V. Joseph, P. Gupta, F. Meshkati, R. Prakash, and P. Tinnakornsrisuphap, "5G industrial networks with CoMP for URLLC and time sensitive network architecture," *IEEE Journal on Selected Areas in Communications*, vol. 37, no. 4, pp. 947–959, Apr. 2019.
200. V. Raghavan, J. H. Ryu, O. H. Koymen, and J. Li, "Towards optimal cooperative beamforming in millimeter wave systems via directional transmissions," *Proceedings of IEEE Wireless Communications and Networking Conference, Austin, TX*, vol. 1, pp. 1581–1586, Apr. 2022.

201. 3GPP TR 38.912 V14.1.0 (2017-06), “Study on New Radio (NR) access technology (Rel. 14),” June 2017.
202. 3GPP TS 38.214 V16.2.0 (2020-07), “Physical layer procedures for data (3GPP TS 38.214 version 16.2.0 Rel. 16),” July 2020.
203. 3GPP TR 38.802 V14.2.0 (2017-09), “Study on New Radio Access Technology Physical Layer Aspects (Rel. 14),” Sept. 2017.
204. T. M. Cover and J. A. Thomas, *Elements of Information Theory*. Wiley Interscience, 1991.
205. S. M. Kay, *Fundamentals of Statistical Signal Processing, Vol. 1: Estimation Theory*. Prentice Hall, 1993.
206. H. V. Poor, *An Introduction to Signal Detection and Estimation*, 2nd ed. Springer, NY, 1994.
207. M. L. McCloud and L. L. Scharf, “Asymptotic analysis of the MMSE multiuser detector for nonorthogonal multipulse modulation,” *IEEE Transactions on Communications*, vol. 49, no. 1, pp. 24–30, Jan. 2001.
208. S. Verdu, *Multiuser Detection*, 1st ed. Cambridge University Press, 1998.
209. V. Raghavan, V. V. Veeravalli, and R. W. Heath, Jr., “Reduced rank signaling in spatially correlated MIMO channels,” *Proceedings of IEEE International Symposium on Information Theory, Nice, France*, vol. 1, pp. 1081–1085, June 2007.
210. R. A. Horn and C. R. Johnson, *Matrix Analysis*. Cambridge University Press, 1985.
211. R. Bhatia, *Matrix Analysis*. Springer-Verlag, 1997.
212. C. R. Rao and M. B. Rao, *Matrix Algebra and its Applications to Statistics and Econometrics*. World Scientific Publishing Co. Pte. Ltd., Singapore, 1998.
213. A. W. Marshall and I. Olkin, *Inequalities: Theory of Majorization and its Applications*. Academic Press, NY, 1979.
214. R. Mathias, “Spectral perturbation bounds for positive definite matrices,” *SIAM Journal on Matrix Analysis and Applications*, vol. 18, no. 4, pp. 959–980, Oct. 1997.
215. D. J. Love, R. W. Heath, Jr., V. K. N. Lau, D. Gesbert, B. D. Rao, and M. Andrews, “An overview of limited feedback in wireless communication systems,” *IEEE Journal on Selected Areas in Communications*, vol. 26, no. 8, pp. 1341–1365, Oct. 2008.
216. R. W. Heath, Jr., N. Gonzalez-Prelcic, S. Rangan, W. Roh, and A. M. Sayeed, “An overview of signal processing techniques for millimeter wave MIMO systems,” *IEEE Journal of Selected Topics in Signal Processing*, vol. 10, no. 3, pp. 436–453, Apr. 2016.
217. F. Khan, *LTE for 4G Mobile Broadband: Air Interface Technologies and Performance*. Cambridge University Press, UK, 2009.
218. 3GPP TS 36.211 V14.2.0 (2017-04), “Evolved Universal Terrestrial Radio Access (E-UTRA); Physical channels and modulation,” Apr. 2017.
219. S. Sesia, I. Toufik, and M. Baker, *LTE – The UMTS Long Term Evolution: From Theory to Practice*, 2nd ed. Wiley, UK, 2013.

220. B. C. Banister and J. R. Zeidler, "Feedback assisted stochastic gradient adaptation of multiantenna transmission," *IEEE Transactions on Wireless Communications*, vol. 4, no. 3, pp. 1121–1134, May 2005.
221. D. J. Love and R. W. Heath, Jr., "Limited feedback diversity techniques for correlated channels," *IEEE Transactions on Vehicular Technology*, vol. 55, no. 2, pp. 718–722, Mar. 2006.
222. P. Xia and G. B. Giannakis, "Design and analysis of transmit beamforming based on limited-rate feedback," *IEEE Transactions on Signal Processing*, vol. 54, no. 5, pp. 1853–1863, May 2006.
223. D. J. Love, R. W. Heath, Jr., W. Santipach, and M. L. Honig, "What is the value of limited feedback for MIMO channels?" *IEEE Communications Magazine*, vol. 42, no. 10, pp. 54–59, Oct. 2004.
224. W. Santipach and M. L. Honig, "Optimization of training and feedback overhead for beamforming over block fading channels," *IEEE Transactions on Information Theory*, vol. 56, no. 12, pp. 6103–6115, Dec. 2010.
225. ——, "Capacity of multi-antenna fading channel with quantized precoding matrix," *IEEE Transactions on Information Theory*, vol. 55, no. 3, pp. 1218–1234, Mar. 2009.
226. V. Raghavan, R. W. Heath, Jr., and A. M. Sayeed, "Systematic codebook designs for quantized beamforming in correlated MIMO channels," *IEEE Journal on Selected Areas in Communications*, vol. 25, no. 7, pp. 1298–1310, Sept. 2007.
227. V. Raghavan and V. V. Veeravalli, "Ensemble properties of RVQ-based limited-feedback beamforming codebooks," *IEEE Transactions on Information Theory*, vol. 59, no. 12, pp. 8224–8249, Dec. 2013.
228. D. Gesbert, H. Bölcskei, D. A. Gore, and A. J. Paulraj, "Outdoor MIMO wireless channels: Models and performance prediction," *IEEE Transactions on Communication*, vol. 50, no. 12, pp. 1926–1934, Dec. 2002.
229. 3GPP TR 38.874 V16.0.0 (2018-12), "Study on integrated access and backhaul (Rel. 16)," Dec. 2018.
230. 3GPP TS 38.106 V17.7.0 (2023-12), "NR Repeater Radio Transmission and Reception (Rel. 17)," Dec. 2023.
231. ZTE, "RP-213592, Study on NR smart repeaters," *3GPP TSG RAN meeting#94e, e-meeting*, Dec. 2021.
232. 3GPP TS 23.304 V18.4.0 (2023-12), "Proximity based Services (ProSe) in the 5G System (5GS) (Rel. 18)," Dec. 2023.
233. G. Liu, F. R. Yu, H. Ji, V. C. M. Leung, and X. Li, "In-band full-duplex relaying: A survey, research issues and challenges," *IEEE Communications Surveys & Tutorials*, vol. 17, no. 2, pp. 500–524, 2nd Quarter 2015.
234. J. N. Laneman, D. N. C. Tse, and G. W. Wornell, "Cooperative diversity in wireless networks: Efficient protocols and outage behavior," *IEEE Transactions on Information Theory*, vol. 50, no. 12, pp. 3062–3080, Dec. 2004.
235. M. R. Souryal and B. R. Vojcic, "Performance of amplify-and-forward and decode-and-forward relaying in Rayleigh fading with turbo codes,"

Proceedings of IEEE International Conference on Acoustics, Speech and Signal Processing, Toulouse, France, vol. 4, pp. 681–684, May 2006.

236. G. Levin and S. Loyka, "Amplify-and-forward versus decode-and-forward relaying: Which is better?" *Proc. IEEE Intern. Zurich Seminar Commun., Zurich, Switzerland*, pp. 1–4, Feb.-Mar. 2012.
237. S. Gopalam, S. V. Hanly, and P. Whiting, "Distributed and local scheduling algorithms for mmWave integrated access and backhaul," *IEEE/ACM Transactions on Networking*, vol. 30, no. 4, pp. 1749–1764, Aug. 2022.
238. ——, "Distributed user association and resource allocation algorithms for three tier HetNets," *IEEE Transactions on Wireless Communications*, vol. 19, no. 12, pp. 7913–7926, Dec. 2020.
239. Lianstar, "RF cellular repeater," Available: [Online]. `https://lianstar.com/en/rf-cellular-repeater.php`.
240. CommScope, "RF repeater systems," Available: [Online]. `https://www.commscope.com/product-type/in-building-cellular-systems/rf-repeater-systems/`.
241. ESRI, "Digital twin," Available: [Online]. `https://www.esri.com/content/dam/esrisites/en-us/media/brochures/digital-twin-technology-resource.pdf`.
242. Facebook, "Terragraph mesh," Available: [Online]. `https://terragraph.com/assets/files/Terragraph_Mesh_Whitepaper-d906f1eb9c3ea7a8c1bbd8552b1f9f2d.pdf`.
243. J. Deygout, "Multiple knife-edge diffraction of microwaves," *IEEE Transactions on Antennas and Propagation*, vol. 14, no. 4, pp. 480–489, July 1966.
244. R. Levie, C. Yapar, G. Kutyniok, and G. Caire, "Radio UNet: Fast radio map estimation with convolutional neural networks," *IEEE Transactions on Wireless Communications*, vol. 20, no. 6, pp. 4001–4015, June 2021.
245. M. Gasse, D. Chetelat, L. Charlin, and A. Lodi, "Learning to branch in MILP solvers," *TTI-C Workshop on Automated Algorithm Design, Chicago*, Aug. 2019.
246. E. Khalil, P. L. Bodic, L. Song, G. Nemhauser, and B. Dilkina, "Learning to branch in mixed integer programming," *Proceedings of AAAI Conference on Artificial Intelligence, Phoenix, AZ*, vol. 30, no. 1, Feb. 2016.
247. M. Lee, G. Yu, and G. Y. Li, "Learning to branch: Accelerating resource allocation in wireless networks," *IEEE Transactions on Vehicular Technology*, vol. 69, no. 1, pp. 958–970, Jan. 2020.
248. Wikipedia, "Set cover problem," Available: [Online]. `https://en.wikipedia.org/wiki/Set_cover_problem`.
249. A. Lozano and S. Rangan, "Spectral vs. energy efficiency in 6G: Impact of the receiver front-end," *IEEE BITS, the Information Theory Magazine*, pp. 1–11, Oct. 2023.
250. P. Skrimponis, N. Hosseinzadeh, A. Khalili, E. Erkip, M. J. W. Rodwell, J. F. Buckwalter, and S. Rangan, "Towards energy efficient mobile wireless receivers above 100 GHz," *IEEE Access*, vol. 9, pp. 20 704–20 716, Nov. 2021.

251. T. Richardson and S. Kudekar, "Design of low-density parity check codes for 5G New Radio," *IEEE Communications Magazine*, vol. 56, no. 3, pp. 28–34, Mar. 2018.
252. T. J. Richardson and R. Urbanke, *Modern Coding Theory*, 1st ed. Cambridge University Press, 2008.
253. T. S. Rappaport, Y. Xing, O. Kanhere, S. Ju, A. Madanayake, S. Mandal, A. Alkhateeb, and G. C. Trichopoulos, "Wireless communications and applications above 100 GHz: Opportunities and challenges for 6G and beyond," *IEEE Access*, vol. 7, pp. 78 729–78 757, June 2019.
254. J. Ma, R. Shrestha, L. Moeller, and D. M. Mittleman, "Channel performance for indoor and outdoor terahertz wireless links," *American Physics Letters Photonics*, vol. 3, no. 5, p. 051601, May 2018.
255. L. Pometcu and R. D'Errico, "Characterization of sub-THz and mmwave propagation channel for indoor scenarios," *Proceedings of European Conference on Antennas and Propagation, London, UK*, vol. 1, pp. 1–4, Apr. 2018.
256. K. Guan, G. Li, T. Kürner, A. F. Molisch, B. Peng, R. He, B. Hui, J. Kim, and Z. Zhong, "On millimeter wave and THz mobile radio channel for smart rail mobility," *IEEE Transactions on Vehicular Technology*, vol. 66, no. 7, pp. 5658–5674, July 2017.
257. A. Afsharinejad, A. Davy, B. Jennings, S. Rasmann, and C. Brennan, "A path-loss model incorporating shadowing for THz band propagation in vegetation," *Proceedings of IEEE Global Telecommunications Conference, San Diego, CA*, vol. 1, pp. 1–6, Dec. 2015.
258. Y. Xing, T. S. Rappaport, and A. Ghosh, "Millimeter wave and sub-THz indoor radio propagation channel measurements, models, and comparisons in an office environment," *IEEE Communications Letters*, vol. 25, no. 10, pp. 3151–3155, Oct. 2021.
259. S. L. H. Nguyen, J. Jarvelainen, A. Karttunen, K. Haneda, and J. Putkonen, "Comparing radio propagation channels between 28 and 140 GHz bands in a shopping mall," *Proceedings of European Conference on Antennas and Propagation, London, UK*, vol. 1, pp. 1–4, Apr. 2018.
260. J. D. Preez, S. Sinha, and K. Sengupta, "SiGe and CMOS Technology for State-of-the-Art Millimeter-Wave Transceivers," *IEEE Access*, vol. 11, pp. 55 596–55 617, 2024.
261. G. M. Rebeiz, "Silicon RFICs for 100-700 GHz communication and radar/sensor systems," *Proceedings of International Conference on Infrared, Millimeter and Terahertz Waves, Delft, Netherlands*, p. 1, Aug.-Sept. 2022.
262. International Telecommunication Union (Radiocommunication Sector), "World Radiocommunication Conference 2023 (WRC-23), Provisional final acts," Nov.-Dec. 2023, Available: [Online]. `https://www.itu.int/dms_pub/itu-r/opb/act/R-ACT-WRC.15-2023-PDF-E.pdf`.
263. S. Kang, M. Mezzavilla, S. Rangan, A. Madanayake, S. B. Venkatakrishnan, G. Hellbourg, M. Ghosh, H. Rahmani, and A. Dhananjay, "Cellular wireless networks in the upper mid-band," *Submitted for publication*, 2023.

264. M. Cai, K. Gao, D. Nie, B. Hochwald, J. N. Laneman, H. Huang, and K. Liu, "Effect of wideband beam squint on codebook design in phased-array wireless systems," *Proceedings of IEEE Global Telecommunications Conference, Washington, DC*, vol. 1, pp. 1–6, Dec. 2016.
265. R. Li, Y. Chen, G. Y. Li, and G. Liu, "Full-duplex cellular networks," *IEEE Communications Magazine*, vol. 55, no. 4, pp. 184–191, Apr. 2017.
266. A. Sabharwal, P. Schniter, D. Guo, D. W. Bliss, S. Rangarajan, and R. Wichman, "In-band full-duplex wireless: Challenges and opportunities," *IEEE Journal on Selected Areas in Communications*, vol. 32, no. 9, pp. 1637–1652, Sept. 2014.
267. I. P. Roberts, J. G. Andrews, H. B. Jain, and S. Vishwanath, "Millimeter-wave full duplex radios: New challenges and techniques," *IEEE Wireless Communications*, vol. 28, no. 1, pp. 36–43, Feb. 2021.
268. Z. Xiao, P. Xia, and X.-G. Xia, "Full-duplex millimeter-wave communication," *IEEE Wireless Communications*, vol. 24, no. 6, pp. 136–143, Dec. 2017.
269. T. Chen, S. Garikapati, A. Nagulu, A. Gaonkar, M. Kohli, I. Kadota, H. Krishnaswami, and G. Zussman, "A survey and quantitative evaluation of integrated circuit-based antenna interfaces and self-interference cancellers for full-duplex," *IEEE Open Journal of the Communications Society*, vol. 2, pp. 1753–1776, July 2021.
270. 3GPP TR 38.858 V18.0.0 (2023-12), "Study on Evolution of NR Duplex Operation (Rel. 18)," Dec. 2023.
271. V. Singh, S. Mondal, A. Gadre, M. Srivastava, J. Paramesh, and S. Kumar, "Millimeter-wave full duplex radios," *Proceedings of International Conference on Mobile Computing and Networking (Mobicom), London, UK*, vol. 2, pp. 1–14, Apr. 2020.
272. M. S. Sim, S. Ahuja, W. Nam, A. Dorosenco, Z. Fan, and T. Luo, "60 GHz mmWave full-duplex transceiver study and over-the-air link verification," *Proceedings of IEEE Global Telecommunications Conference, Rio de Janeiro, Brasil*, vol. 1, pp. 1–6, Dec. 2022.
273. M. Jiang, J. Cezanne, A. Sampath, O. Shental, Q. Wu, O. H. Koymen, A. M. Bedewy, and J. Li, "Wireless fronthaul for 5G and future radio access networks: Challenges and enabling technologies," *IEEE Wireless Communications*, vol. 29, no. 2, pp. 108–114, Apr. 2022.
274. J. Cezanne, M. Jiang, O. Shental, A. M. Bedewy, A. Sampath, O. H. Koymen, and J. Li, "Design of wireless fronthaul With mmWave LOS-MIMO and sample-level coding for O-RAN and beyond 5G systems," *IEEE Open Journal of the Communications Society*, vol. 4, pp. 1893–1912, 2023.
275. D. K. P. Tan, J. He, Y. Li, A. Bayesteh, Y. Chen, P. Zhu, and W. Tong, "Integrated sensing and communication in 6G: Motivations, use cases, requirements, challenges and future directions," *Proceedings of IEEE International Online Symposium on Joint Communications & Sensing (JC&S)*, vol. 1, pp. 1–6, Feb. 2021.

276. S. Saponara, M. S. Greco, and F. Gini, "Radar-on-chip/in-package in autonomous driving vehicles and intelligent transport systems: Opportunities and challenges," *IEEE Signal Processing Magazine*, vol. 36, no. 5, pp. 71–84, Sept. 2019.
277. J. Hasch, E. Topak, R. Schnabel, T. Zwick, R. Weigel, and C. Waldschmidt, "Millimeter-wave technology for automotive radar sensors in the 77 GHz frequency band," *IEEE Transactions on Microwave Theory and Techniques*, vol. 60, no. 3, pp. 845–860, Mar. 2012.
278. S. M. Patole, M. Torlak, D. Wang, and M. Ali, "Automotive radars: A review of signal processing techniques," *IEEE Signal Processing Magazine*, vol. 34, no. 2, pp. 22–35, Mar. 2017.
279. F. Engels, P. Heidenreich, M. Wintermantel, L. Stäcker, M. A. Kadi, and A. M. Zoubir, "Automotive radar signal processing: Research directions and practical challenges," *IEEE Journal of Selected Topics in Signal Processing*, vol. 15, no. 4, pp. 865–878, June 2011.
280. I. Gresham, A. Jenkins, R. Egri, C. Eswarappa, N. Kinayman, N. Jain, R. Anderson, F. Kolak, R. Wohlert, S. Bawell, J. Bennett, and J.-P. Lanteri, "Ultra-wideband radar sensors for short-range vehicular applications," *IEEE Transactions on Microwave Theory and Techniques*, vol. 52, no. 9, pp. 2105–2122, Sept. 2004.
281. G. E. Hinton and H. Bishop, *Deep Learning: Foundations and Concepts*, 1st ed. Springer, 2024.
282. M. Chen, U. Challita, W. Saad, C. Yin, and M. Debbah, "Artificial neural networks-based machine learning for wireless networks: A tutorial," *IEEE Communications Surveys & Tutorials*, vol. 21, no. 4, pp. 3039–3071, Fourth quarter 2019.
283. N. C. Luong, D. T. Hoang, S. Gong, D. Niyato, P. Wang, Y.-C. Liang, and D. I. Kim, "Applications of deep reinforcement learning in communications and networking: A survey," *IEEE Communications Surveys & Tutorials*, vol. 21, no. 4, pp. 3133–3174, Fourth quarter 2019.
284. M. Li, C. Liu, S. V. Hanly, I. B. Collings, and P. Whiting, "Explore and eliminate: Optimized two-stage search for millimeter-wave beam alignment," *IEEE Transactions on Wireless Communications*, vol. 18, no. 9, pp. 4379–4393, Sept. 2020.
285. C. Liu, M. Li, S. V. Hanly, P. Whiting, and I. B. Collings, "Millimeter-wave small cells: Base station discovery, beam alignment, and system design challenges," *IEEE Wireless Communications*, vol. 25, no. 4, pp. 40–46, Aug. 2018.
286. Q. Li, P. Sisk, A. C. Kannan, T. Yoo, T. Luo, G. Shah, B. Manjunath, C. Samarathungage, M. T. Boroujeni, H. Pezeshki, and H. Joshi, "Machine learning based time domain millimeter-wave beam prediction for 5G-Advanced and beyond: Design, analysis, and over-the-air experiments," *IEEE Journal on Selected Areas in Communications*, vol. 41, no. 6, pp. 1787–1809, June 2023.

Index

For Product Safety Concerns and Information please contact our EU representative GPSR@taylorandfrancis.com Taylor & Francis Verlag GmbH, Kaufingerstraße 24, 80331 München, Germany